ALS
Advances in Life Sciences

Impact of Arbuscular Mycorrhizas on Sustainable Agriculture and Natural Ecosystems

Edited by
S. Gianinazzi
H. Schüepp

Springer Basel AG

Editors

Dr. S. Gianinazzi
Laboratoire de Phytoparasitologie
INRA/CNRS, SGAP, INRA
BV 1540
F - 21034 Dijon Cédex
France

Dr. H. Schüepp
Eidgenössische Forschungsanstalt für
Obst-, Wein- und Gartenanbau
CH - 8820 Wädenswil
Switzerland

A CIP catalogue record for this book is available from the Library of Congress,
Washington D.C., USA

Deutsche Bibliothek Cataloging-in-Publication Data
**Impact of arbuscular mycorrhizas on sustainable agriculture
and natural ecosystems** / ed. by S. Gianinazzi ; H. Schüepp. —
Basel ; Boston ; Berlin : Birkhäuser, 1994
  (Advances in life sciences)
  ISBN 978-3-0348-9654-2   ISBN 978-3-0348-8504-1 (eBook)
  DOI 10.1007/978-3-0348-8504-1
NE: Gianinazzi, Silvio [Hrsg.]

© 1994 Springer Basel AG
Originally published by Birkhäuser Verlag in 1994
Softcover reprint of the hardcover 1st edition 1994
Camera-ready copy prepared by the authors
Printed on acid-free paper produced from chlorine-free pulp

9 8 7 6 5 4 3 2 1

This book is dedicated to the memory of

Dr. Justo ARINES (1946-1993)

whose personal qualities and scientific activities within the European
Network COST ACTION 810 will always be very much appreciated

Contents

Preface

This book, prepared by participants of the European network COST ACTION 810 (1989–93) is the outcome of a meeting held in Switzerland (Einsiedeln, September 29 to October 2, 1993) on the "Impact of arbuscular mycorrhizas on sustainable agriculture and natural ecosystems".

COST (Coopération Scientifique et Technique) Networks were created in 1971 by the Commission of European Communities, and later enlarged to include non-European Member States, to promote pre-competitive scientific and technical research in fields of common interest. During the eighties, COST ACTIONS were launched in bio-technological fields, including the network on arbuscular mycorrhizas.

Arbuscular mycorrhizas are a universally found symbiosis between plants and certain soil fungi and essential components of soil-plant systems. They act as a major inter-face by influencing or regulating resource allocation between abiotic and biotic components of the soil-plant system. Arbuscular mycorrhizas are involved in many key ecosystem processes including nutrient cycling and conservation of soil struc-ture, and have been shown to improve plant health through increased protection against abiotic and biotic stresses.

Sustainability can be defined as the successful management of resources to satisfy changing human needs while maintaining or enhancing the quality of the environ-ment and conserving resources. Increasing environmental degradation and instability, due to anthropogenic activities and in particular the increasing fragility of the soil resource, has led to an increased awareness of the need to develop practices resulting in more sustainable natural and agroecosystems.

Because of the fundamental importance of arbuscular mycorrhizas in key eco-system processes, an understanding of the symbiosis is an essential prerequisite for the development of sustainable agricultural systems and of guidelines for the protection of natural ecosystems.

By combining scientific expertise and resources from many European laboratories, the COST ACTION 810 Network has greatly promoted development of a multi-disciplinary approach necessary for progress in the complexity of mycorrhizal research. As a result, considerable advances have been made in (i) the development of biomolecular methods adapted to arbuscular mycorrhizal fungi, (ii) the promotion of innovative ecological research which considers biodiversity, in order to better understand the impact of arbuscular mycorrhizas in sustainable agricultural and natural ecosystems, (iii) the role of arbuscular mycorrhizas in plant development and root morphology, and (iv) the definition of the cellular and molecular basis of plant-fungus exchanges.

Another exciting aspect has been the advance made towards creating a "Banque Européenne des Glomales" (BEG) and the "Taxonomic Expert System" that should stimulate international exchange and greatly facilitate research on arbuscular mycorrhizas.

In this book, well-known specialists review research targeted on the role of arbuscular mycorrhizas in environmental protection and sustainable plant production with view to meeting the growing requirements of a growing society.

Silvio Gianinazzi   Hannes Schüepp
INRA/CNRS, Dijon, France   FAW, Wädenswil, Switzerland

Contributors

David Atkinson SAC, West Mains Road, Edinburgh, EH9 3JG, UK

Luciano Avio Istituto di Microbiologia Agraria, Centro di Studio per la Microbiologia del Suolo C.N.R., Via del Borghetto 80, 56124 Pisa, Italy

Rosario Azcón Estación Esperimental del Zaidin, CSIC, Prof. Abareda 1, Granada, Spain

Conception Azcón-Aguilar Dept. de Microbiología del Suelo y Sistemas Simbióticos, Estación Experimental del Zaidín, CSIC, Prof. Albareda 1, 18008 Granada, Spain

Berta Bago Dept. de Microbiología del Suelo y Sistemas Simbióticos, Estación Experimental del Zaidín, CSIC, Prof. Albareda 1, 18008 Granada, Spain.

Jose Miguel Barea Estación Experimental del Zaidín CSIC Prof. Albareda 1, 18008 Granada, Spain

Graziella Berta Dipartimento di Biologia Vegetale dell'Università di Torino ,Viale Mattioli, 25, 10125 Torino, Italy

Gabor J. Bethlenfalvay US Dept. of Agriculture, Agric. Research Service, Horticultural Crops Research Lab, Corvallis, OR 97330, USA

Paola Bonfante Dipartimento di Biologia Vegetale dell'Università di Torino e Centro di Studio sulla Micologia del Terreno del CNR, Viale Mattioli, 25 10125 Torino, Italy

Anna Silvia Citernesi Istituto di Microbiologia Agraria, Centro di Studio per la Microbiologia del Suolo C.N.R., Via del Borghetto 80, 56124 Pisa, Italy

John C. Dodd International Institute of Biotechnology, P.O. Box 228, Canterbury, Kent, CT2 7YW, UK

Hubert Dulieu Laboratoire de Phytoparasitologie, INRA-CNRS, Station de Génétique et d'Amélioration des Plantes, INRA, BV 1540, F-21034 Dijon cédex, France

Victoria Estaún Institut de Recerca i Tecnologia Agroalimantàries, Centre de Cabrils s/n, E-08348 Cabrils (Barcelona), Spain

Silvio Gianinazzi Laboratoire de Phytoparasitologie, INRA-CNRS, Station de Génétique et d'Amélioration des Plantes, INRA, BV 1540, F-21034 Dijon cédex, France

Vivienne Gianinazzi-Pearson Laboratoire de Phytoparasitologie, INRA-CNRS, Station de Génétique et d'Amélio-ration des Plantes, INRA, BV 1540, F-21034 Dijon cédex, France

Manuela Giovannetti Istituto di Microbiologia Agraria, Centro di Studio per la Microbiologia del Suolo C.N.R., Via del Borghetto 80, 56124 Pisa, Italy

Armelle Gollotte Laboratoire de Phytoparasitologie, INRA-CNRS, Station de Génétique et d'Amélioration des Plantes, INRA, BV 1540, F-21034 Dijon cédex, France

Alexander Hahn Technische Universitat München (Weihenstephan), Lehrstuhl für Botanik, D-85350 Freising, Germany

Kurt Haselwandter Institut für Mikrobiologie, Universität Innsbruck, Technikerstr. 25, A-6020 Innsbruck, Austria

Bertold Hock Technische Universitat München (Weihenstephan), Lehrstuhl für Botanik, D-85350 Freising, Germany

Gisela Höflich Zentrum für Agrarlandschafts- und Landnutzungsforschung, Eberswalder Str. 84, Müncheberg, Germany

Mario Honrubia Departamento de Biología Vegetal. Laboratorio de Micología. Facultad de Biología.Campus de Espinardo. Universidad de Murcia. 30100 Murcia. Spain

John E. Hooker Soil Biology Unit, Department of Land Resources, SAC, Mill of Craibstone, Aberdeen, AB2 9TQ, UK

Maria Jaizme-Vega CITA, Departmento Protección Vegetal, Apartado 60, La Laguna, E-38080, Tenerife, Islas Canarias, Spain

Iver Jakobsen Plant Biology Section, Environmental Science and Technology Department, Risø National Laboratory, DK-4000 Roskilde, Denmark

Peter Jeffries Biological Laboratory, University of Kent, Canterbury, Kent CT2 6NJ, UK

Erik J. Joner Department of Biotechnological Sciences, Agricultural University of Norway, N-1432 Aas, Norway

John Larsen Plant Biology Section, Environmental Science and Technology Department, Risø National Laboratory, DK-4000 Roskilde, Denmark

Corinne Leyval Centre de Pédologie Biologique, C.N.R.S., 17, rue N.D. des Pauvres, B.P.5, F-54501 Vandoeuvre-Les-Nancy Cedex, France

Gigliola Puppi Dipartimento di Biologia Vegetale, Università "La Sapienza", I-00165 Roma, Italy

Soren Rosendahl Botanical Institute, University of Copenhagen, Øster Farimagsgade 2D, DK-1353 Copenhagen, Denmark

Manuel Sánchez-Díaz Departamento de Fisiología Vegetal. Universidad de Navarra. 31080 Pamplona, Spain

Francis E. Sanders Department of Pure and Applied Biology, University of Leeds, Leeds LS2 9JT, UK

Cristiana Sbrana Istituto di Microbiologia Agraria, Centro di Studio per la Microbiologia del Suolo C.N.R., Via del Borghetto 80, 56124 Pisa, Italy

Hannes Schüepp Swiss Federal Research Station, CH–8820 Wädenswil, Switzerland

Diederik van Tuinen Laboratoire de Phytoparasitologie, INRA-CNRS, Station de Génétique et d'Amélioration des Plantes, INRA, BV 1540, F-21034 Dijon cédex, France

Mauritz Vestberg Agricultural Research Centre of Finland, Laukaa Research and Elite Plant Unit, FIN-41340 Laukaa, Finland

Christopher Walker Forestry Commission, Forestry Authority, Northern Research Station, Roslin, Midlothian EH25 9SY, UK

Adolphe Zézé Laboratoire de Phytoparasitologie, INRA-CNRS, Station de Génétique et d'Amélioration des Plantes, INRA, BV 1540, F-21034 Dijon cédex, France

Impact of Arbuscular Mycorrhizas on
Sustainable Agriculture and Natural Ecosystems
S. Gianinazzi and H. Schüepp (eds.)
© 1994 Birkhäuser Verlag Basel/Switzerland

Taxonomy and phylogeny of the *Glomales*

S. Rosendahl, J.C. Dodd[1] and C. Walker[2]

Botanical Institute, University of Copenhagen, Øster Farimagsgade 2D, DK-1353 Copenhagen K, Denmark, [1]*International Institute of Biotechnology, P.O. Box 228, Canterbury, Kent, CT2 7YW, UK,* [2]*Forestry Commission, Forestry Authority, Northern Research Station, Roslin, Midlothian EH25 9SY, UK*

Introduction

The importance of arbuscular mycorrhizal fungi in nutrient uptake by plants has been intensively studied (e.g., Sanders et al., 1974; Gianinazzi-Pearson and Gianinazzi, 1986). The fungal partners in these symbioses are currently placed in the *Zygomycetes* in the order *Glomales*. Little is known about their diversity, especially at the specific- and sub-specific level. The study of diversity among arbuscular mycorrhizal fungi (AMF) is partly hampered by the limited range of useful morphological characters and their inconistent or incomplete use in species descriptions. In efforts to overcome this problem, molecular techniques have been applied. Although isozymes and DNA sequence data can distinguish between fungi at any taxonomic level, the species and genus concept in the *Glomales* are still not well-defined, and further work is needed to establish the level of identification useful in studies of the biology and biodiversity of these ecologically significant fungi.

History

Thaxter (1922) provided the basis for the taxonomy of these fungi, though their trophic status was not known at that time, and both mycorrhizal and non-mycorrhizal fungi were classified together in one genus, *Endogone*. In 1953, the first link was made between *'Endogone'* (= *Glomus mosseae*) spores and the mycorrhizal habit of the fungus (Mosse, 1953). This was followed by the informal description of several 'spore types' (Mosse and Bowen, 1968), seven of which formed arbuscular mycorrhizas. In a monograph detailing the *Endogonaceae* in the Pacific

Northwest (Gerdemann and Trappe, 1974) the family was further defined, and new species described. Five genera were included, namely *Endogone, Glomus, Sclerocystis, Gigaspora* and *Acaulospora.* Since then, two additional genera, *Entrophospora* and *Scutellospora,* have been erected. The genus *Endogone* is not thought to contain members capable of forming arbuscular mycorrhiza, whereas the other six genera contain species forming arbuscular mycorrhiza. More recently, *Endogone* was therefore separated, remaining in its own family and order, and those fungi known or supposed to form arbuscular mycorrhiza were placed in their own order, *Glomales,* with three families: *Glomaceae* (*Glomus* and *Sclerocystis*), *Acaulosporaceae* (*Acaulospora* and *Entrophospora*) and *Gigasporaceae* (*Gigaspora* and *Scutellospora*) (Morton and Benny, 1990). Although many of the approximately 140 currently named species (Walker and Trappe, 1993) were described from field-collected material, and are therefore of unknown mycorrhizal status, it is assumed that they are all capable of forming arbuscules in roots.

Table I. The current taxonomy of the *Glomales* according to Morton and Benny (1990). The family *Endogonaceae* is placed in its own order and is not included.

Order: *Glomales* Morton and Benny

    Suborder: *Glomineae* Morton and Benny

        Family: *Glomaceae* Pirozynski and Dalpé

            Genus: *Glomus* Tulasne and Tulasne
            Genus: *Sclerocystis* (Berkeley and Broome) Almeida and Schenck

        Family: *Acaulosporaceae* Morton and Benny

            Genus: *Acaulospora* (Gerdemann and Trappe) Berch
            Genus: *Entrophospora* Ames and Schneider

    Suborder: *Gigasporineae* Morton and Benny

        Family: *Gigasporaceae* Morton and Benny

            Genus: *Gigaspora* (Gerdemann and Trappe) Walker and Sanders
            Genus: *Scutellospora* Walker and Sanders

Taxonomy at higher levels

The AMF in the *Glomales* were along with the zygosporic *Endogone* initially proposed placed informally in the *Mucorales* (*Zygomycetes*) (Bucholtz, 1912; Moreau, 1953). This was formalised by Benjamin (1979) based tentatively on presumed homology between the *Glomus* spores and the azygospore found in *Mucoraceae*. The incorrectly interpreted observation of zygospores in sporocarps of *Glomus fasciculatum* was used to support this hypothesis (Thaxter, 1922). Structures interpreted as zygospores were formed when spores of *Gigaspora decipiens* were germinated together (Tommerup, 1991), but indications suggest that the resting structures found in the *Glomales*, and used to characterise species, are probably not homologous either within or among the three families in *Glomales*. Even within the currently accepted concept of the genus *Glomus* it is questionable whether the supposed chlamydospores formed by different species represent homologous structures.

Morton and Benny (1990) erected the order *Glomales* to encompass the AMF. They defined the ability to form this type of mycorrhiza to be a key character in the order. While it is beyond doubt that the separation of *Endogonales* and *Glomales* is justified, the latter family contains at least one species, the so-called fine endophyte (*Glomus tenue*) that is probably not closely linked phylogenetically to the other glomalean fungi (Walker, 1992), despite its capacity to form a type of arbuscular mycorrhiza. Another fungus, *Glomus tubiforme* forms ectomycorrhiza. Although, from its spore morphology this fungus can be placed in the genus *Glomus,* when examined with molecular techniques, no product was amplified by the *Glomales*–specific VANS1–primer, indicating that the species probably does not belong in the same group as some other species of *Glomus* (Walker and Simon, unpublished). The arbuscular mycorrhizal status of Glomalean fungi is unquestionably an important character, but because the symbiotic status of many species presently placed in the *Glomales* is unknown its appropriateness as a determinant character of the order can be questioned.

The information gathered from the sequences of the 18S ribosomal genes has proved useful in studies of fungal phylogeny (Bruns et al., 1991). The data on glomalean fungi indicate that they form an ancient group, branching before the *Ascomycetes* and the *Basidiomycetes* (Berbee and Taylor, 1993). However, the study did not clarify the distinction between *Zygomycetes* and *Chytridiomycetes*. This might be due to a polyphyletic nature of the *Zygomycetes*, and the present molecular data alone can neither be used to confirm the placing of the *Glomales* in the *Zygomycetes*, nor to separate it into its own class. Further work on the molecular evolution of species presently placed in the *Zygomycetes* is needed in order to clarify the true affinity of the *Glomales*.

The taxonomic concepts

The concepts used in the taxonomy of the *Glomales* are based on the spore morphology, with the main criteria used for species delimitation being spore size, shape, colour, basal structure (including mode of occlusion), ornamentation, and wall (or wall-layer) structure. The wall structure of the fungi has perhaps been over-stressed during the last decade. Several different wall

4

types have been described in detail (Berch and Koske, 1986; Morton, 1986; Walker, 1983; 1986) and the number of walls and their position have been assigned considerable significance in species designation. The wall characters have been most useful for genera with spores in which several walls or wall layers are present (e.g., *Acaulospora* and *Scutellospora*), but in the genus *Glomus*, usually only one to three walls are present in a single wall group. Wall structure then becomes of less value in species diagnosis. The use of wall structures is likely to be of increasing value for separating taxa at the supra-specific level, since they are useful for ontogenetic studies which should help in determining the different natures of spores within the *Glomales*.

Some problems with the taxonomic concepts used in the *Glomales* have recently been reviewed (Walker, 1992; Bentivenga and Morton, 1993). The uncertainty about the structures that should be considered as the most important as taxonomic characters is a problem in glomalean systematics, but more consistency is likely as species are re-described. The introduction of molecular characters has been useful, but the techniques have not yet changed the current taxonomic concepts (Simon et al., 1993). Nevertheless, there is clear evidence that progress with the polymerase chain reaction (PCR) techniques is being made, both inter- and intra-specific taxa apparently being identifiable through DNA polymorphisms detected through the use of short arbitrary primers (Wyss and Bonfante, 1993).

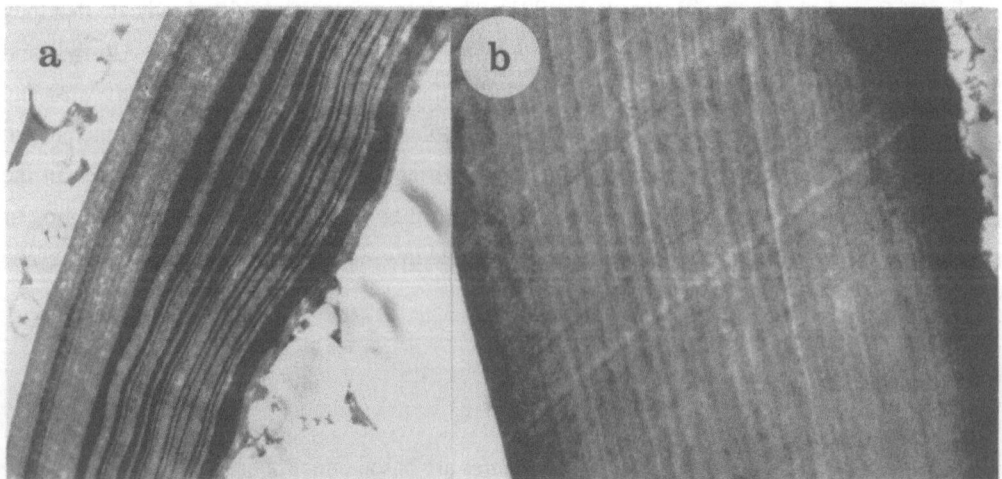

Figure 1. Transmission electron micrographs (TEM) of the structural wall of a) *Glomus intraradices* showing alternating electron-dense and electron-translucent layers and b) *G.geosporum* with a electron-translucent layer of arced microfibrils

Novel approaches in the taxonomy of the *Glomales*

Spore wall characteristics The lack of knowledge on homology of the morphological structures is particularly relevant when identifying wall structures. Comparative electron microscopy is useful in evaluation of such morphological characters. Fig. 1 shows the difference in wall structures between a species fitting the description of *Glomus intraradices* (from Assateague Island, USA) and *G. geosporum* (from Kent, UK). Under the light microscope, the pigmented, laminated wall of the former separates on crushing into several wall layers (up to 5 have been counted), whereas that of the latter often does not split, or may, when mature, split into only two layers. The two walls are evidently of different structure and may not represent homologous structures. TEM reveals the alternating structure of electron-dense and electron-translucent layers (Fig. 1a) in the *G. intraradices*-like spore, but only the electron-translucent structure is seen in *G. geosporum* (Fig. 1b). The electron-translucent layer consists of arced microfibrils (Miller and Jeffries, 1994), as described in *Glomus epigaeum* (Bonfante-Fasolo and Grippiolo, 1984). The electron-dense layers do not appear to have such an organised structure, and these may reflect the lines of fracture observed under the light microscope. Further ultrastructural studies are needed to evaluate the homologies in wall structures. Mosse (1970) showed the great variety of different wall layers in a species of *Acaulospora* when studied with transmission electron microscopy (TEM). Current studies are revealing similar levels of complexity in other species of *Acaulospora* and *Scutellospora* (Dodd, Jeffries and Walker, unpublished).

Molecular characters - Isozymes Molecular techniques are useful in systematics. The techniques based on the PCR and subsequent sequencing of specific DNA and RNA regions have already solved several problems in fungal systematics (Reynolds and Taylor, 1993). The techniques have also been used for AMF, and seem to confirm the current taxonomy at the family level (Simon et al., 1993). For the techniques based on nucleic acids are reviewed in this book (see van Tuinen et al., this volume), in our article we will therefore focus on the use of isozymes in glomalean systematics.

Isozyme variation can be the result of different genetic and epigenetic events. More than one locus for the same enzyme production can be present in the organism. If two loci are detected, these may represent proteins with different functions in the organism. The enzyme malate dehydrogenase is known to have a mitochondrial and a cytoplasmic locus (Harris and Hopkinson, 1976) which are inherited independently. Differences in loci will not usually be found within a species, but some loci may not be detected in certain species. Allelic variation is variation in enzymes produced from a single locus (also termed allozymes). Such variation can result from proteins which have the same function, but with minor differences in amino acid composition. Secondary bands, the result of post transcriptional processes, often occur on gels, and are not related to the genetics of the organism.

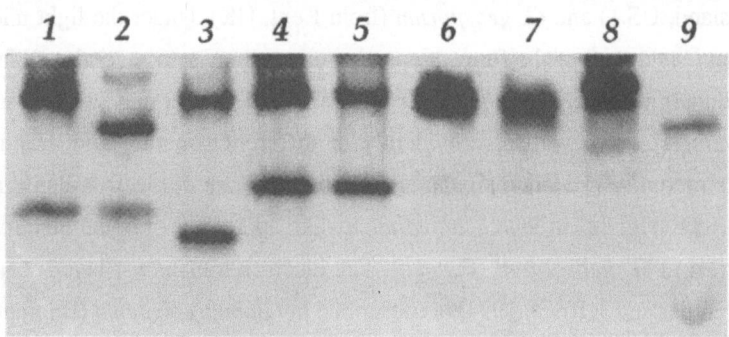

Figure 2. Polyacrylamide gel stained for malate dehydrogenase (MDH). Lane 1 and 2: *Glomus claroideum*; lane 3, 4 and 5: *G. fistulosum*; lane 7: *G. mosseae* from Europe; lane 8: *G. mosseae* from the Philippines and lane 9: *G. geosporum*.

Isozyme characters are easily generated after electrophoretic separation of proteins (see Rosendahl and Sen, 1992), but the interpretation of the results is often controversial. Firstly, the identification of secondary bands is important, as these should not be included in a genetic analysis. Secondly putative loci and alleles must be identified. If the data is to be used for estimation of genetic distance (Rogers, 1986), this interpretation is crucial. In sexually reproducing organisms, the loci and alleles are identified by back-crosses, but the mitotic status of many fungi, including members of the *Glomales*, necessitates an interpretation without benefit of back-crossing. This is done by comparing and contrasting the shape and appearance of the bands, and from studying several isolates of the same species. The gel stained for malate dehydrogenase (Fig. 2) shows two loci in *Glomus claroideum* and *G. fistulosum* (lane 1-5). The two loci are identified on basis of the variation within the loci. The mobility of the band in *G. mosseae* (lane 6 and 7) is similar to the upper locus found in *G. fistulosum*. However, the different appearance of the protein band indicates that the two species have different malate dehydrogenase loci.

Isozyme analysis has been used for taxonomic studies of several species of *Glomus* (Hepper et al., 1988; Rosendahl, 1989). Both studies suffer from failure to identify putative loci and are thus difficult to use for comparing genetic distances. In the study of intraspecific variation in *Glomus mosseae* (Hepper et al., 1989) the morphology of the material was not studied. This species is not well defined, and the results may not therefore represent intraspecific variation as stated by the authors, but may also include an interspecific component. The study on cluster forming species of *Glomus* (Rosendahl, 1989) included *G. intraradices, G. fasciculatum, G. claroideum* and at least three other species. The only morphological species studied from a large number of cultures was *G. intraradices* which showed an intraspecific similarity of 40-50%. The study did not include an interpretation of putative loci and alleles and cannot be used for estimation of genetic distances. Future taxonomic studies including isozymes should address these problems.

Taxonomic problems in *Glomales* that should be given attention

Is the genus Glomus polyphyletic ? The genus *Glomus* is defined by its simple spore type (a presumed chlamydospore) without a specialised hyphal attachment or spore base as found in *Acaulospora* and *Scutellospora*. Moreover, the number of spore walls or wall layers is usually low compared to the other genera. It is possible that an explanation for this lies in a lack of divergence in this ancient mitotic group of fungi, but the few characters used in *Glomus* may cover a polyphyletic origin of the genus by convergent evolution. Walker (1992) suggested that the different occlusions of the subtending hypha found in spores of members of *Glomus* may represent characters that can distinguish fungi at higher taxonomic levels. The endospore-like formation seen in *Glomus maculosum, G. fistulosum* and *G. claroideum* suggests that this spore type may not be homologous with the spore type of *G. mosseae* where the spore content is separated from the hypha by a thin septum formed by the inner laminae of the structural wall.

Isozyme patterns of these groups has been used to study this question (Fig. 2). To discuss the two possible groupings, it is necessary to make a genetic interpretation of the protein bands. Fig.2

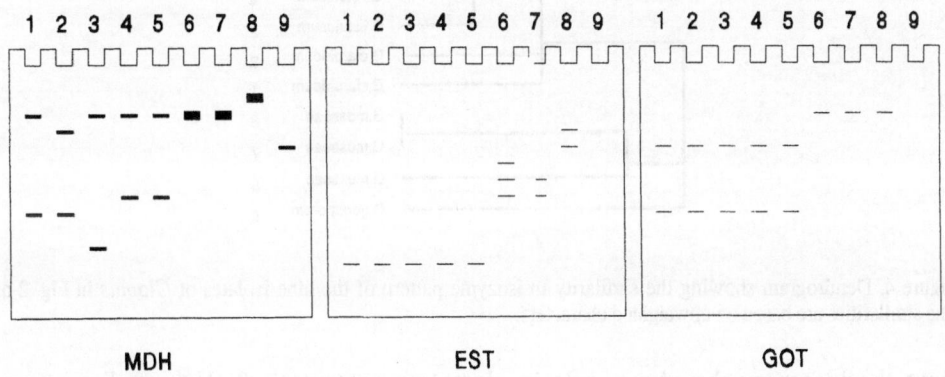

Figure 3. Nine isolates of *Glomus spp.* examined for isozyme pattern on gels stained for malate dehydrogenase (MDH), esterase (EST) and glutamate oxaloacetate transaminase (GOT). The lane numbers refer to the gel shown in Fig. 2.

shows a gel stained for malate dehydrogenase (MDH). Lane 1 and 2 are different strains of *G. claroideum*; 3, 4, and 5 are *G. fistulosum;* 6, 7, and 8 are *G. mosseae* cultures; and lane 9 is *G. geosporum.* The interpretation of the gel is shown in Fig. 3. The pattern of *G. fistulosum* and *G. claroideum* shows two MDH loci. Variation can be found within both loci, with the slowly migrating locus having two alleles, and the faster migrating locus having three. The banding patterns of glutamate oxaloacetate transaminase (GOT) and esterase (EST) are more simple because the loci are monomorphic.

The dendrogram (Fig. 4) shows an unweighted analysis of similarity between the fungi in Fig. 2 and 3. Variation can be seen within *G. fistulosum* and *G. claroideum*, but the two species can clearly be distinguished from *G. mosseae*. A genetic analysis in which the bands are interpreted as either alleles or loci, clearly distinguish the *G. fistulosum* and *G. claroideum* from *G. mosseae* and *G. geosporum*, as the isolates of *G. fistulosum* and *G. claroideum* studied share five enzyme loci, and none of these loci can be found in *G. mosseae* or *G. geosporum*. This should only be regarded as an example, as many more isolates and enzyme loci should be studied, before sound taxonomic conclusions can be drawn. The two taxonomic and systematic questions can be asked: 1) are *G. fistulosum* and *G. claroideum* synonyms?, 2) do *G. mosseae* and *G. fistulosum* with *G. claroideum* represent two independent phylogenetic lineages?

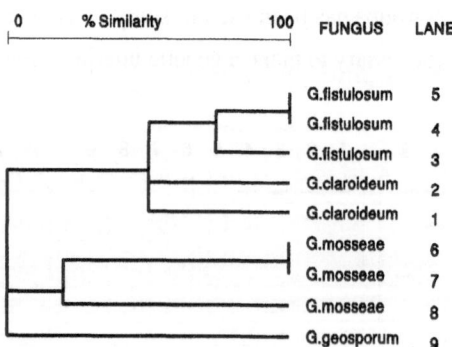

Figure 4. Dendrogram showing the similarity in isozyme pattern of the nine isolates of *Glomus* in Fig. 2 and 3. The similarities are based on unweighted characters.

Does the biogeography play a role in glomalean systematics? New studies of genetic similarities in *Glomus* have shown that based on similarity of putative loci, morphological species are remarkable similar throughout the world (Rosendahl, Dodd, Giovannetti and Walker, unpublished). This supports the hypothesis by Nicolson (1975) and others, that the glomalean fungi spread by continental drift.

Glomus clarum and *G. manihotis* have almost identical spore morphology, spore wall thicknesses and sizes, but differences are observed when MDH loci are compared (Fig. 5). Isolates of *G. manihotis* from the tropics: Brazil, Indonesia and the Philippines (lane 1 and 2), have identical banding patterns but differ from isolates of *G. clarum* coming from the USA and an unknown source (lane 3 and 4). Using genetic analysis the pattern for both groups can be interpreted as allelic variation within two loci. There is a strong case for amalgamating the two species based on morphology alone, but the possibility of keeping them separated based on biogeographical origin supported by evidence from isozyme studies needs further investigation. There are several other possible cases of synonymy of species based on close similarity of spore morphologies (e.g. *Acaulospora elegans* and *A. bireticulata*, and the small-spored *Acaulospora* spp. such as *A. morrowiae*, *A. delicata* and *A. longula*) (Dodd and Walker, unpublished.).

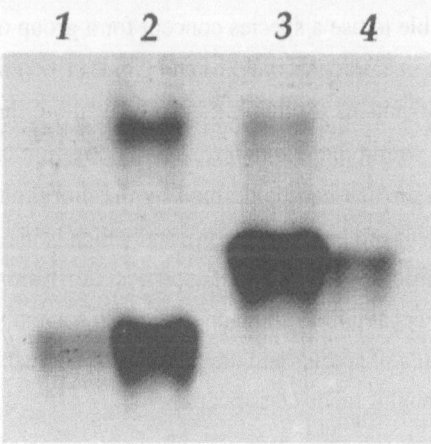

Figure 5. Gel stained for malate dehydrogenase (MDH): Lane 1 and 2 are *Glomus manihotis* and lane 3 and 4 are *G. clarum*.

The brown-spored species of *Gigaspora* differ mainly in minute details of spore ornamentation (Koske and Walker, 1985), but such characteristics may be environmentally plastic. Studies of closely similar morphological species complexes with isozyme techniques could help in clarifying the relationships among them.

Towards a species concept in *Glomales*

A species concept for the glomalean fungi has been proposed (Bentivenga and Morton, 1993) Although the definition includes several terms from evolutionary biology, the practical importance of the concept is limited. The concept is based on regarding the species as reproductively isolated individuals or populations of fungi which is obscure in mitotic fungi. Other attempts have raised the question whether the term species can be used for asexual organisms (Stuessy, 1992), but as probably the majority of fungi do not reproduce sexually (Reynolds and Taylor, 1993), and still maintain an integrity that characterise them as a taxonomic unit, the discussion is purely academic. Moreover, the genetic background for the reproductive barriers in sexually reproducing fungi, is different from what is found in other organisms (Rayner, 1991)

The ecological species concept (Van Valen, 1976) was not found adequate for AMF by Morton (1990) who claimed that the adaptive zone of all AMF is the same and thus without importance in systematics. However, it is known that though AMF may form the same structures inside and outside the roots, the function of the symbiosis may be different. Differences in translocation of phosphorus across the interface have been reported between *Scutellospora* and *Glomus* (Jakobsen et al., 1992). This may indicate that the fungi occupy different niches, and may also explain the co-existence of different morphological types in natural vegetation (Rosendahl et al., 1988).

In conclusion, it is possible to use a species concept for a group of organisms, that describes the variation and similarities in a practical way. Li and Graur (1991) offered a 'taxonomic species concept' as "... a group of organisms resembling each other more than they resemble any other group ...". We suggest that, within the *Glomales,* such a definition can usefully be phrased as a repeatedly recognisable fungus that can be defined by the morphological characteristics of its spores (allowing for differences of gene expression), and which is internally consistent if found in widely separated geographic locations. This species definition will not usually allow generalisations of biological behaviour to be made (though in some instances it might), but it does allow for the comparative study of species and the consequent gradual accumulation of knowledge that might later be used in a phylogenetic classification.

References

Berbee, M.L. and Taylor, J.W. (1993) Ascomycete Relationships: Dating the origin of asexual lineages with 18S ribosomal RNA gene sequence data. in: D.R. Reynolds and J.W. Taylor (eds) *The Fungal holomorph: Mitotic, Meiotic and Pleomorphic Speciation in Fungal Systematics.* CAB International, pp. 67–78.

Berch, S. M. and Koske, R. E. (1986) *Glomus pansihalos,* a new species in the *Endogonaceae, Zygomycetes. Mycologia* 78: 832-836.

Benjamin, R. K. (1979) *Zygomycetes* and their spores. in: B. Kendrik (ed) *The Whole Fungus.* Volume 2. National Museums of Canada, Ottawa. pp. 573- 621

Bentivenga, S. P. and Morton, J. B. (1993) Systematics of glomalean endomycorrhizal fungi: current views and future directions. in: F. Pfleger and R. L. Linderman (eds):*Mycorrhizae and Plant Health.* APS Press. St. Paul, MN.

Bonfante-Fasolo, P. and Grippiolo, R. (1984) Cytochemical and biochemical observations on the cell wall of the spores of *Glomus epigaeum. Protoplasma* 123: 140-151.

Bruns T. D., White T. J. and Taylor, J. W. (1991) Fungal molecular systematics. *Annu. Rev. Ecol. Sys.* 22: 525-564.

Bucholtz, F. (1912) Beiträge zur Kenntnis der Gattung *Endogone* Link. Beih. zum *Botan. Centr.* Abt. 29:147-225.

Gerdemann, J. W. and Trappe, J. M. (1974) The *Endogonaceae* in the Pacific Northwest. *Mycologia Memoir* 5: 1-76.

Gianinazzi-Pearson, V. and Gianinazzi, S. (1986) *Physiological and Genetical Aspects of Mycorrhizae.* INRA, Paris.

Harris, H. and Hopkinson, D. A. (1976) *Handbook of enzyme electrophoresis in human genetics* (with supplements). New York, Oxford American Publishing Co.

Hepper, C. M., Sen, R., Azcon-Aguilar, C. and Grace, C. (1988) Variation in certain isozymes amongst different geographical isolates of the vesicular-arbuscular mycorrhizal fungi *Glomus clarum, Glomus monosporum* and *Glomus mosseae. Soil Biol. and Biochem.* 20: 51-59.

Jakobsen, I., Abbott, L. K. and Robson, A. D. (1992) External hyphae of vesicular-arbuscular mycorrhizal fungi associated with *Trifolium subterraneum* L. *New Phytol.* 120; 371-380.

Koske, R. E. and Walker, C. (1985) Species of *Gigaspora* (*Endogonaceae*) with roughened outer walls. *Mycologia* 77: 702–720.

Li, W.-H. and Graur, D. (1991) *Fundamentals of Molecular Evolution.* Sinauer Associates Inc., Sunderland, Massachusetts.

Miller, A. S. and Jeffries, P. (1994) Ultrastructural observations and a computer model of the helicoidal appearance of the spore wall of *Glomus geosporum. Mycol. Res.* 98: 307-321.

Moreau, F. (1953) Les Champignons. Tome II. Systematique. *Encycl. Mycol.* 23: 941-2120.

Morton, J. B. (1986) Three new species of *Acaulospora* (*Endogonaceae*) from high aluminium, low pH soils in West Virginia. *Mycologia* 78: 641-648.

Morton, J. B. (1990) Species and clones of arbuscular mycorrhizal fungi (*Glomales, Zygomycetes*): Their role in macro- and microevolutionary processes. *Mycotaxon* 37: 493-515.

Morton, J. N. and Benny, G. L. (1990) Revised classification of arbuscular mycorrhizal fungi (*Zygomycetes*): A new order, *Glomales*, two new suborders, *Glomineae* and *Gigasporineae*, and two new families, *Acaulosporaceae*, and *Gigasporaceae*, with an emendation of *Glomaceae. Mycotaxon* 37: 471-491.

Mosse, B. (1953) Fructifications associated with mycorrhizal strawberry roots. *Nature* 171: 974.

Mosse, B. (1970) Honey coloured sessile *Endogone* spores. III. Wall structure. *Arch. Mikrobiol.* 74: 146–154.

Mosse, B. and Bowen, G. D. (1968) A key to the recognition of some *Endogone* spore types. *Trans. Br. Mycol. Soc.* 51: 469-483.

Nicolson, T. H. (1975) Evolution of vesicular-arbuscular mycorrhizas. in: F.E. Sanders, B. Mosse and P.B. Tinker (eds) *Endomycorrhizas*, Acad. Press, London, New York, pp. 25-34.

Rayner, A. D. M. (1991) The challenge of the individualistic mycelium. *Mycologia* 83: 48-71.

Reynolds, D. R. and Taylor, J. W. (eds.)(1993) *The fungal holomorph: Mitotic, meiotic and pleomorphic speciation in fungal systematics.* CAB International, UK., pp. 0-357

Rogers, J. S. (1986) Deriving phylogenetic trees from allele frequencies: A comparison of nine genetic distances. *Syst. Zool.* 35: 297-310.

Rosendahl, S. (1989) Comparisons of spore-cluster forming *Glomus* species (*Endogonaceae*) based on morphological characters and isoenzyme banding patterns. *O. Bot.* 100: 215-223.

Rosendahl, S., Rosendahl, C. N. and Søchting, U. (1988) Distribution of VA mycorrhizal endophytes amongst plant species from a Danish grassland community. *Agric. Ecosys. and Environ.* 29: 329-335.

Rosendahl, S., and Sen, R. (1992) Isozyme analysis of mycorrhizal fungi and their mycorrhizas. *Methods in Microbiology* 24: 169-194.

Sanders, F. E., Mosse, B. and Tinker, P. B. (1974) *Endomycorrhizas.* Academic Press, London.

Simon, L., Bousquet, J., Levesque, R. C. and Lalonde, M. (1993) Origin and diversification of endomycorrhizal fungi and coincidence with vascular land plants. *Nature* 363: 67-69.

Stuessy, T. F. (1992) The systematics of arbuscular mycorrhizal fungi in relation to current approaches to biological classification. *Mycorrhiza* 34: 667-677.

Thaxter, R. (1922) A revision of the *Endogonaceae. Proc. Amer. Acad. Arts. Sci.* 57: 291-351.

Tommerup, I. (1991) The vesicular arbuscular mycorrhizas. *Adv. Plant Pathology* 6: 81-91.

Walker, C. (1983) Taxonomic concepts in the *Endogonaceae*: spore wall characteristics in the species descriptions. *Mycotaxon* 18: 443-455.

12

Walker, C. (1986) Taxonomic concepts in the *Endogonaceae*: II. A fifth morphological wall type in Endogonaceous spores. *Mycotaxon* 25: 95-99.

Walker, C. (1992) Systematics and taxonomy of the arbuscular endomycorrhizal fungi (*Glomales*) - a possible way forward. *Agronomie* 12: 877-897.

Walker, C. and Trappe, J. M. (1993) Names and epithets in the *Glomales* and *Endogonales*. *Mycol. Res.* 97: 339- 344.

Wyss, P. and Bonfante, P. (1993) Amplification of genomic DNA of arbuscular-mycorrhizal (AM) fungi by PCR using short arbitrary primers. *Mycol. Res.* 97: 1351-1357.

Valen, van L. (1976) Ecological species multispecies and oaks. *Taxon* 25: 233-239

Impact of Arbuscular Mycorrhizas on
Sustainable Agriculture and Natural Ecosystems
S. Gianinazzi and H. Schüepp (eds.)
© 1994 Birkhäuser Verlag Basel/Switzerland

Biodiversity and characterization of arbuscular mycorrhizal fungi at the molecular level

D. van Tuinen, H. Dulieu, A. Zézé and V. Gianinazzi-Pearson

Laboratoire de Phytoparasitologie, INRA-CNRS, Station de Génétique et d'Amélioration des Plantes, INRA, BV 1540, F-21034 Dijon cédex, France

Introduction

Biodiversity within a biological group provides the basis for distinguishing members into genera and species according to taxonomic criteria, and between individuals within a species depending on more detailed differences at the genetic level. Diversity between species occurs after a genetic barrier has been created either by a geographic or genetic impedance of gene flow. Divergence can continue by nucleotide substitutions and by mutations in a broader sense (deletions, translocations, duplications), and resulting diversity can be evaluated at the molecular level and used as a phylogenetic character. Diversity at the subspecies level is a function of both mutation rates and gene flow between individuals.

Variability in arbuscular mycorrhizal fungi (AMF) can be appreciated at different levels (Giovannetti and Gianinazzi-Pearson, 1994), but up to date evaluation of their genetic diversity has relied mainly on morphological and structural features of asexual spores (see Rosendahl et al., this volume). Spore characteristics can, however, vary depending on environmental factors and on the physiological status of the fungus. Moreover, it is not possible to differentiate between AMF at the subspecies level, due to the general absence of sexual reproduction and to the recalcitrance of these fungi to grow in the absence of a host plant. For these different reasons, biochemical and molecular tools offer a valuable alternative to resolve taxonomic issues, and to address the question of biodiversity at different taxonomic levels.

Possible sources of genetic diversity in AMF

In the actual state of knowledge, AMF are considered to have originated between the Ordovician and Lower Devonian, 400-500 Myr ago, from a common ancestor in the Glomaceae (Berbee and Taylor, 1992; Pirozynski and Dalpé, 1989; Simon et al., 1993a). Divergence with evolution has led to what are presently recognized as three families, six genera and some 130 species in the order *Glomales* (Morton and Benny, 1990; Walker and Trappe, 1993). Apart from one report (Tommerup, 1988), sexual reproduction is not known for *Glomales* and there is presently no information concerning plasmogamy, karyogamy nor meiosis in these fungi. However, this does not necessarily mean that genetic flux between individuals of the same species is absent or limited, since sexuality or parasexuality could exist but be very simple and limited to the formation of a coenocytic mycelium containing nuclei of different origins. AMF have an aseptate, multinucleate mycelium which originates from an asexual, multinucleate spore. Saprophytic development of the fungus in soil is extremely limited, whereas it is abundant both within root tissues and soil once the symbiosis is established. Vegetative hyphal anastomosis occurs in saprophytic and symbiotic mycelium of AMF (Casana and Bonfante-Fasolo, 1982; Hepper and Mosse, 1975; Tommerup, 1988; unpublished observations) and provides opportunities for somatic reassortment of nuclei, also during growth in the root cortex. It has been suggested that hyphal incompatibility arising between isolates due to geographic distancing could be one mechanism contributing to localized genetic differences between AMF populations (Tommerup, 1988).

As no uninucleate stage is known for AMF, it would be reasonable to assume that hyphae harbour a heterogeneous population of nuclei which would be subject to random genetic drift. If there is no sexual or parasexual reproduction in AMF, variations due to spontaneous allelic mutations in the population of nuclei in any one species would tend to become homogenous during repeated divisions and sporulation. The latter could then be considered simply as a sort of 'packaging' of a sample of nuclei into a limited space. In this case, diversity would become important between species originating from different soils or geographical regions, since the species would have a clonal origin, with random divergence and no gene flow. If sexual or parasexual reproduction occurs, nucleotide substitution frequency will remain the same but exchange of genetic material will be possible, leading to an increasing diversity in genotypes within a same species or group.

In either situation, the theoretical models for origin and scattering of diversity can be based on those of population genetics (Spiess, 1989). In order to clone a unique genotype of the nuclear population, the genetic drift during sporulation has to be taken into consideration.

Estimates of numbers of nuclei in spores of different glomalean fungi give extremely high values of between one and several thousand per spore (see in Giovannetti and Gianinazzi-Pearson, 1994). However, it is not known whether nuclei within a spore originate from a small number which subsequently divides, or whether the entire population migrates directly from the vegetative mycelium. This is pertinent since the possibility of genetic drift would depend on the size of the sample of nuclei. Using a model genotype of one locus and two alleles, p and q respectively, the probability of the frequency of either p or q evolving towards 1 or 0 will be greater if the number of original nuclei in the spore is small. If there is such a random drift phenomenon with sporulation, several reinoculation cycles starting from a single spore will inevitably lead to cloning of one of the single nuclear types (p or q). If, in contrast, the population of nuclei in one spore represents the nuclear diversity of the mycelia from which they have all originated, repeated reinoculation with a single spore is likely to maintain diversity. The analysis of nucleotide sequence divergence through a number of generations starting from a single spore should make it possible to detect an eventual genetic drift and estimate its amplitude.

Estimations of DNA contents of nuclei indicate values from 0.25 pg for *Glomus versiforme* to 0.77 pg for *Gigaspora margarita* (Bianciotto and Bonfante, 1992), but there is no information as to whether the amplitude of such variation could be due to polyploidisation or to amplification of certain DNA sequences (repeated sequences) without changes in the overall information contained within the genome. Estimates of thermal reassociation curves of DNA that has been split and denatured could help to answer this question (Britten and Kohne, 1968).

The genetics of AMF may be relatively simple but it is difficult to analyse. The development of molecular approaches represents a powerful alternative towards understanding biodiversity in these unculturable organisms.

Analysis of diversity in AMF at the molecular level

The study of biodiversity requires tools which provide criteria for defining and resolving biological groups at different taxonomic levels, and different techniques can be applied to analyse genetic diversity depending on the level to be considered. In the case of anamorphic, mitotically reproducing fungi that do not undergo sexual reproduction, like AMF, molecular characters offer extremely interesting possibilities for addressing problems of diversity complexity and phylogenetic relationships.

Isozyme analysis Information on the genetic and nuclear condition of a fungal isolate can be obtained from isozyme analyses. Isozymes are proteins having the same enzymatic activity but coded by different alleles at a same genetic locus (allozymes) or by separate genetic loci. This gives rise to a different tertiary structure of the protein and consequently different electrophoretic mobilities in gels (Micales et al., 1986). Allozyme bands migrate closely together whilst isozymes encoded by different loci occur in different regions of gels.

A number of isozyme systems has been studied in AMF and it has been suggested from analyses of banding patterns that these fungi may be haploid (Rosendahl and Sen, 1992). As the banding patterns represent direct gene products, they can reveal genetic differences between closely related fungi. Isozyme analysis has been applied particularly to members of the genus *Glomus* and, as can be seen in Figure 1, variations in allelic and locus isozyme banding patterns occur between species and isolates. It is interesting to observe an apparent genetic diversity between isolates recognised as *G. mosseae* and of different geographical origin, as previously reported by Hepper et al. (1988a). However, the amplitude of isozyme variability depends very much on the enzyme in question (fig.1), so that unambiguous genetic analyses based on isozyme expression in fact requires testing of a large number of different enzyme systems. The usefulness of isozyme pattern variability as a taxonomic criterion in conjunction with morphological characters is discussed more fully elsewhere (Rosendahl and Sen, 1992; Rosendahl et al., this volume).

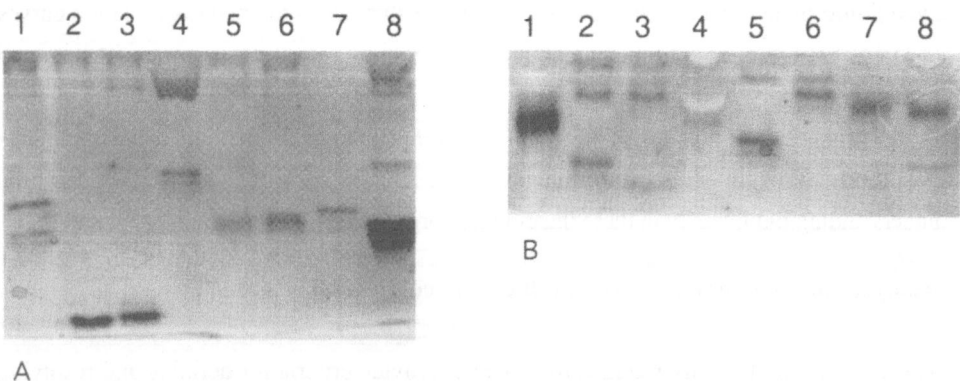

Figure 1. Esterase (A) and malate dehydrogenase (B) isozyme activities of different *Glomus* species. Extracts in lanes from 1) *G. intraradices*, 2) *G. fistulosum*, and 3) *G. claroideum* (DK, S. Rosendahl), 4) *G. geosporum* (UK, J. Dodd), 5) *G. coronatum* (I, M. Giovannetti), 6-7) *G. mosseae* (UK, Rothamsted), 8) *G. mosseae* (DK). Data kindly provided by S. Rosendahl.

Differential expression of alleles or loci at different stages of development or due to dissimilar nuclear conditions can give rise to variations in isozyme banding patterns. For example, Hepper et al. (1986) reported expression of a peptidase locus in symbiotic mycelium, but not in spores, of two *Glomus* species. Nevertheless, some diagnostic isozyme bands are stable under variable growth conditions which makes them potentially useful tools for monitoring otherwise morphologically-indistinguishable AMF in combined infections (Hepper et al., 1986, 1988b; Rosendahl et al., 1989)

Nucleic acid analysis. Methods for analysing diversity at the nucleic acid level essentially screen for substitutions, insertions, deletions or rearrangement of groups of nucleotides. Initially, the molecular techniques used were based on electrophoretic profiles of genomic DNA digested by restriction enzymes and revealed by hybridization with cloned DNA fragments (Restriction Fragment Length Polymorphism: RFLP). Research into diversity at the genomic level in AMF was, until recently, hampered by the problem of obtaining sufficient amounts of digestible DNA required for this type of analysis. However, this field of research is gaining interest with the development of new molecular techniques such as enzymatic amplification of DNA (Saiki et al., 1985; Mullis et al., 1986) or the polymerase chain reaction (PCR), which have completely modified nucleic acid analyses and studies of biodiversity (Gibbons, 1991). With this new technology, it has been possible to have access to the DNA molecule and to the genetic code of organisms for which only minute amounts of DNA are available, and even from fossil species.

Random Amplified DNA Polymorphism. Due to the development of PCR it is now possible to analyse and characterize a species at the DNA level starting from a small amount of material such as a single arbuscular mycorrhizal fungal spore. The Random Amplified Polymorphic DNA (RAPD) method uses short arbitrary primers to amplify DNA fragments by PCR and analyse them electrophoretically (Williams et al., 1990). Depending on the primer used, the banding pattern of DNA fragments varies and can be species-specific. Wyss and Bonfante (1993) used the RAPD approach to detect polymorphism between AMF. They showed that similarity in banding profiles obtained after RAPD amplification was greatest in spores of a same isolate and least between different species. The advantage of this method is that it does not require preliminary knowledge of a DNA sequence, as the DNA fragments are randomly amplified. This method has nevertheless several serious drawbacks. Since the primers used are not specific, the DNA present in any contaminating organism can lead to amplification of a DNA fragment and so give unspecific banding patterns. This is of particular concern for arbuscular mycorrhizal research, as the fungi cannot generally be produced axenically so that bacterial contamination is difficult to avoid. Furthermore, this method cannot be directly used to identify the fungi within roots because of interference by plant DNA. Nevertheless, recent reports have shown this method to be more sensitive than the use of isozymes (Wang et al., 1993) and it can lead to the isolation of specific DNA fragments for which corresponding primers can be generated. Such primers can then be used in combination with PCR-RFLP analysis of variability or, if species-specific, as probes to detect a given fungus.

PCR-RFLP One possible way to overcome unspecific amplification of DNA is to use the technique of PCR-RFLP. In this, a specific DNA fragment is digested, after PCR amplification, by several restriction enzymes. Variations in nucleotide sequences will lead to a different restriction pattern after electrophoresis of the digestion products, if the differences are located within the restriction site of an enzyme. Instead of randomly amplifying a fragment using one arbitrary primer as for RAPDs, two primers specifically flanking the DNA region of interest are used. This requires knowledge of the target DNA sequence, or at least of the primer regions, in order to synthesize corresponding primers. The choice of the target DNA will depend on the presence of conserved nucleotide sequences to allow binding of the primers, and of non-conserved sequences for the detection of variability between genomes. This feature is generally the property of multicopy genes.

One of the main targets for studies of variability and biodiversity in genomic studies are the ribosomal genes. These multicopy genes are made up of three coding regions of different sizes (18S, 5.8S and 25-28S), separated by two non-translated sequences (intra-genic transcribed spacers: ITS). The coding regions have been sufficiently conserved during evolution to enable the design of specific primers for the ribosomal gene. The ITS sequences, separating the 5.8S region from the 18S and 25-28S regions, are very variable and can be used to differentiate between closely related species (White et al., 1990). The ITS regions have been used to identify ectomycorrhizal fungi by PCR-RFLP (Gardes et al., 1991), and to design taxon-specific primers (Gardes and Bruns, 1993). The first gene of AMF to be sequenced was the complete region coding for 18S rRNA subunit from *Glomus intraradices* and *Gigaspora margarita* (Simon et al., 1992). By comparing the sequences obtained with known 18S rRNA sequences from other fungi, these authors were able to generate a primer specific for AMF. This primer simplified subsequent work as the problem of contamination by DNA from other organisms was eliminated. Using this primer in combination with universal primers, almost complete nucleotide sequences for the 18S gene were obtained for twelve different AMF (Simon et al., 1993a). Variability and similarity in these sequences were used to analyse phylogenetic relationships between the fungi, based on the fact that the rate of nucleotide substitution is correlated with divergence between species. The resulting phylogenetic tree was congruent with the classification of the AMF established on morphological diversity. It also concords with the hypothesis that the ancestral form of AMF was *Glomus*-like, originating with the appearance of vascular land plants and subsequently diversifying into the presently recognized different families of *Glomales*. More recently, the technique of Single Strand Conformation Polymorphism (SSCP) has been applied to AMF in combination with taxon-specific primers (Simon et al., 1993b). This technique not only opens the possibility of identifying AMF in root tissues but also of detecting more punctual variations in the genome, since it enables separation of DNA fragments which have undergone point mutations (Hayashi and Yandell, 1993).

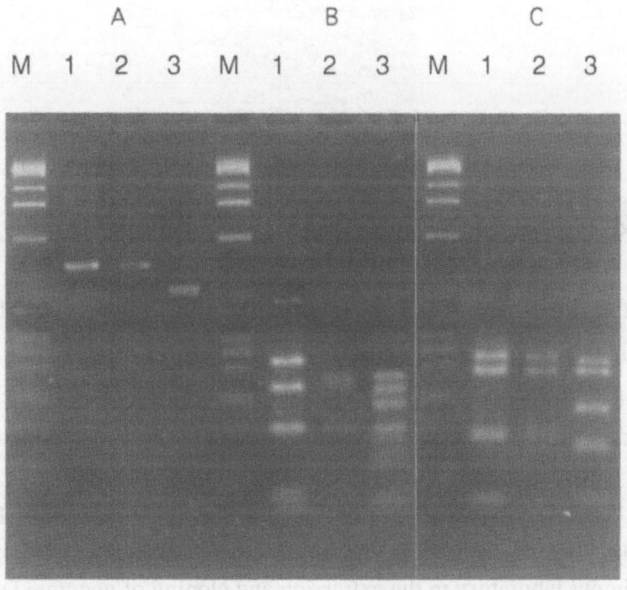

Figure 2. PCR-RFLP analyses of the 5' end of the large rRNA 28S gene subunit of (1) *Glomus mosseae* (F, C. Leyval), (2) *G. mosseae* (UK, Rothamsted) and (3) *G. vesiculiferum*. Fragments were obtained after no digestion (A) and after digestion by Hinf I (B) or RSA I (C). Molecular weight markers (1kb ladder) (M).

The large 25-28S rRNA subunit gene is also suitable for studies of variability and for phylogenetic comparisons. This gene possesses a common conserved core, which in eukaryotes is interspersed with 12 (labelled D1 to D12) divergent and more rapidly evolving domains (Hassouna et al., 1984; Michot and Bachellerie, 1984). The variability of several of these twelve domains has been used for phylogeny studies of *Fusarium* (Guadet et al., 1989). Starting from a single AMF spore, we have analysed by PCR-RFLP an amplified portion of the large subunit rRNA gene, covering the domains D1 and D2 located near the 5' end of the subunit, in two isolates of *Glomus mosseae* and in *G. vesiculiferum*. A variability in length of the amplified fragment was observed between *G. vesiculiferum* and *G. mosseae* (fig.2A). Digestion by different restriction enzymes gave banding patterns which clearly differentiated the two species (fig.2B and C), whereas the two *G. mosseae* isolates could be

20

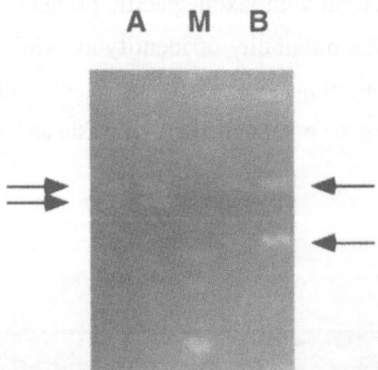

Figure 3. Two single spores of (A) *Glomus mosseae* (F, C. Leyval) and (B) *G. mosseae* (UK, Rothamsted) subjected to PCR-RFLP analyses as described in Figure 2.

distinguished after digestion of the amplified fragment by only one out of nine enzymes tested (fig.3). Although it is necessary to screen a large number of restriction enzymes, these preliminary results indicate that the 28S rRNA subunit gene may be sufficiently variable in AMF to permit differentiation between isolates of species defined on a morphological basis.

DNA cloning: Random cloning of genomic DNA fragments representing sequences of interest can be used to obtain probes to detect variability between genomes. Recent progress has been made in our laboratory in the extraction and cloning of genomic DNA from AMF using a relatively low number of spores (500-2000), and partial genomic libraries of two species have been established in plasmids or phages (Zézé et al., 1994; P. Franken unpublished results). Screening of a partial genomic library of *Scutellospora castanea* has indicated the existence of repeated DNA sequences in moderate or high copy number within the fungal genome. Species-specific and non-specific DNA cloned fragments have been identified and some sequenced with the aim of obtaining primers specific to *Scutellospora castanea* and primers with a broader specificity (unpublished results). The availability of taxon-specific primers will enable the detection of AMF directly in root extracts, whereas the non-specific primers will provide a means for detecting variability between AMF species by PCR-RFLP. For example, amplification products from different AMF obtained with a pair of 20mer primers designed from the sequence of a random clone from *S. castanea* and digested by the restriction enzyme MboI give a banding profile showing differences between *S. castanea* and the other species tested (fig.4). The use of partial genomic libraries to generate specific or non-specific primers or to analyse DNA sequence variability is not as straight forward as the techniques described above as it requires preliminary cloning, screening and sequencing clones of interest, but it does offer the possibility of exploring different parts of the fungal genome.

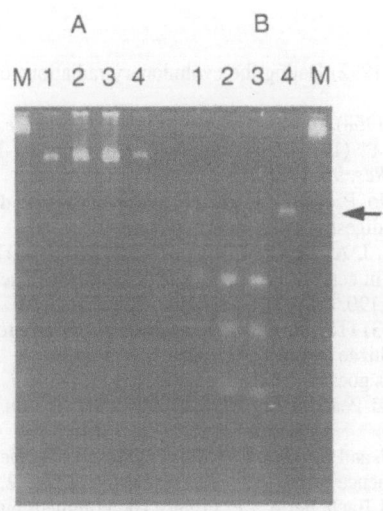

Figure 4. PCR-RFLP analyses after amplification with primers deduced from a clone of a partial genomic library of *Scutellospora castanea*. Fragments from the genomic DNA of *Acaulospora leavis* (1), *Gigaspora rosea* (2), *Glomus caledonicum* (3) and *Scutellospora castanea* (4) were obtained after no digestion [A] or digestion with MboI after amplification [B]. Molecular weight marker (100 bp ladder) (M).

Conclusions

In studies of biodiversity, no one character can give an accurate picture of the amplitude of variation within a given population or taxon, and for AMF there exists no one criterion for taxonomy and phylogeny that is infallible. Different types of characters have to be analysed separately then compared to each other, and it is necessary to be aware that whilst molecular data may be congruent with morphological characters, contradictions may well arise (Kohn, 1992). Furthermore, interpretations of diversity in molecular characters, and in particular DNA sequences, have to bear in mind that each analytical method is focused on a limited portion of the genome and that little is known about how different regions of the fungal genome evolve relative to each other, nor of the pressures that may be selectively affecting change. In conclusion, knowledge of the genetics and the genome complexity of AMF would greatly contribute to a fuller understanding of their biodiversity. In the absence of such information, molecular techniques provide the opportunity of better appreciating genomic variability and of evaluating the magnitude of diversity within this group of obligate symbionts.

Acknowledgement

The authors are grateful to S. Rosendahl and C. Leyval for access to unpublished results.

References

Berbee, M. and Taylor, J.W. (1992) Dating the evolutionary radiations of the true fungi. *Can. J. Bot.* 71: 1114-1127.

Britten, R.J. and Kohne, D.E. (1968) Repeated sequences in DNA. *Science*. 161: 529-540.

Bianciotto, V. and Bonfante, P. (1992) Quantification of the nuclear DNA content of two arbuscular mycorrhizal fungi. *Mycol. Res.* 96: 1071-1076.

Casana, M. and Bonfante-Fasolo, P. (1982) Ife intercellulari ed arbuscoli di *Glomus fasciculatum* (Thaxter) Gerd. et Trappe isolato con digestione enzimatica. *Allionia* 25: 17-25.

Gardes, M., White, T.J., Fortin, J. A., Bruns, T.D. and Taylor, J.W. (1991) Identification of indigenous and introduced symbiotic fungi in ectomycorrhizae by amplification of nuclear and mitochondrial ribosomal DNA. *Can. J. Bot.* 69: 180-190.

Gardes, M. and Bruns, D. (1993) ITS primers with enhanced specificity for basidiomycetes - application to the identification of mycorrhizae and rust. *Mol. Ecol.* 2: 113-118.

Gibbons, A. (1991) Systematics goes molecular. *Science* 251: 872-874.

Giovannetti, M. and Gianinazzi-Pearson, V. (1994) Biodiversity in arbuscular mycorrhizal fungi. *Mycol. Res.* 98 (in press).

Guadet, J., Julien, J., Lafay, J.F. and Brygoo, Y. (1989) Phylogeny of some *Fusarium* species, as determined by large-subunit rRNA sequence comparison. *J. Mol. Evol.* 6: 227-242.

Hassouna, N., Michot,. B. and Bachellerie, J.P. (1984) The complete nucleotide sequence of mouse 28S rRNA gene. Implication for the process of size increase of the large rRNA in higher eukaryotes. *Nucleic Acid Res.* 12: 3563-3583.

Hayashi, K. and Yandell, D.W. (1993) How sensitive is PCR-SSCP? *Hum. Mutat.* 2: 338-346.

Hepper, C.M. and Mosse, B. (1975) Techniques used to study the interaction between Endogone and plant roots. In: Sanders, F.E., Mosse, B. and Tinker, P.B. (eds) *Endomycorrhizas.* Academic Press, London, pp 65-75.

Hepper, C.M., Sen, R. and Maskall, C.S. (1986) Identification of vesicular-arbuscular mycorrhizal fungi in the roots of leek (*Allium porrum* L.) and maize (*Zea mays* L.) on the basis of enzyme mobility during polyacrylamide gel electrophoresis. *New Phytol.* 102: 529-539.

Hepper, C.M., Sen, R., Azcon-Aguilar, C. and Grace, C. (1988a) Variation in certain isozymes amongst different geographical isolates of the vesicular-arbuscular mycorrhizal fungi *Glomus monosporum* and *Glomus mosseae. Soil Biol. Bioch.* 20: 51-59.

Hepper, C.M., Azcon-Aguilar, C., Rosendahl, S. and Sen, R. (1988b) Competition between three species of *Glomus* used as spatially separated introduced and indigenous mycorrhizal inocula for leek (*Allium porrum* L.) *New Phytol.* 110: 207-215.

Kohn, L. (1992) Developing new characters for fungal systematics: an experimental approach for determining the rank of resolution. *Mycologia* 84: 139-153.

Micales, J.A., Bonde, M.R. and Peterson, G.L. (1986) The use of isozyme analysis in fungal taxonomy and genetics. *Mycotaxon* 27: 405-449.

Michot, B. and Bachellerie, J.P. (1984) Secondary structure of mouse 28S rRNA and general models for the folding of the large RNA in eukaryotes. *Nucleic Acid Res.* 12: 4259-4279.

Morton, J.B. and Benny, G.L. (1990) Revised classification of arbuscular mycorrhizal fungi (*Zygomycetes*): a new order, *Glomales*, two new suborders, *Glomineae* and *Gigasporineae*, and two new families, *Acaulosporaceae* and *Gigasporaceae*, with an emendation of *Glomaceae. Mycotaxon* 37: 471-491.

Mullis, K., Faloona, F., Schaarf, S., Saiki, R., Horn, G. and Erlich, H. (1986) Specific enzymatic amplification of DNA in vitro: the polymerase chain reaction. Cold Spring Harbor Symp.*Quant. Biol.* 51: 263-273.

Pirozynski, K.A. and Dalpé, Y. (1989) Geological history of the Glomaceae with the particular reference to mycorrhizal symbiosis. *Symbiosis* 7: 1-36.

Rosendahl, S., Sen, R., Hepper, C.M. and Azcon-Aguilar, C. (1989) Quantification of three vesicular-arbuscualr mycorrhizal fungi (*Glomus* spp) in the roots of leek (*Allium porrum*) on the basis of the activity of diagnostic enzymes after polyacrylamide gel electrophoresis. *Soil Biol. Bioch.* 21: 519-522.

Rosendahl, S. and Sen, R. (1992) Isozyme analysis of mycorrhizal fungi and their mycorrhiza. In: J R Norris, D J Read and A K Varma (eds) *Methods in Micobiology* 24. Academic Press, New York, pp 169-194.

Saiki, R.K., Scharf, S., Faloona, F., Mullis, K.B. and Horn, G.T. (1985) Enzymatic amplification of the beta-globin genomic sequences and restriction site analysis for the diagnosis of sickle cell anemia. *Science* 230: 1350-1354.

Simon, L., Lévesque, R.C. and Lalonde, M. (1992) Rapid quantification by PCR of endomycorrhizal fungi colonizing roots. *PCR Methods and Application* 2: 76-80.

Simon, L., Bousquet, J., Lévesque, R.C. and Lalonde, M. (1993a) Origin and diversification of endomycorrhizal fungi and coincidence with vascular land plants. *Nature* 363: 67-69.

Simon, L., Lévesque, R.C. and Lalonde, M. (1993b) Identification of endomycorrhizal fungi colonizing roots by fluorescent single-strand conformation polymorphism-polymerase chain reaction. *Appl. Environ. Microbiol.* 59: 4211-4215.

Spiess, E.B. (1989). Random genic changes in populations of limited size. In: *Genes in populations.* J. Whiley & Sons, New York, pp 329-360.

Tommerup, I.C. (1988) The vesicular-arbuscular mycorrhizas. *Adv. Plant Path.* 6: 81-91.

Walker, C. and Trappe, J.M. (1993) Names and Epithets in the *Glomales* and Endogonales. *Mycol. Res.* 97: 339-344.

Wang, G., Whittam, T.S., Berg, M. and Berg, D.E. (1993) RAPD (arbitrary primer) PCR is more sensitive than multilocus enzyme electrophoresis for distinguishing related bacterial strains. *Nucleic Acid Res.* 21: 5930-5935.

Williams, J.G.K., Kubelik, A.R., Livak, K.J., Rafalski, J.A. and Tinguey, S.V. (1990) DNA polymorphisms amplified by arbitrary primers are useful as genetic markers. *Nucleic Acid Res.* 18: 6531-6535.

White, T.J., Bruns, T., Lee, S. and Taylor, J. (1990) Amplification and direct sequencing of fungal ribosomal RNA genes for phylogenetics. In: Innis, M.A., Gelfand, D.H., Sninsky, J.J. and White, T.J. (eds) *PCR Protocols, a guide to methods and applications.* Academic Press, San Diego, pp 315-322.

Wyss, P. and Bonfante, P. (1993) Amplification of genomic DNA of arbuscular-mycorrhizal (AM) fungi by PCR using short arbitrary primers. *Mycol. Res.* 97: 1351-1357.

Zézé, A., Dulieu, H. and Gianinazzi-Pearson, V. (1994) DNA cloning and screening of a partial genomic library from an arbuscular mycorrhizal fungus, *Scutellospora castanea. Mycorrhiza* 4 (in press).

Impact of Arbuscular Mycorrhizas on
Sustainable Agriculture and Natural Ecosystems
S. Gianinazzi and H. Schüepp (eds.)
© 1994 Birkhäuser Verlag Basel/Switzerland

Characterization of arbuscular mycorrhizal fungi by immunochemical methods

A. Hahn, V. Gianinazzi-Pearson[1] and B. Hock

Technische Universität München (Weihenstephan), Lehrstuhl für Botanik, D-85350 Freising, Germany
[1]*Laboratoire de Phytoparasitologie, INRA-CNRS, Station de Génétique et d'Amélioration des Plantes, INRA, BV 1540, F-21034 Dijon cédex, France*

Introduction

The application of antibodies (Ab) as molecular probes is based upon their highly selective binding to corresponding antigens. There is a long tradition for the use of immunochemical techniques in microbiology, and although emphasis has been mainly placed on bacteria and viruses, there is an increasing number of examples where Ab have been raised against fungi (Hardham et al., 1986; Straker et al., 1989; Dewey, 1990; Breuil et al., 1992; De Ruiter et al., 1993). In the field of arbuscular mycorrhizal fungi (AMF) immunochemistry represents an extremely useful approach for research into systematic, ultrastructural and ecological aspects pertaining to the fungi forming these highly complex symbioses.

AMF are presently included in the order Glomales which encompasses the genera *Glomus*, *Sclerocystis* (Glomaceae), *Acaulospora*, *Entrophospora* (Acaulosporaceae), *Gigaspora* and *Scutellospora* (Gigasporaceae) (Morton, 1990). Regardless of one's proficiency in recognizing species through spore morphology, the process of identification of AMF in soils is difficult, especially when variations exist within an individual species (Morton, 1988). The microscopic appearance of hyphae is even less reliable as a diagnostic criterion; at best, the morphology of hyphae can be used to differentiate mycorrhizal from other fungi.

Identification of the fungal partner(s) of arbuscular mycorrhizal associations is important for a number of reasons. Commercial production of inoculum requires adequate quality control to assure purity and detection of contaminating fungi. Species, even strains, can differ in their growth promoting activity and ability to infect roots, so that competitive abilities of introduced versus indigenous fungi have to be known to guarantee successful controlled inoculation. Analysis of competition and displacement processes or of geographical and ecological specificity of AMF in plant communities necessitates detailed knowledge of species composition. However, the difficulty linked to identifying distinct fungal species or strains in soil or roots raises the need to develop sensitive detection techniques.

Immunochemistry provides elegant tools for these tasks. The crucial point is the availability of antibodies, which are provided by immunological means. Because of the selective binding of Ab to species-specific antigens, immunochemistry can contribute to the qualitative and quantitative differentiation of fungal isolates. The identification of immunologically specific molecules from AMF is important, not only for taxonomy, but also for the future production of specifically targeted antibodies against different AMF. Ab can also be used to characterize extra- and intracellular surfaces of the fungi, as well as cellular constituents, during interactions with host tissues. Detailed information about surface properties of hyphae and about specific modifications in fungal metabolism is essential in investigations of the host - symbiont dialogue. In this chapter, we briefly discuss immunological procedures to generate efficient probes and give examples of how these have been used for investigations of AMF and processes involved in fungal-plant interactions in symbiotic associations.

Immunochemistry

Immunochemical methods are based on the selective binding properties of antibodies (immunoglobulin glycoproteins) which are produced by a vertebrate in response to injection of a high-molecular-mass foreign substance (antigen). The basic structure of an immunoglobulin, which is made up of a pair of identical heavy and light polypeptide chains held together by disulfide bonds, consists of two main regions named F_{ab} and F_c. (For more details see textbooks on immunology, e.g. Roitt et al., 1991). The F_{ab} region contains the highly variable antibody binding site capable of recognizing and binding to a restricted part of an antigen, the antigenic determinant or epitope. Although there are five distinct classes of antibodies in most higher mammals, immunochemical techniques usually rely upon immunoglobulins of the G class (IgG).

In principle, molecules with a mass of more than 5000 can act as antigens and these are usually proteins, glycoproteins, polysaccharides, lipopolysaccharides or lipids (Mernaugh et al., 1990). The usefulness of an antigen for immunochemistry depends on its biochemical properties, immunogenicity, solubility and stability as well as its location and abundance in the organism. Every antigen has a particular set of epitopes, some of which are shared with other antigens (generic or public epitopes) and some that are not (type-specific or private epitopes). In most instances, an organism will produce more than one Ab per epitope, each differing in affinity towards the epitope. Different systems are employed to screen and estimate specificity of antibodies. Immunofluorescence (IF), immunogold or immunogold-silver labelling, dot immunobinding assays (DIBA) and agglutination tests are used for structural antigens, while precipitation tests, enzyme-linked immunosorbent assays (ELISA) or immunodiffusion tests are more suitable for analysis of soluble antigens.

Crude preparations can be used for the production of antisera but these will yield polyclonal antibodies (pAb) that recognize many different antigenic determinants and require considerable effort to be rendered monospecific. Satisfactory specificities usually require the use of purified cell fractions (proteins, enzymes, carbohydrates) or molecules. In contrast, crude preparations can be used in a monoclonal approach to select for one specific single determinant. However, when monoclonal antibodies (mAb) are raised against crude preparations, it must be kept in mind that strong immune responses towards less specific (shared) epitopes can superimpose a weak response towards a highly specific (private) epitope. In all cases, antigenicity can be increased by chemical fixation (e.g. in polyacrylamide) or by mixing it with an adjuvant before immunization.

There is no consensus about the best source of fungal antigens in mycorrhizal research since nothing is yet known about the nature of species-specific determinants. For identification purposes, for example, the main goal is to narrow the choice to species- or isolate-determining molecules and raise homologous Ab against the relevant antigen. No data have been published yet on the influence of different immunization procedures on the nature of the Ab obtained when AMF antigen is used. It is evident that crushing or homogenizing fungal matter before immunizing will lead to Ab reacting with cytoplasmic, wall or membrane-associated antigens. Consequently, to obtain Ab targeted uniquely against surface antigens of AMF, for example, uncrushed spores have to be used as immunogen.

An alternative approach is to prepare antigens from a heterologous source, as long as homology with AMF antigens is great enough. Since AMF are unculturable organisms, such well-defined heterologous probes (from other organisms) are proving useful tools for investigating molecular modifications in the fungi during symbiotic interactions.

Polyclonal versus monoclonal antibodies

Conventional antisera, obtained by injection of antigen into a vertebrate (e.g. rabbit, sheep, goat, horse) and subsequent bleeding, are a heterogeneous mixture of Ab because they are polyclonal due to the fact that multiple epitopes on the antigen induce the proliferation of several responding clones of Ab-producing cells. Consequently, although an antiserum is unique in its specific composition of polyclonal antibodies (pAb) and therefore in binding properties and performance, different antisera (even from the same animal) must be separately assessed for their suitability in any particular immunochemical assay.

Hybridoma technology has led to the production of monoclonal antibodies (mAb) which are directed against one single epitope and characterized by a uniform specificity and affinity towards an antigen. In this way, mAb identify single molecules as well individual regions in a same molecule. The procedure for mAb production, which is well standardized, involves fusion of antibody-secreting cells (spleen or lymphonodes) from the immunized animal (usually rat or mouse) with a suitable proliferating murine myeloma cell line (Köhler and Milstein, 1975;Harlow and Lane, 1988). Hybridoma cultures secreting Ab with the required specification are subsequently isolated and cloned to give individual cell lines which incessantly produce a single type of antibody and which may be stored for extensive periods of time in liquid nitrogen. Crude supernatants of cell cultures may be used for assays but purification of the mAb is preferable for longer storage and standardisation.

Crude antiserum or mAb can be used but it is preferable to purify the immunoglobulins by affinity chromatography with protein A (Harlow and Lane, 1988). Specific pAb can be further purified by affinity chromatography against the antigen when this is available in sufficient amounts. The serum is passed over the antigen bound to a solid matrix, which retains the specific Ab. After washing to remove unspecifically bound Ab, the pure specific Ab can be eluted from the matrix. To our knowledge, this has not been attempted yet with antigens of AMF, presumably due to the lack of sufficient amounts of pure antigen. Both pAb and mAb have advantages and disadvantages and these are summarized in Table I:

Table I: Comparison of the monoclonal and the polyclonal approach for the production of Ab against AMF antigens.

Ab properties	polyclonal Ab	monoclonal Ab
supply	limited and variable	unlimited production possible
uniformity	changing properties with different sera and bleedings	constant properties
classes and subclasses	typical spectrum	one defined isotype
affinity	mixture of Ab with different affinities	uniformly high or low, can be selected by test format
cross-reactivities	occur as result of different specificities and low affinity interactions	depending on individual Ab
demands on the antigen	high purity required for specific antisera	impure antigens or mixtures can be used
taxonomic specificity	not below species level	strain/isolate level
technology	easy	difficult
cost	low	high

Advantages of pAb include their less expensive, easy production, a possibly higher titer and higher affinity for most antigens, and consequently a stronger signal than mAb due to the greater number of antigens recognized (Fig. 1). The problems encountered with pAb are linked in particular to the possibility of cross-reactivities due to their variable specificity or low affinity. This is illustrated in Fig. 1 which presents a model with AMF spores serving as antigen A and carrying four different types of antigenic determinants. If another antigen (B) is probed, which carries a common antigenic determinant, the antiserum will react with antigen B because it contains antibodies toward the common antigen. This is the basis of cross-reactivity due to sharing of determinants by two different antigens (X1). A different kind of cross-reactivity can exist if the properties of different antibody binding sites are sufficiently similar, though not identical, and if the selectivity of an antibody is not absolute (X2). Although cross-reactivities are prohibitive to single species identification, both types of cross-reactivity may be used for serogrouping.

The problems raised by pAb can be overcome by the use of hybridoma technology. The benefits derived from this technology include the virtually unlimited supply of specified Ab, fewer problems with unspecific binding and a system that, within limits, allows a selective screening for those mAb that have the desired specificities and affinities towards a given antigen. However, as already pointed out, the strength of the signal obtained with mAb is usually lower than with pAb, due to the higher frequency of cross-reacting epitopes recognized by pAb.

Polyclonal and monoclonal antibodies have been raised against spore material (both soluble fractions and wall epitopes) and extraradical mycelium of a few species of AMF. The Ab are generally checked for cross-reactivity against species of fungi from the Glomales and other fungal taxa using immunodiffusion, ELISA, DIBA and immunocytochemical techniques. The latter are the most discriminating since antigen concentration exposed to the Ab is not defined. Aspecificity of pAb has been their main drawback in immunochemical investigations of AMF. However, a considerable improvement in their specificity has recently been obtained by using a purified precipitate of the soluble protein fraction from spores as immunogen (Cordier et al., 1994). This gave antibodies that only recognize *Gigaspora* and that discriminate between species in this genus. The problem associated with mAb appears to be the relatively low chances of raising them against AMF antigens. Although a number of mAb have been obtained by this method (Hahn et al., 1993, 1994), low fusion rates of spleen cells and myelomas have been experienced, and low percentages (<20%) of Ab-producing hybridomas obtained. Out of six fusions of murine spleen cells immunized with different isolates of *Glomus*, only four cell lines of hybridomas expressed mAb with a variety of specificities against the antigens. Similarly, Wright et al. (1987) obtained only two mAb-secreting cell lines from two spleens using soluble AMF antigen. This difficulty in obtaining mAb has been confirmed for non-mycorrhizal fungal antigens. For example, out of

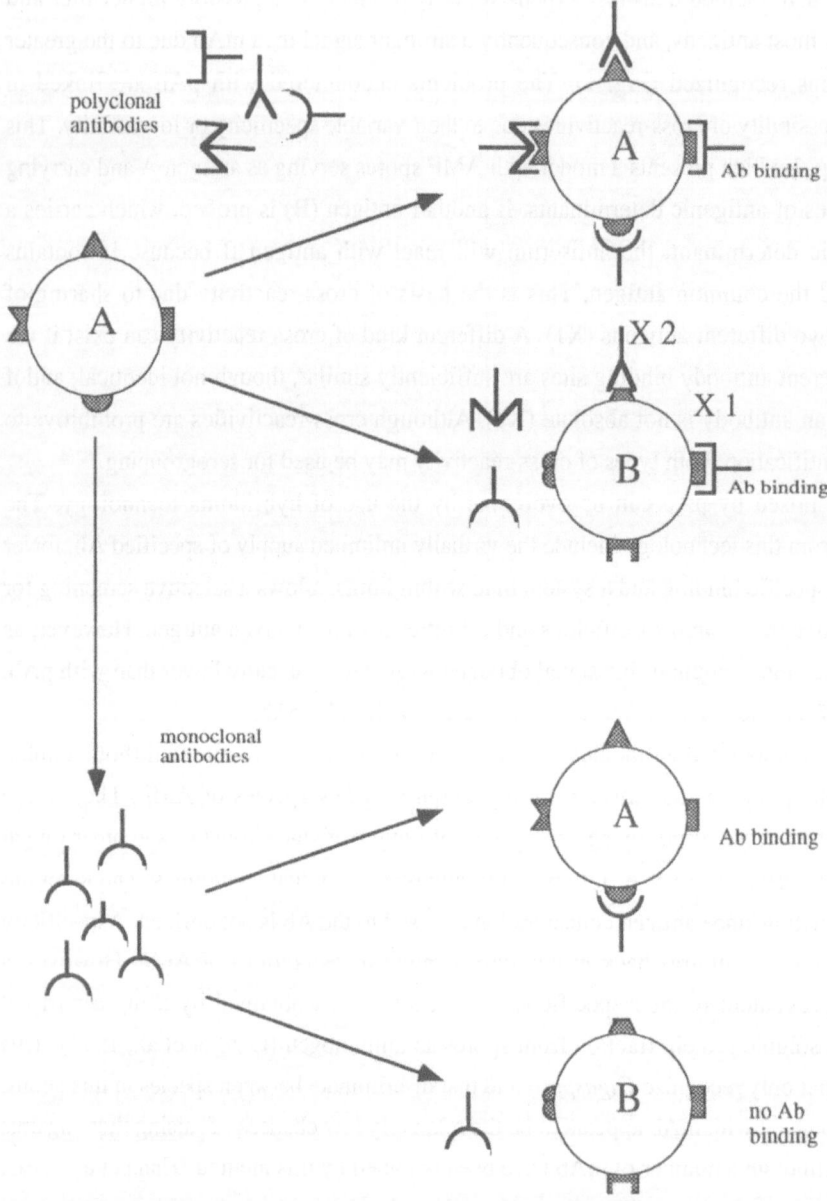

polyclonal
antibodies

A

monoclonal
antibodies

A   Ab binding

X 2

X 1

B   Ab binding

A   Ab binding

B   no Ab
binding

Figure 1: Immunolabelling of AMF spores with polyclonal and monoclonal Ab; Explanation of cross reactivities (modified from Hahn et al. 1994); X 1: cross reaction due to shared antigens; X 2: cross reaction due to lack of specificity.

16 fusions of murine spleens immunized with the pathogen *Pseudocercosporella herpotrichoides*, only three cell lines could be derived that expressed mAb which specifically recognized the antigen (Dewey, 1990).

Serogrouping of AMF

In all cases where pAb have been raised against AMF antigens, some cross-reactivity has been reported. However, a careful selection of cross-reacting pAb may be used for serogrouping of organisms. Although it is clear that a great number of Ab with a wide spectrum of cross-reactivities is necessary to draw hard and fast conclusions, we have attempted to apply this to fungi within the Glomales using published results obtained with polyclonal antisera (Table II).

Table II: Survey of reactivity of different polyclonal Ab raised against AMF. Test systems used included ELISA, DIBA and IF; for comparison, all reactions are expressed as: - no reaction; (+) ambiguous reaction; + strong reaction.

Reference[a]	2)	2)	3)	3)	3)	4)	5)	5)	3)	3)	3)	1)
Acaulospora laevis	+	(+)	+	+	+		(+)	+	-	-	-	-
Gigaspora calospora	+	+	+	+	+				-	-	-	-
Gi. margarita			+	+	+	(+)*	+	(+)	-	-	-	
Gi. sp. (WUM 6)	+	+										
Glomus caledonium	(+)								+	+	+	
G. monosporum	(+)											
G. etunicatum					-							
G. intraradices						-						
G. mosseae									+	+	+	+
G. clarum			-	+	-				+		+	
G. fasciculatum				-	+				+	+	+	
G. macrocarpum			-	-	-				+	+	+	
G. tenue				+	-				-		-	
G. epigaeum (versiforme)												+
G. deserticola												+
Sclerocystis dussii			-	-	-				+	+	+	
Scutellospora sp.						-	-					
non-mycorrhizal fungi	-	-	-						-	-	-	-

[a]1) Kough et al., 1983; 2) Wilson et al., 1983; 3) Aldwell et al., 1985; 4) Friese and Allen, 1991; 5) Sanders et al., 1992.
*) strong reaction with two of three isolates tested.

The antisera produced by Kough et al. (1983) were directed against extracted cell wall preparations of spores of *Glomus versiforme*, then termed *G. epigaeum*. The resulting pAb showed inter-species cross-reactions to spore surfaces of two other species, *G. mosseae* and *G. deserticola*, but no cross-reactions with other Glomales or non-mycorrhizal fungi in immunofluorescence tests. Wilson et al. (1983) developed an antiserum against cell wall material from the germinating hyphae of a *Gigaspora* species. In this case, immunofluorescent reactions with non-mycorrhizal fungi occurred which could be avoided by diluting the antiserum. Cross-reactions were strongest with *Acaulospora laevis* and *Gigaspora calospora*, while reactions with two isolates of *Glomus* were negligible. The cross-reaction with *Acaulospora* was diminished by immunodepletion of the antiserum with *Acaulospora* antigen. Strongest signals with *Gigaspora* were detected on the inside of the hyphae suggesting that the majority of the antigenic components were localized there. Similar observations have been made with mAb against spores of AMF (Hahn et al., 1993), suggesting a greater immunogenicity of cell components associated with inner wall or outer membrane structures. Aldwell et al. (1985) confirmed these close immunochemical similarities between the genera *Glomus* and *Sclerocystis* on the one hand and *Acaulospora* and *Gigaspora* on the other, by ELISA of six polyclonal antisera raised against different Glomales. More recently, Sanders et al. (1992) attempted to identify the antigenic components responsible for the serological affinity between *A. laevis* and *Gigaspora margarita*, previously observed by Aldwell et al. (1983). They showed that this was related to the presence of glycoconjugates in the antigen preparation from the soluble fraction of spore extracts. The serogrouping illustrated in Table II concords with results from 18S rDNA sequence data analysis, and which suggest that *Acaulospora* and *Entrophospora* are sister groups of *Scutellospora* and *Gigaspora*, with *Glomus* as an outgroup (Simon et al., 1993). However, it is incongruent with analyses of structural characters and cell wall chemistry, which indicate that phylogenetic relationships between the Acaulosporaceae and Gigasporaceae are not so close (Morton, 1990; Gianinazzi-Pearson et al., 1994), underlying the complexity of criteria necessary in resolving the taxonomic issues for these fungi.

Immunodetection and identification of AMF in soil and roots

Field studies often require information on the amount of fungal inoculum in soil, its competition with the indigenous microorganisms and its ability to colonize host plants. The selectivity of Ab can be utilized to assay for AMF in soil and roots. However, immunochemical quantification in complex systems like soil is severely hampered by cross-reactivities or unspecific binding to soil particles. Friese and Allen (1991) using antisera against *Gigaspora margarita* coupled to a fluorescent stain were able to differentiate the spores of the homologous isolate from those of other genera of AMF and even from some heterologous isolates of *Gigaspora margarita* (cf. Table II).

Immunodetection was recently employed by Weinbaum et al. (1992) to track *Acaulospora* and *Scutellospora* species under field conditions. They were able to show that these vary in different regions during one growing season. This approach opens the possibilty of following the survival of introduced AMF in the field in competition with indigenous populations.

Important modifications occur in the structure and macromolecular organization of AMF walls between extraradical and intraradical hyphae (Bonfante-Fasolo et al., 1990) and these must be accompanied by changes in the nature and properties of surface molecules. Consequently, pAb obtained against surface components will probably be less adapted to detecting the fungi within root tissues. Aldwell and Hall (1986) used pAb raised against whole extraradical hyphae and indirect ELISA to monitor competition between two genera of AMF in double-inoculation green-house experiments. They were able to conclude that *G. mosseae* was more competitive than *A. laevis* in colonizing *Trifolium repens* and in displacing populations of other inoculated AMF. Soluble spore components as immunogens have proved more successful for direct immunolocalization of fungi in roots, where antigen frequency cannot be controlled to increase Ab specificity. pAb obtained in this way have been used to serologically detect and localize mycelium of *Gigaspora margarita* in roots by immunocytochemistry (Gianinazzi and Gianinazzi-Pearson, 1992). Since the corresponding antigens were only localized within metabolically active hyphae, it was also possible to use Ab to assess the presence of living infection by ELISA (Ravolanirina and Gianinazzi-Pearson, 1992). However, Ab did not specifically discriminate between several

Table III: Cross-reactions of mAb against spores of different isolates of *Glomus*; + strong reaction; (+) ambiguous reaction; - no reaction.

mAb (immunogen)	A5B1 (*G. etunicatum*)	H6A12 (*G. mosseae*)	D12F11 (*G. scintillans*)
G. etunicatum S329	+	-	-
G. etunicatum Whs	+	-	-
G. fasciculatum	-	-	-
G. globisporum S328	-	-	-
G. intraradices	-	-	-
G. macrocarpum	-	-	+
G. mosseae I	-	+	+
G. mosseae Whs	-	+	(+)
G. scintillans	-	-	+
non-mycorrh. organisms (8)	-	-	-

34

Glomales and, as already mentioned, specificity has since been considerably improved by targeting protein antigens (Cordier et al., 1994). For example, pAb have now been obtained against *Gigaspora rosea* which not only distinguish living fungal hyphae (Fig. 3) but also between hyphae of *Gigaspora* species and other AMF that may be present within roots (unpublished results). It is now possible to envisage the identification and purification of species-specific protein immunogens, for example by Western blot analyses, in order to raise monospecific pAb. The number of mAb raised against AMF antigens is limited. Table III gives an overview of the reactivity of mAb raised against three different *Glomus* species by Hahn et al. (1993).

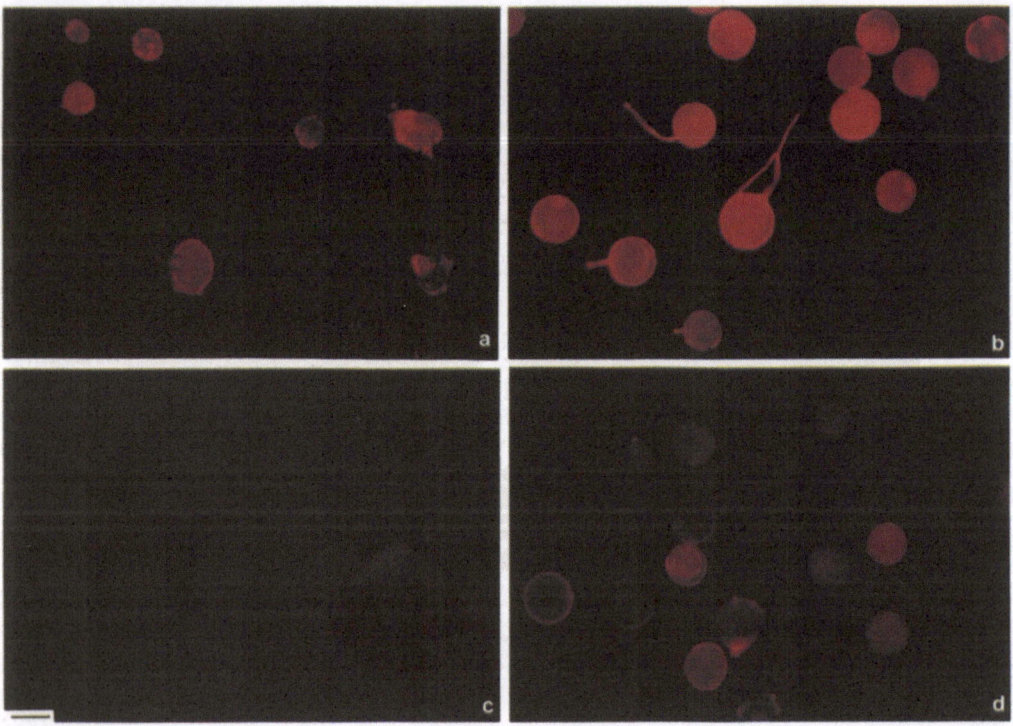

Figure 2: Immunofluorescent labelling of spores of *Glomus* species: a) *G. etunicatum* Whs; b) *G. etunicatum* isolate S329; c) *G. mosseae*; d) *G. etunicatum* isolate S329. a-c) The mAb A5B1 as primary Ab selectively labels *G. etunicatum*, but not *G. mosseae*; d) unspecific murine IgG as primary Ab (control) yields a low fluorescence signal. scale bar = 100 um.

Immunofluorescence of fresh spores and immunogold-silver staining of semi-thin sections of spore walls were used to screen for interesting mAb. One mAb raised against surface antigens of an isolate of *Glomus etunicatum* was able to selectively bind to spore walls of its homologous antigen from two different sources while six other species of *Glomus* were not labelled. The immunofluorescence photographs in figure 2 illustrate the specificity of this mAb which only gives a fluorescent signal with its homologous species.

Employing hybridoma technology, Wright et al. (1987) were able to produce mAb against *G. occultum* that distinguished by ELISA this species from 15 other species of *Glomus* and 29 species of AMF in total. The epitope recognized by the mAb was a soluble protein associated with the cell membrane or some particulate fractions in spores and hyphae, again underlying the interest of this type of cell component for specificity. The mAb could differentiate in the intensity of their reaction among isolates of *G. occultum* from various different sources, an observation which has been confirmed by Oramas-Shirey and Morton (1990). This indicates that antigenic properties may even vary at the strain level, resulting either from different epitope densities in the spores or from a conformational change altering the affinity of the mAb for the antigen. Wright and Morton (1989) used a modified DIBA technique to specifically stain fungal matter with these mAb on a nitrocellulose "print" of the crushed root. In this way, *G. occultum*, which is not stainable with conventional dyes, could be selectively detected and quantified in whole root systems.

Immunocytochemical characterization of structural and functional cell components in AMF

Monoclonal and polyclonal antibodies can be used to determine modifications in fungal cell components related to infection development and activity. This involves the immunocytochemical localization of the corresponding antigen(s) at the cellular level by electron microscopy (Gianinazzi and Gianinazzi-Pearson, 1992). This type of immunoanalysis of AMF is for the moment limited to the use of heterologous antibodies, since purification of sufficiently high amounts of homologous antigen is not presently feasible for these unculturable organisms. Using commercially available pAb and mAb probes for such an approach, Gianinazzi-Pearson et al. (1994) detected ß(1-3) glucan polymers in inner spore and hyphal walls of fungi belonging to the *Glomaceae* (Figure 4) and *Acaulosporaceae*, which is an unusual feature for *Zygomycetes*. In contrast, these structural polysaccharides were not found in members of the *Gigasporineae*. More detailed studies of *G. mosseae* and *A. laevis* during mycorrhizal development have shown that immunolabelling frequency for ß(1-3) glucans progressively decreases as hyphae colonize host tissues, and that no antigen can be detected in walls of arbuscules (Lemoine, Gollotte and Gianinazzi-Pearson, unpublished results). These observations provide further evidence of the molecular simplification of fungal walls with arbuscular mycorrhizal establishment, previously suggested from biochemical and cytochemical analyses (Bonfante-Fasolo et al., 1990).

Heterologous antibodies are also useful for physiological studies, and especially for determining relationships between synthesis and activity of molecules. They have recently been used to study vacuolar alkaline phosphatase which is a characteristic enzyme of AMF (Gianinazzi and Gianinazzi 1994). The active enzyme is virtually abent from germ tubes growing out from

36

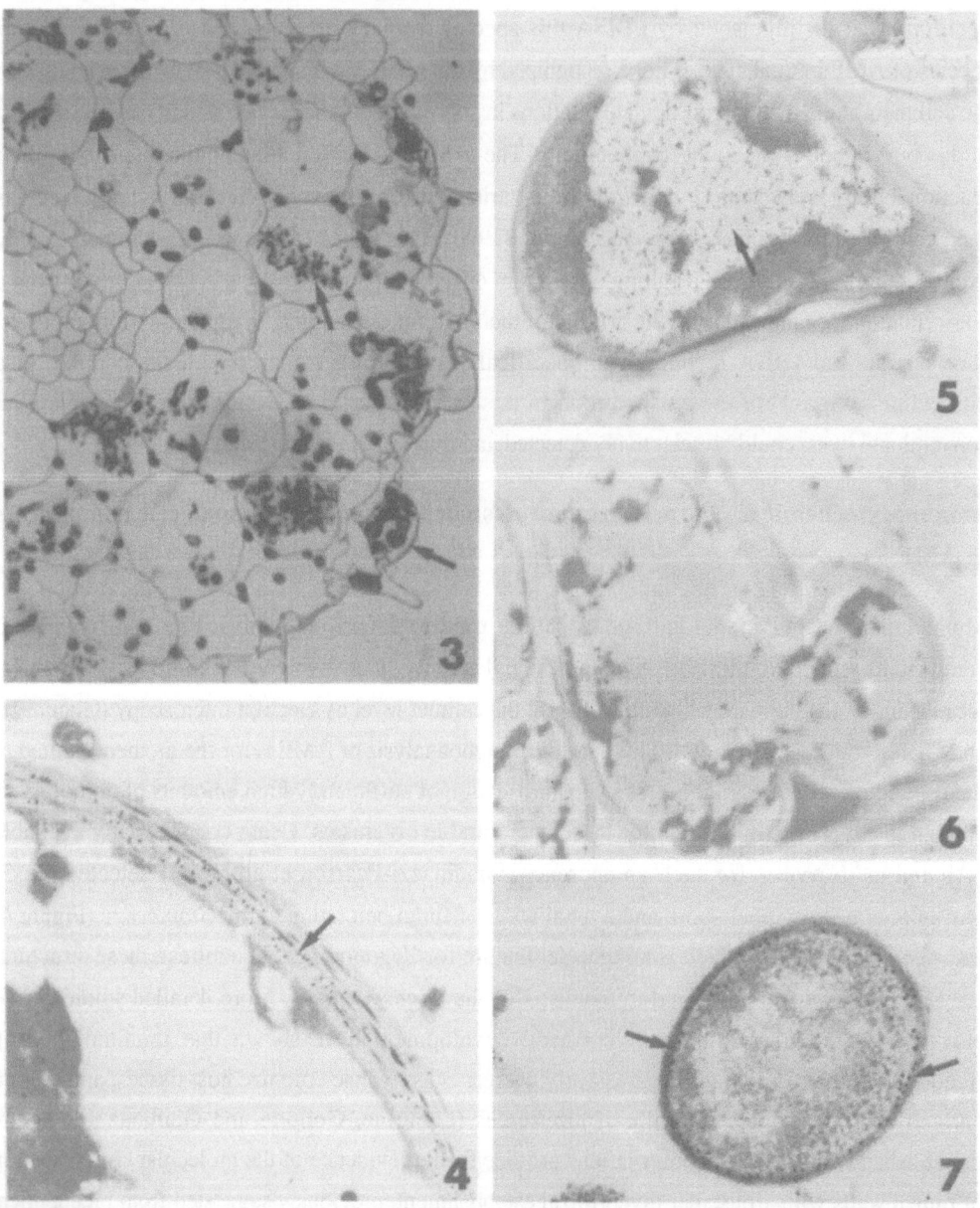

Figures 3 to 7: Immunogold localisation of AMF antigens: 3) detection of *Gigaspora rosea* hyphae in leek roots after silver enhancement, 4) ß(1-3) glucans in the wall of an extraradical hypha, 5) vacuolar alkaline phosphatase in an active intercellular hypha and 6) insignificant detection in a senescing hypha, 7) plasmamembrane ATPase in a very young arbuscule branch.

spores and is rare in abortive entry points formed on myc⁻ resistant roots. It can, on the contrary be detected throughout the living fungal mycelium as infection develops (Tisserant et al., 1993), suggesting that it is influenced by host factors. In order to determine whether *de novo* enzyme synthesis and/or activation is responsible for this change in enzyme behaviour, heterologous pAb raised against *E. coli* alkaline phosphatase have been used to localize the enzyme in AMF (Tisserant 1993). Alkaline phosphatases are enzymes with highly conserved domains and immunochemical tests like DIBA showed that pAb detected antigens only in mycorrhizal roots or enzymically extracted mycelium. The pAb localize antigenic determinants specifically within the fungal vacuole of metabolically active intraradical mycelium (Fig. 5), that is the site of enzyme activity, but not in ageing hyphae (Fig. 6). However, practically no antigen can be detected in vacuoles of germ tubes, suggesting that the host somehow induces increased synthesis of AMF alkaline phosphatase which is then activated in the symbiosis. Such an approach could be applied to other fungal enzyme systems; for example, plasmamembrane ATPase activity only occurs at tips of germ tubes but the enzyme is active in both intercellular and arbuscular hypae in roots (see Gianinazzi-Pearson and Gianinazzi, 1994). Preliminary studies indicate that yeast ATPase pAb cross-react with the AMF enzyme (Fig. 7).

Conclusions

Both pAb and mAb raised against surface antigens or soluble components of spores of *Glomales* can provide powerful tools for investigating taxonomic, ecological and physiological aspects of AMF alone or in mycorrhizal symbiosis. In the first two instances, immunochemical data may be used in two ways: either for identification of isolates which requires highly specific Ab, or to establish relationships between species by serogrouping using Ab with a wide spectrum of crossreactivities. It is possible to increase antibody specificity to AMF by identifying antigens responsible for cross-reactions, purifying specific antigens or by the use of well-defined cell fractions such as soluble proteins or carbohydrates. An alternative to the preparation of antibodies from target fungal antigens is the use of heterologous antibodies, as long as they are well characterized and their homology is great enough. These are opening the possibility of identifying cellular modifications and investigating protein expression of AMF that are linked to symbiotic interactions and mycorrhiza functioning.

Acknowledgements
We thank C. Cordier, M.C. Lemoine, A. Gollotte and B. Tisserant for access to unpublished data, and T. Giersch for valuable discussions.

38

References

Aldwell, F. E. B. and Hall, I. R. (1986) Monitoring spread of *Glomus mosseae* through soil infested with *Acaulospora laevis* using serological and morphological techniques. *Trans. Br. Mycol. Soc.* 87: 131-134.

Aldwell, F. E. B., Hall, I. R. and Smith, J. M. B. (1983) Enzyme-linked immunosorbent assay (ELISA) to identify endomycorrhizal fungi. *Soil Biol. Biochem.*. 15: 377-378.

Aldwell, F. E. B., Hall, I. R. and Smith, J. M. B. (1985) Enzyme-linked immunosorbent assay as an aid to taxonomy of the *Endogonaceae. Trans. Br. Mycol. Soc.* 84: 399-402.

Bonfante-Fasolo, P., Faccio, A., Perotto, S. and Schubert, A. (1990) Correlation between chitin distribution and cell wall morphology in the mycorrhizal fungus *Glomus versiforme. Mycol. Res.* 94: 157-165.

Breuil, C., Luck, B. T., Rossignol, L., Little, J., Echeverri, C. J., Banerjee, S. and Brown. D, L. (1992) Monoclonal antibodies to *Gliocladium roseum*, a potential biological control fungus of sap-staining fungi in wood. *J. Gen. Microbiol* .138: 2311-2319.

Cordier, C., Gianinazzi-Pearson, V. and Gianinazzi, S. (1994) Immunodétection de champignons endomycorhiziens à arbuscules *in planta. Acta Bot. Gallica* (in press).

De Ruiter, G. A., van Bruggen-van der Lugt, A. W., Bos, W., Notermans, S. H. W., Rombouts, F. M. and Hofstra, H. (1993) The production and partial characterization of a monoclonal IgG antibody specific for moulds belonging to the order Mucorales. *J. Gen. Microbiol.* 139: 1557-1564.

Dewey, F. M. (1990) The use of monoclonal antibodies to detect plant invading fungi. In: A. Schots (ed) *Monoclonal Antibodies in Agriculture-Proceedings of the Symposium "Perspectives for Monoclonal Antibodies in Agriculture".* Laboratorium for Monoclonal Antibodies, Wageningen, Niederlande-Pudoc, pp 21-25.

Friese, C. F. and Allen, M. F. (1991) Tracking the fates of exotic and local VA mycorrhizal fungi: methods and patterns. *Agric. Ecosys. Environ.* 34: 87-96.

Gianinazzi, S. and Gianinazzi-Pearson, V. (1992) Cytology, histochemistry and immunocytochemistry as tools for studying structure and function in endomycorrhizas. In: J.R. Norris, D.J. Read and A.K. Varma (eds) *Methods in Microbiology* 24: 109-139.

Gianinazzi-Pearson, V. and Gianinazzi, S. (1978) Enzymatic studies on the metabolism of vesicular-arbuscular mycorrhiza. II soluble alkaline phosphatase specific to the mycorrhizal infection in onion roots. *Physiol. Plant Pathol.* 12: 45-53.

Gianinazzi-Pearson, V. and Gianinazzi, S. (1994) Proteins and protein activities in endomycorrhizal symbioses. In: B. Hock and A.K. Varma (eds) *Mycorrhiza: Function, Molecular Biology and Biotechnology.* Springer, Heidelberg (in press).

Gianinazzi-Pearson, V., Lemoine, M. C., Arnould, C.and Morton, J. B. (1994) Localization of ß(1-3) glucans in spore and hyphal walls of fungi in the Glomales: taxonomic and phylogenetic implications. *Mycologia* (in press).

Hahn, A., Bonfante, P., Horn, K., Pausch, F. and Hock, B. (1993) Production of monoclonal antibodies against surface antigens of spores from arbuscular mycorrhizal fungi by an improved immunization and screening procedure. *Mycorrhiza* 4: 69-78.

Hahn, A., Horn, K. and Hock, B. (1994) Serological properties of mycorrhizae. In: B. Hock and A.K. Varma (eds) *Mycorrhiza: Function, Molecular Biology and Biotechnology.* Springer-Verlag, Heidelberg, in press.

Hardham, A. R., Suzaki, E. and Perkin, J. L. (1986) Monoclonal antibodies to isolate-, species-, and genus-specific components on the surface of zoospores and cysts of the fungus *Phytophthora cimmamomi. Can. J. Bot* .64: 311-321.

Harlow, E. and Lane, D. (1988) *Antibodies: A Laboratory Manual.* Cold Spring Harbour Laboratory, New York.

Köhler, G. and Milstein. C, (1975) Continuous cultures of fused cells secreting antibody of predefined specifity. *Nature* 256: 495-497.

Kough, J., Malajczuk, N. and Linderman, R. G. (1983) Use of the indirect immunofluorescent technique to study the vesicular-arbuscular fungus *Glomus epigaeum* and other *Glomus* species. *New Phytol* .94: 57-62.

Mernaugh, R. L., Mernaugh, G. R. and Kovacs G R (1990) The immune response: antigens, antibodies, antigen-antibody interactions. In: R. Hampton, E. Ball and S. De Boer (ed) *Immunochemical Methods for the Detection of Viral and Bacterial Plant Pathogens.* APS Press, St. Paul, pp 3-14.

Morton, J. B. (1988) Taxonomy of VA mycorrhizal fungi: classification, nomenclature, and identification. *Mycotaxon* 32: 267-324.

Morton, J. B. (1990) Evolutionary relationships among arbuscular mycorrhizal fungi in the Endogonaceae. *Mycologia* 82: 192 -207.

Oramas-Shirey, M. and Morton, J. B. (1990) Immunological stability among different geographic isolates of the arbuscular mycorrhizal fungus *Glomus occultum. Abstracts of the 90th Annual Meeting of the American Society for Microbiology:* 311 (Q-138).

Ravolanirina, F. and Gianinazzi-Pearson, V. (1992) VA endomycorrhization of microplants of grapevine: an immunological approach. In: J.N. Wolf (ed) *Interactions between Plants and Microorganisms*. Regional Seminar, International Foundation for Science, Stockholm, pp. 186-198.

Roitt, I. M. (1991) *Essential immunology*. 7th ed., Blackwell Scientific Publ., Oxford.

Sanders, I. R., Ravolanirina, F., Gianinazzi-Pearson, V., Gianinazzi, S. and Lemoine, M. C. (1992) Detection of specific antigens in the vesicular-arbuscular mycorrhizal fungi *Gigaspora margarita* and *Acaulospora laevis* using polyclonal antibodies to soluble spore fractions. *Mycol. Res.* 96: 477-480.

Simon, L, Bousquet, J., Lévesque, C. and Lalonde, M. (1993) Origin and diversification of endomycorrhizal fungi and coincidence with vascular land plants. *Nature* 363: 67-69

Straker, C. J., Gianinazzi-Pearson, V., Gianinazzi, S., Cleyet-Marel, J. C. and Bousquet, N. (1989) Electrophoretic and immunological studies on acid phosphatase from a mycorrhizal fungus of Erica hispidula. *New Phytol* .111: 215-221.

Tisserant, B. (1993) VA endomycorrhization of woody plants: root architecture and functional activity of the endomycorrhizal symbiosis. *Thesis*, University of Dijon, F: 136 pages.

Tisserant, B., Gianinazzi-Pearson, V., Gianinazzi, S. and Gollotte, A. (1993) In planta histochemical staining of fungal alkaline phosphatase activity for analysis of efficient arbuscular mycorrhizal infection. *Mycol. Res.* 97: 245-250.

Weinbaum, B. S., Allen, M. F. and Friese, C. F. (1992) Tracking the fate of *Acaulospora* and *Scutellospora* using fluorescent antibodies. *Newsletter of the MSA* 43: 54.

Wilson, I. M., Trinick, M. J. and Parker, C. A. (1983) The identification of vesicular-arbuscular mycorrhizal fungi using immunofluorescence. *Soil Biol. Biochem.* 15: 439-445.

Wright, S. F. and Morton, J. B. (1989) Detection of vesicular-arbuscular mycorrhizal fungus colonization of roots by using a dot-immunoblot assay. *Appl. Environm. Microbiol.* 55: 761-763.

Wright, S. F., Morton, J. B. and Sworobuk, J. E. (1987) Identification of a vesicular-arbuscular mycorrhizal fungus by using monoclonal antibodies in an enzyme-linked immunosorbent assay. *Appl. Environ. Microbiology* 53: 2222-2225.

European Bank of Glomales – An essential tool for efficient international and interdisciplinary collaboration

J.C. Dodd[1], V. Gianinazzi-Pearson[2], S. Rosendahl[3] and C. Walker[4]

[1]*The International Institute of Biotechnology, P.O. Box 228, Canterbury, Kent, CT2 7YW, U.K.*
[2]*Lab de Phytoparasitologie INRA/CNRS, Station de Génétique et d'Amélioration des Plantes, INRA, BV 1540, 21034 Dijon Cedex, France*
[3]*University of Copenhagen, Dept of Mycology, Oster Farimagsgade 2D, Copenhagen 1353, Denmark*
[4]*The Forestry Authority, Research Division, Northern Research Station, Roslin, Midlothian EH25 9SY, Scotland.*

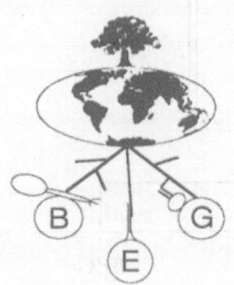

Arbuscular mycorrhizal fungi play a major role in the nutrient acquisition of most plants in terrestrial ecosystems. This group of mutualistic symbionts is currently placed within the Zygomycetes in the order Glomales. Research with these fungi has chiefly occurred over the past 30 or more years and their important role within agricultural, horticultural and forestry systems recognised. However, knowledge of the biodiversity of these fungi is limited due to the difficulty in culturing them away from the host. To help overcome these problems an initiative arising from a European collaborative network (COST 8.10) has resulted in the development of an infrastructure for a European Bank of Glomales (BEG). We are now in a position to launch this pan-european entity which will help focus research on these fungi within Europe and possibly worldwide. This paper outlines the aims, activities, structure, registration procedures and contact points for the BEG.

Acknowledgments

We would like to acknowledge the efforts of J.P. Masson (DG XII) in his support of the BEG idea. Sarah Dines (International Institute of Biotechnology) receives our sincere thanks for her efforts in helping design the logo and layout of the launch.

What is the BEG?

Collaboration between countries within the COST 8.10 initiative (1989-93) led to the idea of establishing a working group for the study of the systematics and taxonomy of the Glomales (Arbuscular mycorrhizal fungi - AMF). It soon became obvious that difficulties in identification were being caused by synonymy of species names and inadequate species descriptions. Large amounts of funds are being spent worldwide on research into isolates of AMF whose identity is often confused or unknown. From this evolved an idea for the establishment of a database for AMF within the framework of COST 8.10. A database and registration scheme have been developed whereby samples of cultures can be sent and registered within the BEG following authentication. This is seen as a prerequisite to the achievement of a unified approach to research with arbuscular mycorrhizas in Europe and possibly worldwide.

Aims of the BEG

- Establishment of a database and registration scheme.
- Research into systematics and taxonomy.
- Establishment of a centralised collection in Europe for the conservation of biodiversity of AMF.
- Quality assurance of inoculum used in experimentation within European Research.

The BEG Structure

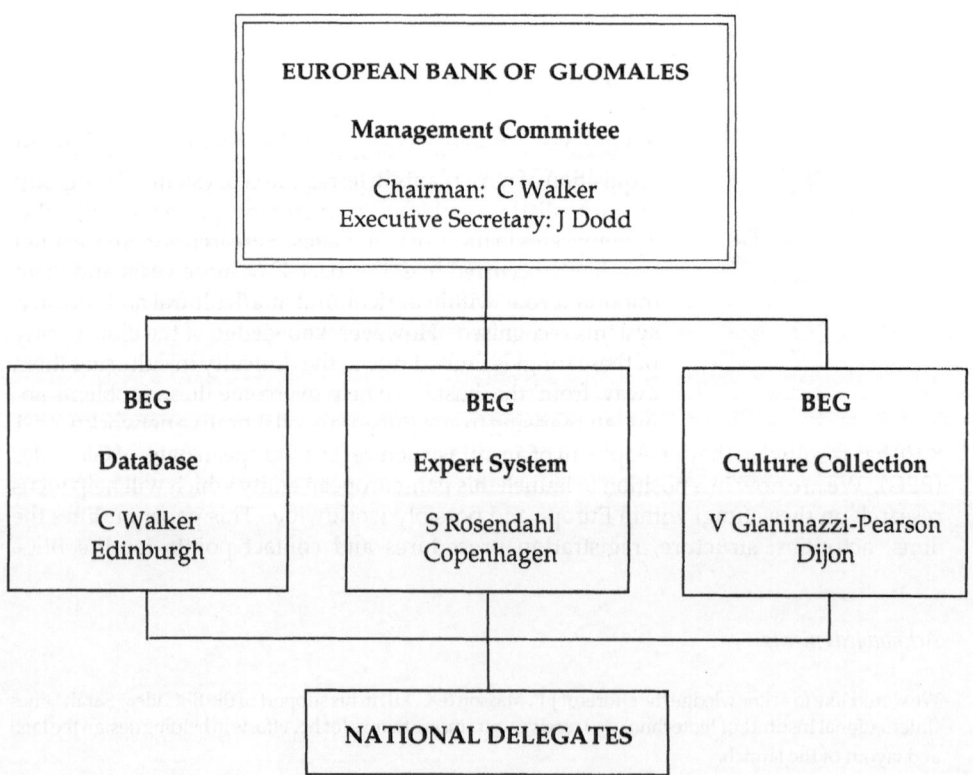

EUROPEAN BANK OF GLOMALES

Management Committee

Chairman: C Walker
Executive Secretary: J Dodd

BEG	BEG	BEG
Database	Expert System	Culture Collection
C Walker Edinburgh	S Rosendahl Copenhagen	V Gianinazzi-Pearson Dijon

NATIONAL DELEGATES

BEG Activities

• Registration of cultures of AMF from anywhere in the world in the BEG database to be maintained at the Forestry Authority, Scotland. These will be given a BEG number in the database that can be used when articles are published within European Journals. This will facilitate review of articles and allow the research work for that particular isolate to be collated within the database.

• A living Culture Collection is being established at INRA, Dijon, France for up to 100 isolates. Initially cultures from European scientists will be targeted but isolates from other ecosystems throughout the world will be gratefully received.

• The development of a prototype Expert System is underway to aid identification of AMF involving collaboration between several European groups and the Expert Centre for Taxonomic Identification, University of Amsterdam and centred at the University of Copenhagen, Denmark.

• Future developments within the BEG will be communicated via a BEG Newsletter and interested parties can register for inclusion on a mailing list at The International Institute of Biotechnology, Kent, UK.

How do I register cultures with the BEG?

• Living pot-culture material (approx. 200-300g of colonised substrate) can be submitted to the BEG, accompanied by a soil import licence if being sent from outside the UK. The licence can be obtained from C. Walker at the Forestry Authority, Scotland. The culture will be authenticated (alteration of the name previously used by the donor may occur) and the culture allocated a BEG number. This information will be passed to the donor who should use that number in future publications (users of registered cultures will be encouraged to submit voucher specimens for authentication at the time of submission of articles to scientific journals). When living material is sent for registration the donor can state if he wishes the material to be cultured or be used for identification only. No cultures will be cultured or passed onto a third party without the expressed permission of the donor! In some cases, however, it may prove necessary to culture the sample to gain more spores for proper identification of the isolate and the donor will be asked accordingly.

• At present the BEG committee are willing to provide small samples (50g) of two BEG cultures for scientific investigation only (Non commercial use!). These are:

Acaulospora longula (Walker)	- BEG 8
Acaulospora laevis (LPA1)	- BEG 13
Glomus claroideum (Rosendahl)	- BEG 14
Glomus geosporum (Kent)	- BEG 11
Glomus mosseae (LPA5)	- BEG 12

The material can be requested via the Culture Collection, INRA, Dijon and relevant information on culture conditions and suitability for experimental design (pH of soil, temperature etc.) will be sent. As the bank develops we will reevaluate this facility and hopefully include more isolates.

BEG Contacts

Chairman of the BEG: Dr. C. Walker, The Foresty Authority, Research Division, Northern Research Station, Roslin, Midlothian EH25 9SY, Scotland. Fax: 314455124 (e-mail YFCW15@CASTLE.ED.AC.UK)
Should be contacted if samples are to be sent to the BEG for registration and authentication.

Executive Secretary: Dr. J. C. Dodd, The International Institute of Biotechnology, University of Kent at Canterbury, P.O. Box 228, Canterbury, Kent CT2 7YW, UK. Fax: 227463482 (e-mail JCD@UKC.AC.UK).
Should be contacted for registration on list for future newsletters on the BEG.

Head of Expert System: Dr. S. Rosendahl, University of Copenhagen, Dept. of Mycology, Oster Farimagsgade 2D, Copenhagen 1353, Denmark. Fax: 4535322321 (e-mail SPOROL@VM.UNI-C.DK).
Should be contacted for information regarding the development of the Expert System.

Head of Culture Collection: Dr. V. Gianinazzi-Pearson, Lab. de Phytoparasitologie INRA/CNRS, Station de Génétique et d'Amélioration des Plantes, INRA, BV 1540, 21034 Dijon Cedex, France. Fax: 3380633263 (e-mail GIANINA@EPOISSES.INRA.FR).
Should be contacted for sample of AMF from the Culture Collection.

National Delegates

Contact can also be made through any of the following National Delegates.

M. Bodmer, Swiss Federal Research Station for Fruit-Growing, Viticulture and Horticulture, Wädenswil 8820, Switzerland. Fax: 417806341 (e-mail BODMERM@RZVAX.ETHZ.CH).
A. Camprubi, Depto. de Patologia Vegetal, IRTA-Centro Investicacion Agraria, Carratera de Cabrils S/N, Cabrils, Barcelona 08348, Spain. Fax: 3437533954.
J. C. Dodd (see above)
V. Gianinazzi-Pearson (see above)
M. Giovannetti, Instituto di Microbiologia Agraria, Centro di Studio per la Microbiologia del Suolo, Via del Borghetto 80, 56124 Pisa, Italy.
D. Mitchell, Dept. of Botany, University College Dublin, Belfield, Dublin 4, Eire. Fax: 35317061153.
S. Rosendahl (see above)
M. Vestberg, Agricultural Research Centre of Finland, Laukaa Research and Elite Plant Unit, Juntula, Laukaa 41340, Finland. Fax: 35841633835. (e-mail @CC.HELSINKI.FI:KES2@MTTK).
H. Von Alten, Institut für Pflanzenkrankheiten und Pflanzenschutz der Universität Hannover, Herrenhäuser Str. 2, 30419 Hannover, Germany. Fax: 5117623015.

LA BANQUE EUROPÉENE DES GLOMALES (BEG)

(European Bank of Glomales)

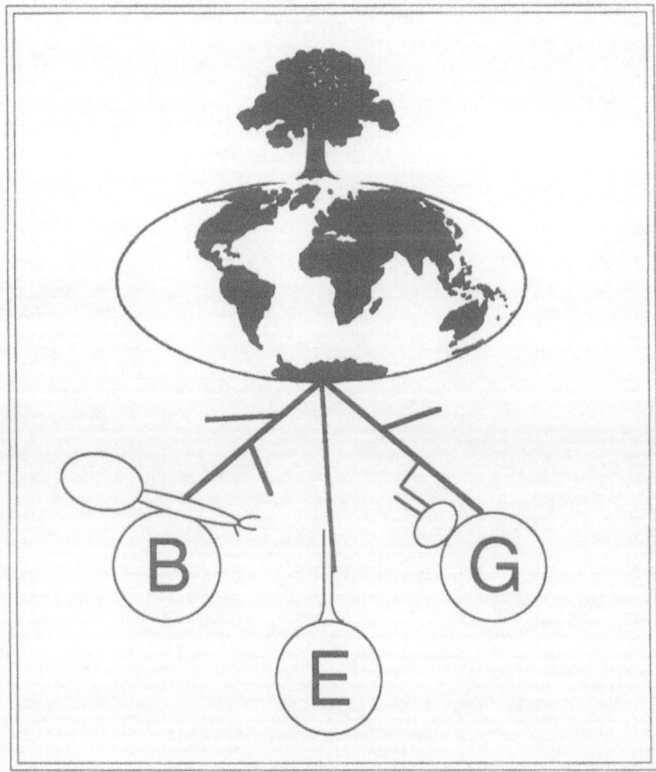

Mission Statement

To enhance, through pan-european collaboration, efficiency of research into the conservation and biodiversity of endomycorrhizal fungi, given their critical role in the survival of plants in most ecosystems.

BEG Committee
25th May 1993

*"The study of plants without their mycorrhizas
is the study of artefacts"*

*"The majority of plants, strictly speaking,
do not have roots; they have mycorrhizas"*

Impact of Arbuscular Mycorrhizas on
Sustainable Agriculture and Natural Ecosystems
S. Gianinazzi and H. Schüepp (eds.)
© 1994 Birkhäuser Verlag Basel/Switzerland

Physiological characteristics of the host plant promoting an undisturbed functioning of the mycorrhizal symbiosis

C. Azcón-Aguilar and B. Bago

Dept. de Microbiología del Suelo y Sistemas Simbióticos, Estación Experimental del Zaidín, CSIC, Prof. Albareda 1, 18008 Granada, Spain

Introduction

The term 'mycorrhiza' refers to the association between fungi and roots, or any other organ of higher plants involved in plant nutrient uptake from soil. This association is usually considered a mutualistic symbiosis because of the highly interdependent, and commonly beneficial, relationships established between both partners, in which the host plant receives mineral nutrients via fungal mycelium (mycotrophism), while the heterotrophic fungus obtains carbon compounds from the host photosynthesis (Harley and Smith, 1983; Harley, 1989).

During the process of mycorrhiza formation, in which the plant 'accepts' the fungal colonization with an almost complete absence of rejection reaction, a series of root-fungus interactions occur giving way to the integration of both organisms in a common 'supra-organism'. Despite the scarcity of experimental information, it can be assumed that the establishment of the symbiosis must be the result of a continuous molecular 'dialogue' between plant and fungus, as carried out through the exchange of both recognition and acceptance signals. The result of this dialogue will finally depend on the genome expression of both partners, the plant and the fungus involved in the symbiosis (Smith and Gianinazzi-Pearson, 1988).

Concerning arbuscular mycorrhizal associations, the most ubiquitous of all mycorrhizal types, the integration of both symbionts takes place at different levels: (i) structural, defined by arbuscule formation, a process by which the fungus modifies certain cells to form a dual host/fungal plasmalemma structure; (ii) physiological and biochemical, where the integration is mainly expressed by changes in the nutrient uptake properties of the roots when colonized by the fungus. These changes can be seen both in the tissue characteristics and in the biochemical processes (which include a rearrangement of particular enzymatic activities) involved in nutrient uptake, as well as in the effectiveness of the process itself. The establishment of a series of complex interfaces between fungus and plant, which are supposed to be involved in nutrient exchange between symbionts, and the concomitant metabolic and enzymatic modifications occurring, are indicative of the high degree of physiological integration of the macro and micro components of this symbiosis (Smith and Gianinazzi-Pearson, 1988).

Mycorrhizal symbiosis is regulated/controlled by a series of mechanisms in the plant and modulated by the environmental conditions. Particularly, it has been established that some physiological properties of the host plant are responsible determinants for suitable mycorrhiza formation and functioning. The analysis of these plant characteristics is the objective of the present review study.

Interdependence relationships between arbuscular mycorrhizal fungi and plants

Although arbuscular mycorrhizal fungi (AMF) have a certain, but limited, saprophytic ability (Warner and Mosse, 1980; Azcón-Aguilar and Barea, 1985), it seems that they depend on the establishment of the mycorrhizal association to accomplish their life cycle (Burggraaf and Beringer, 1987; Smith and Gianinazzi-Pearson, 1988). Consequently, AMF are considered to be both ecologically and physiologically obligate symbionts, since it has not yet been possible to grow them on synthetic media (Hepper, 1984; Azcón-Aguilar and Barea, 1994).

Regarding plant evolution on earth, there are evidences from fossil records suggesting that arbuscular mycorrhizal associations played a key role in the colonization of land by ancient plants (Stubblefield and Taylor, 1988; Pirozynski and Dalpé, 1989), a hypothesis that has been recently reinforced by phylogenetic studies based on modern molecular biology approaches (Simon et al., 1993). Since those early stages of plant (and mycorrhiza) appearance, both of the symbionts have developed a mutualistic dependency on each other. However, for the plant it could be inferred that the tendency of evolution has been toward non-mycotrophy (Brundrett, 1991). In fact, the dependence that most plants exhibit on arbuscular mycorrhizas for growth is not as strong as that of the AMF on plant. Actually, some plants have became independent of mycorrhizal fungi for optimal development (non mycotrophic plants). Some other plant species benefit from mycorrhizal associations only under certain conditions, usually when growing in soils having a very low fertility level, thus, these species are considered facultatively mycotrophs or facultatively mycorrhizal. These plant species have developed regulatory systems that allow them to establish the mycorrhizal association only in the environmental conditions such that they can benefit from the symbiosis (Koide and Li, 1990). Finally, a third group of plants, known as obligatorily mycotrophs, exhibits a strong dependence on mycorrhizal associations for growth and development and, ultimately, for surviving to reproductive maturity, in almost all environmental conditions. Such plant species are, in any case, able to control both the extent of fungal colonization in their root system, and the fungal activity, once the mycorrhiza have been established.

In some cases, however, plant growth depressions have also been described as a result of mycorrhiza formation (Bethlenfalvay and Pacovsky, 1983; Koide, 1985). These results suggest that the regulatory mechanisms carried out by the host plant were overcome by the mycorrhizal fungi. This can be attributed to very aggressive fungi, able to escape from the control mechanisms of the plant, or to physiological alterations in the plant that disable it to control the symbiosis. In any case, as we will see later, temporal growth depressions can be tolerated by the host plant if the result of the symbiosis were positive for plant growth in a long term.

Plants can control the mycorrhizal symbiosis at two different levels. Firstly, by preventing mycorrhiza formation, and secondly, once mycorrhiza is established, by depressing fungal activity and/or limiting fungal spread in the root tissues.

Physiological characteristics of the plant involved in mycorrhiza formation

Different features of the host plant, depending mainly (but not exclusively) on the morphological and physiological characteristics of the root system, condition its ability to acquire nutrients and, consequently, its capability to form mycorrhiza and the benefit the host obtains from its mycorrhizal status. Indirectly, factors affecting mycorrhiza formation can also affect the functioning of the symbiosis.

In general, plants able to effectively absorb relatively immobile available soil resources, such as phosphorus, are less dependent on mycorrhiza formation (Brundrett, 1991). It is known that the ability of the plant's root system to absorb nutrients having a very limited mobility in the soil solution is positively correlated to its surface area, which, in turn, is a consequence of the architecture of the root system (branching pattern, length and diameter of the roots, length and number of root hairs, etc..). Thus, plants with a sparse branching frequency, low number of lateral roots and few (and/or short) root hairs appear to be more dependent on mycorrhiza to acquire mineral nutrient uptake (Brundrett, 1991).

However, the geometry of the root system is not the only characteristic involved in nutrient uptake and, in some way, in mycorrhizal dependency. The activity of the root system, its growth rate and plasticity (the ability to respond quickly to localized or temporal changes in soil conditions, or in nutrient levels in soil profiles) are also important features conditioning the ability of the plant to cope with soil conditions which are not conducive to nutrient uptake and, consequently, its necessity to rely on mycorrhiza establishment to overcome these problems.

In addition to the activity and geometry of the root system, other important characteristics in determining the mycorrhizal dependency of the plant will be the nutrient uptake efficiency of the root system (mainly for phosphorus), as well as the plant's internal requirements for the nutrient and the nutrient utilization efficiency. Plants with a slow growth rate usually have less demand for nutrients, thus they can be more tolerant to low nutrient soils and less dependent on mycorrhiza.

Mechanisms of plant resistance to mycorrhiza formation The mechanisms by which non mycotrophic plants inhibit arbuscular mycorrhiza formation are not fully understood. There are probably different mechanisms involved in different groups of non host plants. In certain cases, root exudates from non mycotrophic species have been shown to inhibit AMF spore germination (El-Atrach et al., 1989). In most cases, however, hyphae from AMF can grow around roots of non mycorrhizal plants without revealing any inhibitory effect from root exudates (Glenn et al., 1988; Giovannetti et al., 1993 a). These hyphae, however, are not stimulated to grow and branch as usually occurs in the rhizosphere of host plants (Elias and Safir, 1987; Glenn et al., 1988; Bécard and Piché, 1989; Giovannetti et al., 1993 a and b). It seems that root exudates from non host plants lack some diffusable stimulant of AMF, produced by host plants, which is able to elicit hyphal branching and proliferation around roots, facilitating, consequently, the contact of the hyphae with the root surface (Glenn et al., 1988; Giovannetti et al., 1993 a and b).

In addition to these facts, there is evidence that if one hypha contacts the root surface of a non host species, swellings similar to appressoria can be formed. However, further penetration of root tissues is not usually observed, or, in some cases, a very small number of aborted unit infections can be found, restricted to intercellular hyphae, but without intracellular colonization. This suggests that inhibition of arbuscular mycorrhizal development in non host roots is related to intrinsic factors of the plant (Ocampo et al., 1980; Glenn et al., 1988; Giovannetti et al., 1993 a). There are studies about the influence of intergeneric grafts between host and non host plants on the formation of arbuscular mycorrhizas in which lupin shoots reduced mycorrhizal colonization of mycotrophic pea rootstocks and completely inhibited arbuscule formation. These results suggest that resistance to AMF colonization is, at least partially, determined by mobile factors produced in the shoots of the non mycotrophic plant (Gianinazzi-Pearson and Gianinazzi, 1992).

In leguminous plants a striking correlation has been described between the ability to effectively nodulate with *Rhizobium* and to form arbuscular mycorrhiza, suggesting that certain steps in the establishment of both types of symbiosis are under a common genetic control (Duc et al., 1989; Bradbury et al., 1993). It is interesting to note that the non nodulating alfalfa mutants, which did not develop normal arbuscular mycorrhizal associations, had significantly more appressoria, and bigger, than the mycorrhizal genotype (Bradbury et al., 1993). This means that early recognition events between the plant and the AMF are taking place, as indicated by the high number of appressoria formed, but further colonization of the root by the fungus does not occur, probably because other recognition events are blocked. It has been suggested that appressoria formation on the roots of these mutants may elicit reactions related to wall structure that are usually associated with the activation of defence responses to pathogens, which are not normally induced, or very weakly, by AMF (Gollotte et al., 1993). The higher number, and the larger size, of the appressoria formed in the non mycorrhizal alfalfa plants, in relation to the mycorrhizal ones, can be interpreted as an attempt to colonize the roots by the AMF.

It can be concluded that mycorrhiza formation is a complex process involving a set of developmental steps. In each of them there are signals that have to be recognized by the plant to allow the fungus to proceed to the next step in colonizing the root. These signals are often effective only at a defined developmental or physiological stage and under certain environmental conditions, whereas they are inactive at others. The lack of recognition at any of these developmental steps would halt fungal growth and, consequently, mycorrhiza formation.

Studies on the effect of AMF on the growth and development of micropropagated, mycotrophic plants showed that AMF inoculation immediately after the *in vitro* phase, in which roots formed are not completely functional, induced, in some cases, either aborted infection or very intense infection development, apparently outside the control of the host (Azcón-Aguilar et al., 1994). It seemed as if some roots (or plantlets) were not mature enough to control the growth of the AMF in the root tissues and, consequently, to establish the symbiosis. Thus, they reacted against the fungus (abortive infection) and, if they failed to prevent colonization, it appeared rather disorganized and uncontrolled. This effect seemed to be a consequence of the physiological status of the plantlets and/or their roots. Consequently, it can be deduced that a certain physiological maturity of the plant is required for the establishment of a functional mycorrhizal symbiosis.

Physiological characteristics of the plant involved in mycorrhizal functioning

As it has been already pointed out, AMF depend for their growth and functioning on the supply of carbon compounds by the host plant as derived from photosynthesis. This means that the normal functioning of the mycorrhizal symbiosis is, in turn, dependent on a number of physiological characteristics of the host plant, of which photosynthetic activity is one of the most important. In this context, it is also relevant to consider the source-to-sink relationships regulating carbon allocation to the heterotrophic component of the mycorrhizal symbiosis.

Regulation mediated through carbohydrate levels In general, it has been well established that formation and functioning of arbuscular mycorrhizas are regulated by the root carbohydrate level, which, in turn, is influenced by the mineral nutrition of the plant and its photosynthetic activity.

Actually, AMF need carbon compounds for the building up of their biomass and the maintenance of their metabolic rates, but they also need energy to take up nutrients from soil and translocate them to the plant. Therefore, the more effective the fungus is in nutrient uptake, the higher the demand for carbohydrates they must obtain from the host plant. A reduction in root carbohydrate availability will consequently reduce mycorrhizal functioning.

The net benefit of the mycorrhizal symbiosis depends on the benefit derived from the uptake of mineral nutrients, against the cost of maintaining the mycosymbiont, in terms of the carbon supplied by the plant. In any case, if photosynthesis is limited by phosphorus, which is usually the case when phosphorus is limiting growth, the greater carbon expenditure below-ground by mycorrhizal plants, could somehow be compensated for by the general trend of higher carbon assimilation (Kucey and Paul, 1982; Snellgrove et al., 1986; Brown and Bethlenfalvay, 1988). This trend can be explained not only in terms of phosphorus acquisition, but also by the increase in the photosynthate sink strength induced by the mycorrhiza. In fact, it has been described an enhancement of carbon assimilation rate independent from any phosphate-mediated effect (Brown and Bethlenfalvay, 1988).

If mycorrhizal induced growth depressions occur, they can be originated by reductions in the levels of carbohydrates available for plant growth, due to the carbon utilization by the AMF, when these can not compensate such drainage by enhancing carbon assimilation. Growth depressions induced by mycorrhiza can also be explained by other mechanisms, such as phosphorus competition in conditions of very limited phosphorus availability. However, the imbalance between both carbon expenditure and increased carbon assimilation in the plant induced by the AMF is what is most common.

The host plant would be expected to restrict mycorrhizal formation and functioning in cases where the cost of the symbiosis is higher than the benefit. That is the case when photosynthesis is depressed by factors other than phosphorus, for example light (Tester et al., 1986), or when carbon supply is limited as a consequence of grazing (Trent et al., 1988). In fact, the decrease in the proportion of mycorrhizal root length as a consequence of low photon irradiance was explained as a reduction in the formation of entry points by the AMF (Tester et al., 1986). Under low light intensity mycorrhizal colonization and/or activity can be depressed and, consequently, growth responses reduced or eliminated.

According to Koide (1993), however, even if the host plant is able to regulate arbuscular mycorrhizal fungal activity, it would not be necessary for mycorrhiza to promote growth continuously in order to be maintained by the plant. Temporal growth depressions could be tolerated by the plant if the net consequence of the symbiosis were positive for plant development in a long-term scale. This is important, from the ecological point of view, because soil resources in natural communities are subjected to temporary changes, depending on moisture content and temperature. Fluctuations in the level of available phosphate can be important according to environmental conditions. Thus, it could be useful for the plant to maintain the symbiosis during periods of no net benefit, just for having it already established when soil resources become less available.

Therefore, it is reasonable to expect that the host plant would reduce the colonization level by AMF under conditions of no net benefit, as it normally happens, but without completely eliminating the symbiosis. When the conditions change and the mycorrhizal symbiosis becomes more beneficial, the contrary (to allow colonization to increase) would also be expected.

This is most probable to occur in plants with less developed mechanisms for acquiring and utilizing phosphorus efficiently, which rely more on mycorrhiza for optimal growth. It would be expected, consequently, that plants with more efficient mechanisms for phosphorus acquisition and utilization are also more efficient in regulating mycorrhizal functioning.

It has been suggested that host genotypes can regulate carbon expenditure by the mycorrizal symbiosis, and that those genotypes less dependent on mycorrhiza for optimal growth have more accurate regulation mechanisms (Graham and Eissenstat, 1994)

Regulation mediated through phosphorus The activity of the mycorrhizal fungi has also been shown to be decreased by high phosphorus levels inside roots (Koide and Li, 1990). In fact, there is much experimental evidence showing that mycorrhizal colonization is reduced in plants having a high level of phosphorus (Barea, 1991). It is important, however, to consider, when talking about reduction in the percentage of the mycorrhizal infection of roots, that this reduction may actually be the result of a decrease in fungal growth and spread rate in the root, or of an increase in the growth rate of the root itself (Amijee et al., 1989). High levels of phosphorus in the plant actually reduce mycorrhizal colonization of the root system, even in the absence of root length increases, and this reduction is also reflected in the density of arbuscular development (Smith and Gianinazzi-Pearson, 1988), the amount of external mycelium (Abbott et al., 1984) and the number of entry points (Amijee et al., 1989).

These effects have been attributed to reductions in the exudation rate in the rhizospheric soil, or in the cellular spaces in the root cortex, of possible substrates necessary for fungal growth, such as soluble carbohydrates and amino acids. Elias and Safir (1987) observed that hyphal growth of an AMF was stimulated by exudates from *Trifolium repens,* but only if the plants had been grown under phosphate limitation.

In some plants these reductions in root exudation were mediated through decreases in the permeability of root membranes induced by the high phosphorus content of the plant (Ratnayake et al., 1978; Grahan et al., 1981; Cooper, 1984). In other plants, however, an effect of phosphorus on the permeability of the root membranes was not observed, and the reduction in root exudation was associated only with a decrease in the concentration of soluble carbohydrates in root (Jasper et al., 1979; Same et al., 1983; Thomson et al., 1986).

The latter authors, in fact, correlate increased phosphorus levels in the plants with decreased concentrations of soluble carbohydrates in the roots, and suggest that the mechanism by which the host plant controls the spread of the fungus inside the root is through the supply of carbon compounds (Thomson et al., 1990). This has been suggested by numerous experiments and seems to be corroborated by the fact that reductions in mycorrhizal activity induced by high phosphorus concentrations were more severe at suboptimal light levels (Son and Smith, 1988).

More recent studies, however, find that increased phosphorus supply to leek plants results in progressively increased levels of soluble carbohydrates in the roots. Moreover, in conditions in which increased concentrations of phosphorus inhibited mycorrhizal colonization, the levels of soluble carbohydrates in the root were at a maximun (Amijee et al., 1993). These authors consequently exclude the hypothesis that greater concentrations of soluble carbohydrates in roots favour mycorrhizal establishment.

When high levels of phosphorus were added to leek plants, these authors found a reduction in the growth rate of the external mycelium, in the formation of new entry points and also a high proportion of abortive entry points (Amijee et al., 1993). They concluded that the delay in the formation of new entry points from the external mycelium is the rate-limiting step for AMF colonization of the root in the presence of high levels of P, and suggested that the delay was, at least partially, due to an increased resistance of root tissues to the penetration by AMF, as it commonly happens in roots of non mycorrhizal plants. The reduction in the growth rate of the external mycelium and, consequently, in the formation of new entry points could also be explained if the mycelium were not invigorated to grow and extend along the root as a result of the lack, or diminution, of growing substrates translocated to it from the host plant through the internal mycelium (Sanders and Sheikh, 1983).

In any case, and whatever the mechanism involved, it seems clear from experimental evidence that the host plant controls the spread of the AMF inside the root tissues and limits the level of root colonization by controlling the growth rate of the fungus inside or outside the roots.

It has been suggested that arbuscular mycorrhizal symbioses are, in some way, self-regulatory (Hayman, 1983), and because one of the main effects of mycorrhizal colonization is the increase in phosphorus uptake by the host plant, phosphorus has been involved in this self-regulation mechanism. This basically consist in that, under conditions where phosphorus is not the limiting factor for plant growth, mycorrhizal colonization tends to be inhibited. However, when phosphorus is limiting plant growth, extensive colonization of the root system is allowed. This regulatory mechanism seems to be mediated by physiological alterations of the host plant, and more precisely of the root system (Koide and Li, 1990).

Nevertheless, arbuscular mycorrhizal symbiosis appears to exert a regulatory mechanism to overcome temporal changes in the availability of P in soil. This is based on the loading and unloading of polyphosphate granules. In fact, synthesis and breakdown of polyphosphate granules in vacuoles may constitute a self-regulating mechanism in mycorrhizal functioning (Smith and Gianinazzi-Pearson, 1988), since it allows the storage of phosphorus in an osmotically inactive form when the nutrient is in excess, and its transfer to the host when needed. Therefore, AMF can potentially be maintained over a range of soil P concentrations (Barea et al., 1993).

When phosphorus availability in soil is extremely low, the development of the mycorrhizal symbiosis can also be reduced. This has been observed in several instances (Koide and Li, 1990; Abbott et al., 1984), being consistent with the idea that arbuscular mycorrhizal development is well controlled by the host plant, and that the colonization of the root by the AMF is limited under any condition preventing that mycorrhizal colonization could promote growth. Therefore, even when phosphorus is limiting growth, mycorrhizal colonization will be low if it can not significantly improve the phosphate nutrition of the host plant.

In situations of extremely low phosphate availability in soil, it seems that the root, rather than the shoot, is the responsible for the regulation of mycorrhizal infection (Koide and Li, 1990) through some physiological alteration.

Regulation mediated through general plant nutrition Colonization of roots by AMF is also regulated by the general status of the nutrition of the plant and the nutrient availability in soil. In general, the proportion of root length colonized by AMF increases with decreasing nutrient availability. High levels of nitrogen can increase root colonization (Hepper, 1983), probably by making phosphorus the limiting factor for plant growth and, consequently, by making the plant more receptive to mycorrhiza formation. It is difficult, however, to generalize the effect of nutrients different from phosphorus, because their influence is going to depend on the general balance between them.

In general, the plant would be expected to reduce mycorrhizal colonization when nutrients other than P limit plant growth. However, it would also be expected to increase root colonization when the supply of other nutrients required for growth is sufficient to make phosphorus the limiting factor.

Regulation of mycorrhizal colonization under stress conditions Under stress conditions, plants may undergo a series of temporary or permanent changes in their metabolism and physiology which can induce disruption in the functioning of arbuscular mycorrhizal associations. When these stresses affect photosynthetic activity, reductions in the amount of carbon compounds affect plant growth and usually depress root activity through changes in the translocation pattern of photosynthate to the root and/or reductions in the levels of available photosynthate (Andersen and Rygiewicz, 1991). In these conditions, mycorrhizal functioning is disrupted and competition for carbon compounds between roots and AMF can occur.

In general, photosynthesis and C allocation processes are sensitive indicators of physiological stress. In situations where the photosynthetic efficiency is reduced by foliar stress, more carbohydrates can be retained in leaves for repair mechanisms and less is available for translocation to roots and AMF. Because AMF act as a strong physiological sink for carbon compounds (Koch and Johnson, 1984; Douds et al., 1988) they probably suffer at the expense of other metabolic sinks, the consequences of the reduction in available carbohydrates. In fact, decreases in the mycorrhizal colonization level have been observed in different stress situations such as, for instance, when light supply is limited (Bethlenfalvay and Pacovsky, 1983; Tester et al., 1986), when growing at very high (Haugen and Smith, 1992) or low temperatures (Hetrick and Bloom, 1984), or when exposed to ozone (McCool and Menge, 1983).

A similar situation was found when plants were supplied with phenmedipham. This herbicide is known to inhibit photosynthesis and when sprayed on alfalfa and sorghum leaves, root concentration of total and reducing sugars decreased (Ocampo and Barea, 1985). Although the percentage of mycorrhizal root length was not affected by herbicide application, fungal metabolism, determined by using a vital stain, was reduced after 48 h application (Ocampo and Barea, 1985). These results suggest that, even in situations in which no changes in root colonization levels are detected by using the most common, non-vital, staining procedures, the plant can be constantly regulating fungal activity, and also that the fungi have to modulate their metabolism, according to environmental changes or stresses.

However, apart from the decreased colonization degree of the root system by AMF, which means that they reduce their growth rate by the limited supply of carbon compounds, competition for available carbohydrates may induce growth depressions in the host plant. This suggests that AMF are strong competitors for carbon compounds and, under these stress situations affecting photosynthetic activity, the plant, although able to reduce fungal growth, fails to inhibit it completely. Consequently, it suffers the strong competition imposed on it by the fungi, which can finally induce a growth depression.

A similar situation can be found when the plant is affected by stresses of biotic type (shoot pathogens), in which arbuscular mycorrhiza can increase the severity of the stress by means of competition for available carbohydrates in the stressed plants.

In relation to the carbon cost of the symbiosis, it has been suggested that there is an optimal level of mycorrhizal colonization above which there is no increase in phosphorus uptake, although the plant continues to supply carbon compounds for the AMF metabolism (Douds et al., 1988). This assumption is based on experiments in which 'half mycorrhizal' plants (using a split-root system) had the same phosphorus content as mycorrhizal ones, but allocated half of the carbon compounds to the root system (Douds et al., 1988). However, since mycorrhizal plants were bigger than half mycorrhizal ones, although total P content in leaves was similar, it must be assumed that the higher carbon cost of the mycorrhiza was compensated, and even overcome, by other physiological effects of the symbiosis aside from phosphorus uptake.

Conclusions

The plant features relevant to mycorrhiza formation mainly concern the morphological and physiological characteristics of the root systems, and those related in promoting mycorrhizal functioning, are obviously those involved in the feeding and growth regulation of its heterotrophic fungal partner. Therefore, the photosynthesis related processes, and the subsequent source-to-sink relationships for carbon compounds allocation, are critical to maintain a functional and efficient biotrophic association.

A number of plant characteristics related to the degree of dependency on arbuscular mycorrhiza establishment for a given plant species to develop suitably, are able to modulated mycorrhizal functioning in such host plant. Thus, nutrient demand imperatives and nutrient use efficiency are critical to assure mycorrhizal operativity. Additionally, cultivar-associated factors, such as a high relative growth rate, are also desirable plant traits fovouring mycorrhizal functioning. In fact, recommendations for breeding programmes, that would take mycorrhizas into account, must consider these plant features.

The use of myc- mutants, grafting experiments, and comparative studies involving cultivars in the same species, are current research tools to ascertain plant involvement in arbuscular mycorrhizal development and control. Molecular biology and immunological approaches are now being applied to identify the origin (plant or fungus) of key compounds controlling arbuscular mycorrhizal functioning. Thus, it is likely feasible to improve in the nearest future the knowledge concerning the contribution of both symbionts for a suitable mutualistic development in arbuscular mycorrhizal associations.

Acknowledgements

The authors thank Dr. J. M. Barea, Department of Soil Microbiology and Symbiotic Systems, Estación Experimental del Zaidín, Granada, for his comments and suggestions. This study was supported by CICYT-Spain (Project AGR 91 - 0605-C02-01).

58

References

Abbott, L.K., Robson, A.D. and De Boer, G. (1984) The effect of phosphorus on the formation of hyphae in soil by the vesicular-arbuscular mycorrhizal fungus *Glomus fasciculatum. New Phytol.* 97: 437-446.

Amijee, F., Stribley, D.P. and Tinker, P.B. (1993) The development of endomycorrhizal root systems. VIII. Effects of soil phosphorus and fungal colonization on the concentration of soluble carbohydrates in roots. *New Phytol.* 123: 297-306.

Amijee, F., Tinker, P.B. and Stribley, D.P. (1989) The development of endomycorrhizal root systems. VII. A detailed study of effects of soil phosphorus on colonization. *New Phytol.* 111: 435-446.

Andersen, C.P., and Rygiewicz, P.T. (1991) Stress interactions and mycorrhizal plant response: Understanding carbon allocation priorities. *Environm. Poll.* 73: 217-244.

Azcón-Aguilar, C. and Barea, J. M. (1985) Effect of soil micro-organisms on formation of vesicular-arbuscular mycorrhizas. *Trans. Brit. Mycol. Soc.* 84: 536-537.

Azcón-Aguilar, C. and Barea, J.M. (1994) Saprophytic growth of arbuscular mycorrhizal fungi. In: B. Hock and A. Varma (eds) *Mycorrhiza: Structure, Function, Molecular Biology and Biotechnology*, Springer-Verlag, Heidelberg, (in press).

Azcón-Aguilar , C., Encina, C. L., Azcón, R. and Barea, J. M. (1994) Effect of arbuscular mycorrhiza on growth and development of *Annona cherimola* micropropagated plants. *Agric. Sci. Finl.* (In press).

Barea J.M. (1991) Vesicular-arbuscular mycorrhizae as modifiers of soil fertility. In: B.A. Stewart (ed) *Advances in Soil Science, Volume 15*, Springer-Verlag, New York, pp. 1-39.

Barea J.M., Azcón, R. and Azcón-Aguilar, C. (1993) Mycorrhiza and crops. In: D.S. Ingram, and P.H. Williams (eds) *Advances in Plant Pathology, Volume 9*, Academic Press, London, pp. 167-189.

Bécard, G. and Piché, Y. (1989) Fungal growth stimulation by CO_2 and root exudates in vesicular-arbuscular mycorrhizal symbiosis. *Appl. and Environm. Microbiol.* 55: 2320-2325.

Bethlenfalvay, G.J., and Pacovsky, R. S. (1983) Light effects in mycorrhizal soybeans. *Plant Physiol.* 73: 969-972.

Bradbury, S.M., Peterson, R.L. and Bowley, S.R. (1993) Further evidence for a correlation between nodulation genotypes in alfalfa (*Medicago sativa* L.) and mycorrhiza formation. *New Phytol.* 124: 665-673.

Brown, M.S. and Bethlenfalvay, G.J. (1988) The *Glycine-Glomus-Rhizobium* symbiosis. VII. Photosynthetic nutrient-use efficiency in nodulated mycorrhizal soybeans. *Plant Physiol.* 86: 1292-1297.

Brundrett, M. (1991) Mycorrhizas iN natural ecosystems. *Adv. Ecol. Res.* 21: 171-313.

Burggraaf, A.J. and Beringer, J.E. (1989) Absence of nuclear DNA synthesis in vesicular arbuscular mycorrhizal fungi during in vitro development. *New Phytol.* 1: 25-33.

Cooper, K.M. (1984) Physiology of VA mycorrhizal associations. In: C.L. Powell and D.J. Bagyaraj (eds)*VA mycorrhiza.*, CRC press, Boca Raton, Florida, pp. 155-186.

Douds, Jr. D.D., Johnson, R. and Kock, K.E. (1988) Carbon cost of the fungal symbiont relative to net leaf P accumulation in a split-root VA mycorrhizal symbiosis. *Plant Physiol.* 86: 491-496.

Duc, G., Trouvelot, A., Gianinazzi-Pearson, V. and Gianinazzi, S. (1989) First report of non-mycorrhizal plant mutants (Myc⁻) obtained in pea (*Pisum sativum* L.) and fababean (*Vicia faba* L.). *Plant Science* 60: 215-222.

El-Atrach, F., Vierheilig, H. and Ocampo, J.A. (1989) Influence of non-host plants on vesicular-arbuscular mycorrhizal infection of host plants and on spore germination. *Soil Biol. Biochem.* 21: 161-163.

Elias, K.S. and Safir, G.R. (1987) Hyphal elongation of *Glomus fasciculatus* in response to root exudates. *Appl. Environm. Microbiol.* 53: 1928-1933.

Gianinazzi-Pearson, V., and Gianinazzi, S. (1992) Influence of intergeneric grafts between host and non-host legumes on formation of vesicular-arbuscular mycorrhiza. *New Phytol.* 120: 505-508.

Giovannetti, M., Avio, L., Sbrana, C. and Citernesi, A.S. (1993a) Factors affecting appressorium development in the vesicular-arbuscular mycorrhizal fungus *Glomus mosseae* (Nicol. & Gerd.) Gerd. & Trappe. *New Phytol.* 123: 115-122.

Giovannetti, M., Sbrana, C., Avio, L., Citernesi, A. S. and Logi, C. (1993b) Differential hyphal morphogenesis in arbuscular mycorrhizal fungi during pre-infection stages. *New Phytol.* 125: 587-593.

Glenn, M.G., Chew, F.S. and Williams, P.H. (1988) Influence of glucosinolate content of *Brassica* (Cruciferae) roots on growth of vesicular-arbuscular mycorrhizal fungi. *New Phytol.* 110: 217-225.

Gollotte, A., Gianinazzi-Pearson, V., Giovannetti, M., Sbrana, C., Avio, L. and Gianinazzi, S. (1993) Cellular localization and cytochemical probing of resistance reactions to arbuscular mycorrhizal fungi in a 'locus a' myc⁻ mutant of *Pisum sativum* L. *Planta* 191: 112-122.

Graham, J.H. and Eissenstat, D.M. (1994) Host genotype and the formation and function of VA mycorrhizae. *Plant and Soil.* 159: 179-185.

Graham, J.H., Leonard, R.T. and Menge, J.A. (1981) Membrane mediated decrease in root exudation responsible for phosphorus inhibition of vesicular-arbuscular mycorrhiza formation. *Plant Physiol.* 68: 548-552.

Harley, J.L. (1989) The significance of mycorrhiza. *Mycology Research* 92: 129-139.

Harley, J.L. and Smith, S. E. (1983) *Mycorrhizal Symbiosis*. Academic Press, New York.

Haugen, L.M. and Smith, S.E. (1992) The effect of high temperature and fallow period on infection of mung bean and cashew roots by the vesicular-arbuscular mycorrhizal fungus *Glomus intraradices*. *Plant and Soil* 145: 71-80.

Hayman, D.S. (1983) The physiology of vesicular arbuscular endomycorrhizal symbiosis. *Can. J. Bot.* 61: 944-963.

Hepper, C.M. (1983) Effect of nitrate and phosphate on the vesicular-arbuscular mycorrhizal infection of lettuce. *New Phytol.* 92: 389-399.

Hepper, C.M. (1984) Isolation and culture of VA mycorrhizal (VAM) fungi. In: C.L. Powell and D.J. Bagyaraj (eds)*VA mycorrhiza.*, CRC press, Boca Raton, Florida, pp. 95-112.

Hetrick, B.A. and Bloom, J. (1984) The influence of temperature on colonization of winter wheat by vesicular-arbuscular mycorrhizal fungi. *Mycologia* 76, 953-956.

Jasper, D.A., Robson, A.D. and Abbott, L. K. (1979) Phosphorus and the formation of vesicular arbuscular mycorrhizas. *Soil Biol. Biochem.* 11: 501-505.

Koch, K.E. and Johnson, C.R. (1984) Photosynthate partitioning in split-root citrus seedlings with mycorrhizal and nonmycorrhizal root systems. *Plant Physiol.* 75: 26-30.

Koide, R.T. (1985) The nature of growth depressions in sunflower caused by vesicular-arbuscular mycorrhizal infection. *New Phytol.* 99: 449-462.

Koide, R.T. (1993) Physiology of the mycorrhizal plant. In: D.S. Ingram and P.H. Williams (eds) *Advances in Plant Pathology, Volume 9*, Academic Press, London, pp. 33-54.

Koide, R.T. and Li, M. (1990) On host regulation of the vesicular-arbuscular mycorrhizal symbiosis. *New Phytol.* 114: 59-74.

Kucey, R.M.N. and Paul, E.A. (1982) Carbon flow, photosynthesis and N_2 fixation in mycorrhizal and nodulated fababeans (*Vicia faba* L.). *Soil Biol. Biochemi.* 14: 407-412.

McCool, P. and Menge, J.A. (1983) Influence of ozone on carbon partitioning in tomato: Potential role of carbon flow in regulation of the mycorrhizal symbiosis under conditions of stress. *New Phytol.* 94: 241-247.

Ocampo, J.A. and Barea, J.M. (1985) Effect of carbamate herbicides on VA mycorrhizal infection and plant growth. *Plant and Soil* 85: 375-383.

Ocampo, J.A., Martin, J. and Hayman, D.S. (1980) Influence of plant interactions on vesicular-arbuscular mycorrhizal infections. I. Host and non-host plants grown together. *New Phytol.* 84: 27-35.

Pirozynski, K.A. and Dalpé, Y. (1989) Geological history of the Glomaceae with particular reference to mycorrhizal symbiosis. *Symbiosis* 7: 1-36.

Ratnayake, M., Leonard, R.T. and Menge, J.A. (1978) Root exudation in relation to supply of phosphorus and its possible relevance to mycorrhiza formation. *New Phytol.* 81: 533-552.

Same, B. ., Robson, A.D. and Abbott, L.K. (1983) Phosphorus, soluble carbohydrates and endomycorrhizal infection. *Soil Biol. Biochem.* 15: 593-597.

Sanders, F.E. and Sheikh, N.A. (1983) The development of vesicular-arbuscular mycorrhizal infection in plant root systems. *Plant and Soil* 71: 223-246.

Simon, L., Bousquet, J., Lévesque, R.C. and Lalonde, M. (1993) Origin and diversification of endomycorrhizal fungi and coincidence with vascular land plants. *Nature* 363: 67-69.

Smith, S.E. and Gianinazzi-Pearson, V. (1988) Physiological interactions between symbionts in vesicular-arbuscular mycorrhizal plants. *Ann. Rev. Plant Physiol. Plant Mol. Biol.* 39: 221-244.

Snellgrove, R.C., Stribley, D.P., Tinker, P.B. and Lawlor, D.W. (1986) The effect of vesicular-arbuscular mycorrhizal infection on photosynthesis and carbon distribution in leek plants. In: V. Gianinazzi-Pearson and S. Gianinazzi (eds) *Physiological and Genetical Aspects of Mycorrhizae*, INRA, Paris, pp. 421-424.

Son, C.L. and Smith, S.E. (1988) Mycorrhizal growth responses: interactions between photon irradiance and phosphorus nutrition. *New Phytol.* 108: 305-314.

Stubblefield, S.P. and Taylor, T.N. (1988) Recent advances in palaeomycology. *New Phytol.* 108: 3-25.

Tester, M., Smith, S.E., Smith, F.A. and Walker, N.A. (1986) Effects of photon irradiance on the growth of shoots and roots, on the rate of initiation of mycorrhizal infection and on the growth of infection units in *Trifolium subterraneum* L. *New Phytol.* 103: 375-390.

Thomson, B.D., Robson, A.D. and Abbott, L.K. (1986) Effects of phosphorus on the formation of mycorrhizas by *Gigaspora calospora* and *Glomus fasciculatum* in relation to root carbohydrates. *New Phytol.* 103: 751-765.

Thomson, B.D., Robson, A.D. and Abbott, L.K. (1990) Mycorrhizas formed by *Gigaspora calospora* and *Glomus fasciculatum* on subterranean clover in relation to soluble carbohydrate concentrations in roots. *New Phytol.* 144: 217-225.

60

Trent, J.D., Wallace, L.L., Svejcar, T.J. and Christiansen, S. (1988) Effect of grazing on growth, carbohydrate pools, and mycorrhizae in winter wheat. *Can. J. Plant Sci.* 68: 115-120.

Warner, A. and Mosse, B. (1980) Independent spread of vesicular-arbuscular mycorrhizal fungi in soil. *Trans. Brit. Mycol. Soc.* 74: 407-410.

REGULATION OF MYCORRHIZAL FUNCTIONING

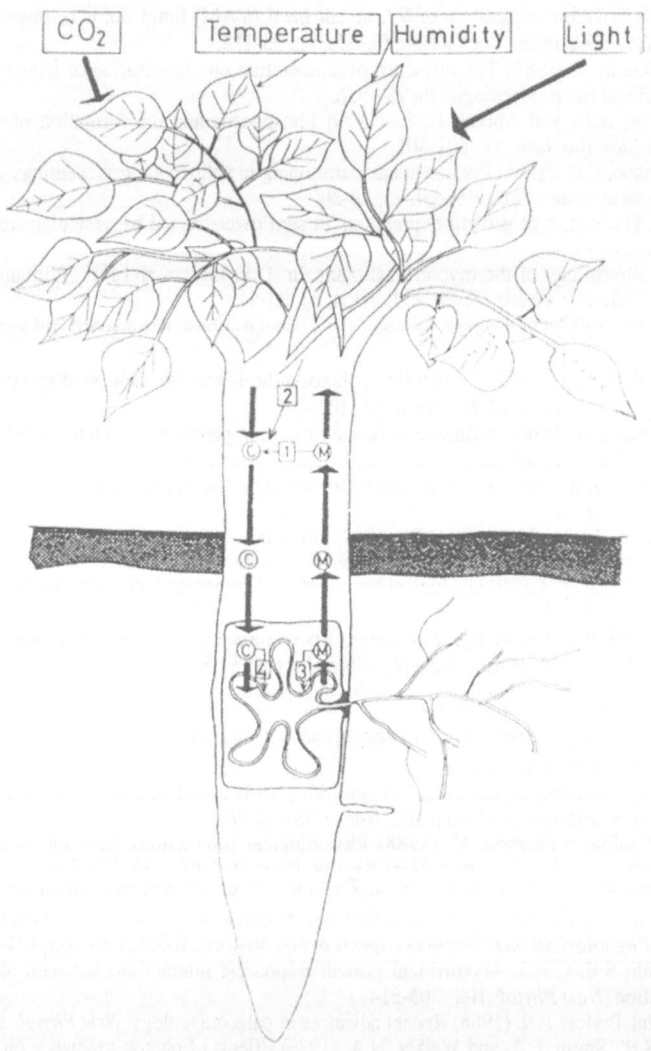

Plant-fungus interactions in mycorrhizas are regulated by a number of feed-back control mechanisms. Among these: (1) = Effect of mineral nutrients (M) on the allocation of carbohydrate (C) produced in plant photosynthesis. This is critical to control the shoot-to-root ratio; (2) Effect of photosynthesis-regulating factors on the amount of carbon available to be transferred to the heterotrophic part of the system (root + mycorrhiza); (3) Effect of mineral nutrient concentrations in root cells on arbuscule development; and (4) effect of carbon compounds on arbuscule functioning.

Recognition and infection process, basis for host specificity of arbuscular mycorrhizal fungi

M. Giovannetti, C. Sbrana, A.S. Citernesi, L. Avio, A. Gollotte[1], V. Gianinazzi-Pearson[1] and S. Gianinazzi[1]

Istituto di Microbiologia Agraria, Centro di Studio per la Microbiologia del Suolo C.N.R., Via del Borghetto 80, 56124 Pisa, Italy
[1]Laboratoire de Phytoparasitologie, INRA-CNRS, Station de Génétique et d'Amélioration des Plantes, INRA, BV 1540, F-21034 Dijon cédex, France

Introduction

Arbuscular mycorrhizal fungi (AMF) are obligate biotrophs, which derive nutrients from living cells of host plants (Lewis, 1973). The process of infection of host roots is characterized by distinct stages involving a number of complex morphogenetic changes in the fungi: spore germination, hyphal differentiation, appressorium formation, root penetration, intercellular growth and arbuscule formation.

Spore germination and some independent hyphal growth of AMF can occur in the absence of roots, and this saprophytic phase is crucial to their life cycle since survival depends on the ability to rapidly and efficiently infect suitable hosts. Appressoria, which are generally considered as evidence of fungal recognition of a potential host plant (Staples and Macko, 1980), can be formed as early as 36 hours after the beginning of the interaction of an AMF with a host root (Giovannetti and Citernesi, 1993). However, the dialogue between the two partners of the symbiosis must initiate well before with host plant signals, probably in root exudates, being perceived by the fungal symbiont and inducing differential hyphal morphogenesis and growth prior to contact with the root surface and the appressorium development (Hepper, 1984; Giovannetti et al., 1993b). AMF are not only capable of recognizing hosts at an early stage, but they are also able to discriminate between host and non-host plants (Glenn et al., 1985; Giovannetti et al., 1993a, b; Schreiner and Koide, 1993) indicating that in spite of their wide host range, this group of obligate biotrophs does show some sort of host specificity. Once the host has triggered these initial steps in the infection process, the AMF undergo a sequence of events which lead to their establishment in a new ecological niche, the root.

Knowledge of the mechanisms regulating the different steps in the formation of arbuscular mycorrhiza is essential to a thorough understanding of the symbiotic association. Here we review what is known about the early stages of the infection process by AMF, with particular emphasis on recognition responses which are prerequisites to mycorrhiza establishment and therefore central to the proliferation and survival of these fungi.

Pre-infection events

A. Spore germination and hyphal growth *In vitro* and *in vivo* studies have shown that spore germination, the first event in the life cycle of AMF, is influenced by spore dormancy, environmental conditions such as pH, temperature, water potential, nutrient content and soil microbial activities (Mosse, 1959; Green et al., 1976; Hepper, 1984; Siqueira et al., 1985; Mayo et al., 1986). More recently, evidence has emerged that plant metabolites in root exudates may play an important role in the regulation of spore germination and that some flavonoid compounds may be involved in stimulatory effects (Gianinazzi-Pearson et al., 1989; Tsai and Phillips, 1991; Nairet al., 1991; Siqueira et al., 1991; Bécard et al., 1992; Kape et al., 1992b). Host roots positively influence spore germination and this effect appears to be specific since neither roots nor root exudates from non-hosts show any effect (Powell, 1976; Daniels and Trappe, 1980; Tommerup, 1985; Glenn et al., 1985; Gemma and Koske, 1988; El-Atrachet al., 1989; Gianinazzi-Pearson et al. 1989; Avio et al., 1990). However, such differences between host and non-host effects on spore germination have not been observed in root systems that are genetically transformed with *Agrobacterium rhizogenes* (Bécard and Piché, 1990; Schreiner and Koide, 1993), but the reason for this is not known.

After germination, hyphae follow a forward, linear growth pattern, with regular, perpendicular branchings. It has been calculated for *Glomus mosseae,* for example, that the average growth rate of hyphae in the absence of host root factors can reach 1.65 μm/min, that hyphal density is 3.5-3.8 mm/mm^2, and that hyphal growth is arrested after 12 days culture (Giovannetti et al., 1993b). Although *Gigaspora margarita* has much larger spores, hyphae formed from germinating spores behave in much the same way (Bécard and Piché, 1989). More extensive hyphal growth from spores both *in vitro* and *in vivo* has been reported in response to root exudates from host plants, before physical contact between the symbionts (Mosse, 1962; Mosse and Hepper, 1975; Graham, 1982; Hepper, 1984; Bécard and Piché, 1989; Gianinazzi-Pearson et al., 1989; Giovannetti et al., 1993b). As for spore germination, this growth-promoting effect is host-specific. Roots of non-host species neither enhance nor inhibit hyphal growth of AMF, showing that such plants must lack necessary eliciting factors (Glenn et al., 1985 and 1988; Gianinazzi-Pearson et al., 1989; Avio et al., 1990; Vierheilig and Ocampo, 1990; Giovannetti et al., 1993a).

There is some evidence for the role of certain flavonoids and volatiles in the stimulation of hyphal growth in AMF, although effects are concentration-dependent (Carr et al., 1985; Gianinazzi-Pearson et al., 1989; Tsai and Phillips, 1991; Nair et al., 1991; Bécard et al., 1992; Morandi et al., 1992). Flavonoids may have a regulatory role in processes essential to initial infection events, as in other plant-microbe interactions (Phillips and Tsai, 1992). However, it is not yet evident whether they, or other host factors, act as signal molecules in the mycorrhizal associations or whether they are simply a nutrient source for AMF. Chemotropism of hyphae towards host roots has been shown for the AMF *Gigaspora gigantea,* and it has been suggested that volatiles could be involved in this directional growth (Koske, 1982; Gemma and Koske, 1988). As for other biotrophic organisms (Currier and Strobel, 1986; Dixon and Lamb, 1990), volatiles could represent an important factor for AMF to locate host roots because of their rapid, long-distance diffusion in the soil.

Nevertheless, even in the presence of host root factors, hyphal growth from spores is arrested after 20-24 days indicating that although AMF are able to respond to the stimuli coming from host roots, they cannot perform the successive stages of their life cycle if further events of the symbiosis do not take place (Bécard and Piché, 1989; Giovannetti et al., 1993a).

The reasons for this inability for prolonged independent growth remain unknown. It was originally suggested that lack of DNA synthesis and nuclear proliferation could be the cause (Burggraaf and Beringer 1989), but this has been refuted by recent observations on nuclear migration and replication of nuclear DNA in hyphae growing out from spores (Bianciotto and Bonfante, 1992; Bécard and Pfeffer, 1993), so that more complex mechanisms must be responsible.

B. Differential hyphal morphogenesis Some authors have observed that root exudates of host plants not only elicit hyphal proliferation, but also exert a morphogenetic effect on AMF. In fact, hyphal branching increases in the presence of host roots or their exudates (Graham, 1982; Elias and Safir, 1987; Mosse, 1988), and hyphal tips branch profusely within a few mm of host roots (Mosse and Hepper, 1975; van Nuffelen and Schenck, 1984; Bécard and Fortin, 1988). It has been proposed that this kind of hyphal morphology could have assimilatory functions or, alternatively, represent a change necessary for the fungus to become infective.

64

More detailed investigations of the early events in AMF infection, using an *in vivo* system where mycelium is separated from roots by a permeable membrane, have shown that differential hyphal morphogenesis, before appressorium formation, is a more complex process (Giovannetti et al., 1993b). When AMF grow over the membrane overlying host roots, individual hyphae abandon their original linear growth pattern of relatively regular branching, spacing and apical dominance, to adopt an irregular behaviour with reduced inter-hyphal spacing, frequent septa and repeatedly altered directional growth (Fig. 1). The modified hyphae grow at a rate of 3.4 μm/min, proliferate profusely, giving rise to a thick hyphal network where hyphal density is 4-fold and hyphal branching 7-fold that of mycelium growing in the absence of roots (Giovannetti et al., 1993b).

This differential morphogenesis appears to be in response to a stimulus which is host-specific.[1] No change in growth pattern has been observed when AMF are challenged with roots either of non-mycorrhizal plants such as *Brassica, Spinacia, Beta,* or of non-hosts such as *Pinus* or *Abies* (ectomycorrhizal), *Arbutus* (arbutoid mycorrhizal), *Vaccinium* (ericoid mycorrhizal) (Giovannetti et al., 1993a; Giovannetti et al., 1994). It can therefore be concluded that AMF are able to discriminate between roots of host plants and all other plant species, through the recognition of specific signals emitted by host roots.

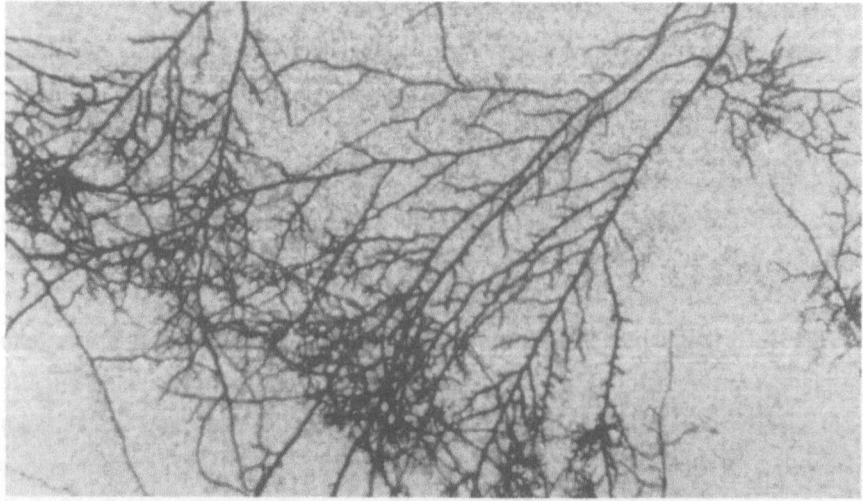

Figure 1. Light micrograph showing differential morphogenesis in hyphae of an AMF elicited by the roots of a host plant growing underneath a millipore membrane (x 80).

The significance of this differential hyphal morphogenesis in AMF lies in its role as a precise recognition response, the earliest detected so far, where the extensive hyphal branching probably has the function of searching for suitable sites for appressorium formation. Moreover, the occurrence of these initial morphogenetic events suggests that the cytoskeleton may be implicated in mediating early fungal response to host root signals (Åström et al., 1994), as reported for other biotrophic fungi (Tucker et al., 1986; Kwon and Hoch, 1991).The nature of the host signals first recognized by AMF is not known, and further research is necessary into the host factor(s) involved in order to determine whether the morphogenetic activity of root exudates is due to a combination of compounds or to a single one, and what is the chemical nature and effective concentration. It can, however, be concluded from the studies mentioned so far that all plant species forming arbuscular mycorrhiza must release chemical signals with a similar activity. These must be lacking in the rhizosphere of all other plants, whether they be non-mycorrhizal or hosts to other types of mycorrhizal fungi, and this could be the basis of host specificity. Research into the AMF genes specifically activated during this stage of infection could provide some clues as determinants essential to hyphal growth, appressorium differentiation and to the establishment of subsequent interactions between the mycorrhizal symbionts.

C. Appressorium formation Appressoria of AMF are inflated, multinucleate structures which vary in their number, shape and size, depending on the fungal species (Garriock et al., 1989; Fortuna et al., 1992) (Fig. 2).

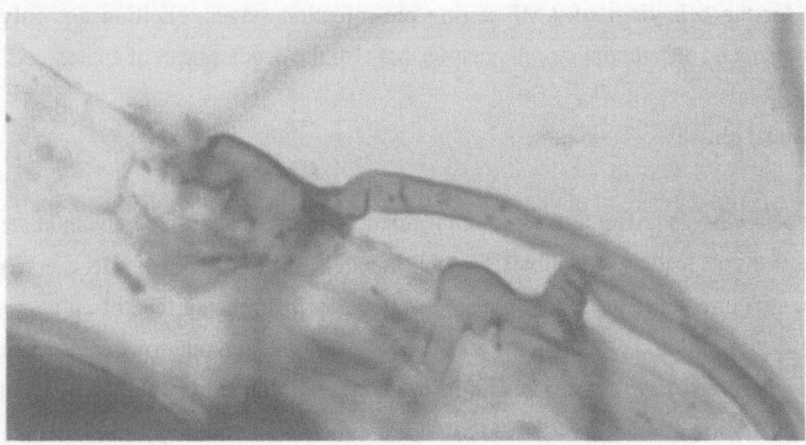

Figure 2. Light micrograph showing appressoria formed by an AMF on the surface of host roots (x 500)

Appressorium formation is generally considered the most decisive event in fungal recognition and infection of a host (Staples and Macko, 1980). These infection structures are in fact only developed by AMF on the surface of host roots, and not on roots of non-mycorrhizal plants or on plants which are hosts of ecto-, arbutoid and ericoid mycorrhizae (Tommerup, 1984; Glenn et al., 1985; 1988; Bedmar and Ocampo, 1986; Avio et al., 1990; Giovannetti et al., 1993a, 1994). Furthermore, appressoria are formed regardless of the outcome of the interaction or of the formation of a functional symbiosis, as illustrated in the case of mycorrhiza-resistant myc⁻ pea mutants and an alfalfa genotype (Duc et al., 1989; Gianinazzi-Pearson et al., 1991; Bradbury et al., 1993a; b). A time-course study of appressorium formation by *G. mosseae* has shown that this fungus is capable of differentiating infection structures within 36 hours after the beginning of interactions with the hosts *Helianthus annuus* and *Ocimum basilicum* (Giovannetti and Citernesi, 1993).

Nothing is known about the phenomenon of hyphal adhesion by AMF to host roots nor the nature of signals inducing appressorium formation. The former could be mediated by compounds occurring in root mucilages, or on root and hyphal surfaces. In investigations of factors affecting appressorium formation, it has been found that even in the presence of host root exudates, thigmotropic stimuli such as nylon, silk, cellulose, polyamide or glass threads do not elicit differentiation of these infection structures (Giovannetti et al., 1993a). However, specific topographical signals may be implicated in mediating appressorium formation on host roots, since hyphal growth is frequently orientated along epidermal grooves and appressoria are formed over clinal and anticlinal wall junctions between adjacent epidermal cells (Garriock et al., 1989; Giovannetti et al., 1993a), as in other biotrophic fungi (see for example Hoch et al., 1987; Teruhne et al., 1993). Consequently, root factors from host plants are once more decisive for the behaviour of AMF during pre-infection stages, eliciting not only hyphal chemotropism and differential morphogenesis, but also the development of appressoria.

Penetration phase

A. Susceptible hosts After appressorium formation on the root surface, the next step in the sequence of events leading to symbiosis establishment is that of growth of infection hyphae out from this structure. Several infection pegs can develop from the surface of an appressorium adhering to root cells and often two are able to successfully infect, outgrowing the others which progressively retract their cytoplasm and form septa (Garriock et al., 1989; Giovannetti et al., 1993a). Penetration of outer root tissues can occur either intercellularly, or intracellularly with the formation of simple or large coiled hyphae in epidermal and hypodermal cells.

Information about mechanisms by which AMF penetrate cells underlying appressoria is lacking. Localized changes in host wall texture have occasionally been observed during root penetration, suggesting cell wall degrading activities at this stage (Jacquelinet-Jeanmouginet al., 1987), and biochemical studies have shown that AMF possess limited cellulase and pectinase activities (García-Romera et al., 1991; García-Garrido et al.,1992). However, a precise role of such enzymes in the early stages of root penetration has to be demonstrated, and the question as to the mechanism/s by which AMF penetrate host cells is still largely unanswered.

Little is known about early plant reactions to the development of penetration hyphae. There are occasional reports of epidermal or hypodermal cells forming wall thickenings at the point of penetration, but no such response has been observed in the majority of host plants. Only slight increases in chitinase and peroxidase activities, but not in phenolics nor phytoalexin accumulation, have been found in the earliest period of arbuscular mycorrhizal establishment and neither callose nor pathogenesis-related PRb1 protein have been detected at penetration points (Spanu and Bonfante, 1988; Spanu et al., 1989; Morandi et al., 1984; Gollotte et al., 1993, 1994). It would therefore appear that in arbuscular mycorrhiza, plant defence responses are either only weakly elicited or are suppressed during the penetration phase of the plant-fungus interaction. Such low expression of defence mechanisms is maintained throughout the fully compatible, functional symbiosis (Gianinazzi, 1991).

B. Resistant mutants Mycorrhizal resistant plants have been identified among non nodulating (nod⁻) pea mutants, where the development of the fungus is stopped at the stage of appressorium formation on the root surface. These mutants, called myc⁻ in analogy to the nod⁻ character (Duc et al., 1989; Gianinazzi-Pearson et al., 1991), provide an interesting experimental model for identifying plant or fungal genes involved in the early events of AMF infection for the following reasons: 1. they are chemically induced mutants so that, presumably, only a small number of bases in a gene is affected; 2. they are genetically stable; 3. they keep their myc⁻ phenotype even when changing inoculation conditions, contrary to reportedly myc⁻ alfalfa genotypes which become infected by indigenous endophytes (Bradbury et al., 1991, 1993); 4. they belong to four groups of complementation, suggesting that at least four plant genes are involved in the early steps of AMF infection; 5. contrary to resistance to pathogens, the myc⁻ character is under recessive genetic control; 6. up to now all myc⁻ plants have a nod⁻ phenotype.

Ultrastructural analyses of myc⁻ mutants have shown that, contrary to host plants (myc⁺), there is elicitation of important wall thickenings in epidermal cells in contact with the appressoria of an AMF (Gianinazzi-Pearson et al., 1991). These wall structures resemble paramural deposits formed in incompatible plant-pathogen interactions (Bracker and Littlefield, 1973), and defence associated compounds are likewise located in them such as phenolics, ß-1,3 glucans (callose) and pathogenesis-related PRb1 protein (Gollotte et al., 1993, 1994, and unpublished results). Furthermore, an arabinogalactanprotein, suggested to be involved in cell-to-cell recognition events and eventually in defence responses, preferentially accumulates in the

68

myc⁻ paramural deposits (Gollotte et al., unpublished results). These observations clearly demonstrate that, although only weak resistance responses occur in myc$^+$ plants, an AMF is able to induce a strong defence reaction in myc⁻ mutant pea plants so that root penetration cannot occur.

In mycorrhizal myc$^+$ plants, specific genes involved in symbiosis establishment may play a regulatory role towards defence genes. Consequently, when symbiosis genes are altered in their function, as in the myc⁻ mutants, defence reactions against an invading AMF are no longer suppressed (Gianinazzi-Pearson et al., 1991; Gollotte et al., 1993). All root-infecting pathogens tested so far give similar infection patterns in myc$^+$ or in myc⁻ plants, whether these be fungi (*Aphanomyces euteiches, Rhizoctonia* species, *Chalara elegans*) (unpublished results), bacteria (*Agrobacterium tumefaciens*) or nematodes (*Meloidogyne* species) (Gollotte et al., 1993). It seems therefore, that the myc⁻ mutants are only resistant to symbiotic microorganisms which establish compatible interactions with myc$^+$ plants (AMF and *Rhizobium*). A similar type of defence response can be elicited in *Rhizobium*-legume interactions, for example, by exopolysaccharide-deficient bacterial mutants in pseudonodules of alfalfa (Niehaus et al., 1993). These observations, together with the constant coincidence of the myc⁻ and nod⁻ characters in the mutants and their specificity towards AMF and *Rhizobium*, suggest that common mechanisms may control some early infection events in both types of symbiosis.

Conclusions

Arbuscular mycorrhizal symbiosis is established through a multi-step process consisting of a cascade of recognition events, and leading to complete morphological and physiological integration of the two partners (Gianinazzi-Pearson, 1984). There are at least three stages in the sequence of recognition processes at which plant-fungus interactions can be halted. AMF spores germinate and hyphae branch in all directions, until they meet signals diffusing from roots of host plants. These act as cues to AMF so that hyphae begin to proliferate, increasing the chances of coming into contact with host roots. The fungi also respond by modifications in their hyphal morphogenesis which culminate in appressorium formation. Such differential hyphal morphogenesis does not occur in the absence of host roots, so that the infection process is not initiated. After appressorium formation, root penetration by the fungus depends on the genome of host plants. Where this is altered, as in mycorrhiza-resistant plant mutants, penetration is hindered by plant defence reactions and infection is inhibited, whereas a successful infection develops and a functional symbiosis is established in unmodified hosts.

In conclusion, the molecular dialogue established in the rhizosphere between arbuscular fungi and host plants leads to an initial selective recognition, but further gene activation, both in the fungus and in the plant, is necessary for the development of a meaningful symbiosis.

Future research will disclose the full significance of pre-infection events for AMF. These biotrophic organisms must have evolved an efficient system for locating their hosts, which in the end ensures not only their own survival, but also the fitness and maintenance of the plant ecosystem they live in.

References

Åström, H., Giovannetti, M., and Raudaskoski, M. (1994) Cytoskeletal components in the arbuscular mycorrhizal fungus *Glomus mosseae*. *Mol. Pl. Micr. Int.* 7: 309-312.

Avio, L., Sbrana, C., and Giovannetti, M. (1990) The response of different species of *Lupinus* to VAM endophytes. *Symbiosis* 9: 321-323.

Bedmar, E.J. and Ocampo, J.A. (1986) Susceptibilidad da distintas variedades de guisante, veza y lupino a la infeccion por *Glomus mosseae*. *Anales de Edafologia y Agrobiologia* 45: 231-238.

Bécard, G., Douds, D.D., and Pfeffer, P.E. (1992) Extensive in vitro hyphal growth of vesicular-arbuscular mycorrhizal fungi in the presence of CO_2 and flavonols. *Appl. and Environ. Microbiol.* 58: 821-825.

Bécard, G. and Fortin, J.A. (1988) Early events of vesicular-arbuscular mycorriza formation on Ri T-DNA transformed roots. *New Phytol.* 108: 211-218.

Bécard, G. and Pfeffer, P.E. (1993) Status of nuclear division in arbuscular mycorrhizal fungi during in vitro development. *Protoplasma* 174: 62-68.

Bécard, G. and Piché, Y. (1989) Fungal growth stimulation by CO_2 and root exudates in vesicular-arbuscular mycorrhizal symbiosis. *Appl. and Environ. Microbiol.* 55: 2320-2325.

Bécard, G. and Piché, Y. (1990) Physiological factors determining vesicular-arbuscular mycorrhizal formation in host and non-host Ri T-DNA transformed roots. *Can. J. Bot.* 68: 1260-1264.

Bianciotto, V. and Bonfante-Fasolo, P. (1992) Quantification of the nuclear DNA content of two arbuscular mycorrhizal fungi. *Mycol. Res.* 96: 1071-1076.

Bracker, C.E. and Littlefield, L.J. (1973) Structure of host-pathogen interfacies. In: R.J.W. Byrde and C.V. Cutting (eds) *Fungal pathogenicity and plant's response*. Academic Press, London, New York, pp.159-317.

Bradbury, S.M., Peterson, R.L., and Bowley, S.R. (1991) Interactions between three alfalfa nodulation genotypes and two *Glomus* species. *New Phytol.* 119: 115-120.

Bradbury, S.M., Peterson, R.L., and Bowley, S.R. (1993a) Further evidence for a correlation between nodulation genotypes in alfalfa (*Medicago sativa* L) and mycorrhiza formation. *New Phytol.* 124: 665-673.

Bradbury, S.M., Peterson, R.L., and Bowley, S.R. (1993b) Colonization of three alfalfa (*Medicago sativa* L.) nodulation fenotypes by indigenous vesicular-arbuscular mycorrhizal fungi. *Symbiosis* 15: 207-215.

Burggraaf, J.P. and Beringer, J.E. (1989) Absence of nuclear DNA synthesis in vesicular-arbuscular mycorrhizal fungi during in vitro development. *New Phytol.* 111: 25-33.

Carr, G.R., Hinkley, M.A., Le Tacon, F., Hepper, C.M., Jones, M.G.K., and Thomas, E. (1985) Improved hyphal growth of two species of vesicular-arbuscular mycorrhizal fungi in the presence of suspension-cultured plant cells. *New Phytol.* 101: 417-426.

Currier, W.W. and Strobel, G.A. (1986) Chemotaxis of *Rhizobium* spp. to plant root exudates. *Plant Physiol.* 59: 820-823.

Daniels, B.A. and Trappe, J.M. (1980) Factors affecting spore germination of the vesicular-arbuscular mycorrhizal fungus, *Glomus epigaeus*. *Mycologia* 72: 457-471.

Dixon, R.A. and Lamb, C.J. (1990) Molecular communication in interactions between plants and microbial pathogens. *Ann. Revi. Plant Phys and Plant Mol. Biol.* 41: 339-367.

Duc, G., Trouvelot, A., Gianinazzi-Pearson, V., and Gianinazzi, S. (1989) First reports of non-mycorrhizal plant mutants (Myc-) obtained in pea (*Pisum sativum*) and fababean (*Vicia faba*). *Plant Science* 60: 215-222.

El-Atrach, F., Vierheilig, H., and Ocampo, J.A. (1989) Influence of non-host plants on vesicular-arbuscular mycorrhizal infection of host plants and on spore germination. *Soil Biol. Biochem.* 21: 161-163.

Elias, K.E. and Safir, G.R. (1987) Hyphal elongation of *Glomus fasciculatus* in response to root exudates. *Appl. Environm. Microbiol.* 53: 1928-1933.

Fortuna, P., Citernesi, A.S., Morini, S., Giovannetti, M., and Loreti, F. (1992) Infectivity and effectiveness of different species of arbuscular mycorrhizal fungi in micropropagated plants of Mr.S. 2/5 plum rootstocks. *Agronomie* 12: 825-829.

García-Garrido, J.M., García-Romera, I., and Ocampo, J.A. (1992) Cellulase production by the vesicular-arbuscular mycorrhizal fungus *Glomus mosseae* (Nicol. and Gerd.) Gerd. and Trappe. *New Phytol.* 121(2): 221-226.

García-Romera, I., García-Garrido, J.M., Martinez-Molina, E., and Ocampo, J.A. (1991) Production of pectolytic enzymes in lettuce root colonized by *Glomus mosseae. Soil Biol. Biochem.* 23(6): 597-601.

Garriock, M.L., Peterson, R.L., and Ackerley, C.A. (1989) Early stages in colonization of *Allium porrum* (leek) roots by the vesicular-arbuscular mycorrhizal fungus, *Glomus versiforme. New Phytol.* 112: 85-92.

Gemma, J.N. and Koske, R.E. (1988) Pre-infection interactions between roots and the mycorrhizal fungus *Gigaspora gigantea*: chemotropism of germ-tubes and root growth response. *Trans. Brit. Myc.Soc.* 91: 123-132.

Gianinazzi, S., (1991) Vesicular-arbuscular (endo) mycorrhizas: cellular, biochemical and genetic aspects. *Agric. Ecosyst. Environm.* 35: 105-119.

Gianinazzi-Pearson, V. (1984) Host-fungus specificity, recognition and compatibility in mycorrhizae. In: D.I.S. Verma and I. Hohn (eds) *Plant gene research basic knowledge and application: gene involved in microbe-plant interactions.* Springer-Verlag, New York, Vienna, pp.225-253.

Gianinazzi-Pearson, V., Branzanti, B., and Gianinazzi, S. (1989) In vitro enhancement of spore germination and early hyphal growth of a vesicular-arbuscular mycorrhizal fungus by host root exudates and plant flavonoids. *Symbiosis* 7: 243-255.

Gianinazzi-Pearson, V., Gianinazzi, S., Guillemin, J.P., Trouvelot, A., and Duc, G. (1991) Genetic and cellular analysis of resistance to vesicular-arbuscular (VA) mycorrhizal fungi in pea mutants. In: H. Hennecke and D.I.S. VermA (eds) *Advances in molecular genetics of plant-microbe interactions.* Kluwer Academic Publishers, pp.336-342.

Giovannetti, M., Avio, L., Sbrana, C., and Citernesi, A.S. (1993a) Factors affecting appressorium development in the vesicular- arbuscular mycorrhizal fungus *Glomus mosseae* (Nicol. & Gerd.) Gerd. & Trappe. *New Phytol.* 123: 114-122.

Giovannetti, M. and Citernesi, A.S. (1993) Time-course of appressorium formation on host plants by arbuscular mycorrhizal fungi. *Mycol. Res.* 97: 1140-1142.

Giovannetti, M., Sbrana, C., Avio, L., Citernesi, A.S., and Logi, C. (1993b) Differential hyphal morphogenesis in arbuscular mycorrhizal fungi during pre-infection stages. *New Phytol.* 125: 587-594.

Giovannetti, M., Sbrana, C., and Logi, C. (1994) Early processes involved in host recognition by arbuscular mycorrhizal fungi. *New Phytol.* In press.

Glenn, M.G., Chew, F.S., and Williams, P.H. (1985) Hyphal penetration of *Brassica* (*Cruciferae*) roots by a vesicular-arbuscular mycorrhizal fungus. *New Phytol.* 99: 463-472.

Glenn, M.G., Chew, F.S., and Williams, P.H. (1988) Influence of glucosinolate content of *Brassica* (*Cruciferae*) roots on growth of vesicular-arbuscular mycorrhizal fungi. *New Phytol.* 110: 217-225.

Gollotte, A., Gianinazzi-Pearson, V., and Gianinazzi, S. (1994) Etude immunocytochimique des interfaces plant- champignon endomycorhizien à arbuscules chez des pois isogéniques myc[+] ou resistant á l'endomycorhiza-tion (myc[-]). *Acta Botanica Gallica.* In press.

Gollotte, A., Gianinazzi-Pearson, V., Giovannetti, M., Sbrana, C., Avio, L., and Gianinazzi, S. (1993) Cellular localization and cytochemical probing of resistance reactions to arbuscular mycorrhizal fungi in a 'locus a' myc[-] mutant of *Pisum sativum* L. *Planta* 191: 112-122.

Graham, J.H. (1982) Effect of citrus exudates on germination of chlamydospores of the vesicular-arbuscular mycorrhizal fungus, *Glomus epigaeum. Mycologia* 74: 831-835.

Green, N.E., Graham, J.H., and Schenck, N.C. (1976) The influence of pH on the germination of vesicular-arbuscular mycorrhizal spores. *Mycologia* 68: 929-934.

Hepper, C.M. (1984) Regulation of spore germination of the vesicular-arbuscular mycorrhizal fungus *Acaulospora laevis* by soil pH. *Trans. Brit. Mycol. Soc.* 83: 154-156.

Hoch, H.C., Staples, R.C., Whitebread, B., Comeau, J., and Wolf, E.D. (1987) Signaling for growth orientation and cell differentiation by surface topography in *Uromyces*. *Science* 235: 1659-1662.

Jacquelinet-Jeanmougin, J., Gianinazzi-Pearson, V., and Gianinazzi, S. (1987) Endomycorrhizas in the Gentianaceae. II. Ultrastructural aspectes of symbiont relationships in *Gentiana lutea* L. *Symbiosis* 3: 269-286.

Kape, R., Wex, K., Parniske, M., Görge, E., Wetzel, A., and Werner, D. (1992) Legume root metabolites and VA-Mycorrhiza development. *J. Pl. Phys.* 141: 54-60.

Koske, R.E. (1982) Evidence for a volatile attractant from plant roots affecting germ tubes of a VA mycorrhizal fungus. *Trans. Brit. Mycol. Soc.* 79: 305-310.

Kwon, Y.H. and Hoch, H.C. (1991) Initiation of appressorium formation in *Uromyces appendiculatus*: organization of the apex, and the responses involving microtubules and apical vesicles. *Can. J. Bot.* 69: 2560-2573.

Lewis, D.H. (1973) Concepts in fungal nutrition and the origin of biotrophy. *Biological Reviews* 48: 261-278.

Mayo, K., Davis, R.E., and Motta, J. (1986) Stimulation of germination of spores of *Glomus versiforme* by spore-associated bacteria. *Mycologia* 78: 426-431.

Morandi, D., Bailey, J.A., and Gianinazzi-Pearson, V. (1984) Isoflavonoid accumulation in soybean roots infected with vesicular-arbuscular mycorrhizal fungi. *Physiol. Plant Path.* 24: 357-364.

Morandi, D., Branzanti, B., and Gianinazzi-Pearson, V. (1992) Effect of some plant flavonoids on in vitro behaviour of an arbuscular mycorrhizal fungus. *Agronomie* 12: 811-816.

Mosse, B. (1959) The regular germination of resting spores and some observations on the growth requirements of an *Endogone sp.* causing vesicular- arbuscular mycorrhiza. *Trans. Brit. Mycol. Soc.* 42: 273-286.

Mosse, B. (1962) The establishment of vesicular-arbuscular mycorrhiza under aseptic conditions. *J. Gen. Microbiol.* 27: 509-520.

Mosse, B. (1988) Some studies relating to "independent" growth of vesicular-arbuscular endophytes. *Can. J. Bot.*66: 2533-2540.

Mosse, B. and Hepper, C.M. (1975) Vesicular-arbuscular mycorrhizal infections in root organ cultures.*Physiol.Plant Path.* 5: 215-223.

Nair, M.G., Safir, G.R., and Siqueira, J.O. (1991) Isolation and identification of vesicular-arbuscular mycorrhiza-stimulatory compounds from clover (*Trifolium repens*) roots. *Appl. Environm. Microbiol.* 57: 434-439.

Niehaus, K., Kapp, D., and Pühler, A. (1993) Plant defence and delayed infection of alfalfa pseudonodules induced by an exopolysaccharide (EPSI)-deficient *Rhizobium meliloti* mutant. *Planta* 190: 415-425.

Phillips, D.A. and Tsai, S.M. (1992) Flavonoids as plant signals to rhizosphere microbes. *Mycorrhiza* 1: 55-58.

Powell, C.L. (1976) Development of mycorrhizal infections from *Endogone* spores and infected root fragments.*Trans. Brit. Mycol. Soc.* 66: 439-445.

Schreiner, R.P. and Koide, R.T. (1993) Stimulation of vesicular-arbuscular mycorrhizal fungi by mycotrophicand nonmycotrophic plant root systems. *Appl. Environm. Microbiol.* 59: 2750-2752.

Siqueira, J.O., Safir, G.R., and Nair, M.G. (1991) Stimulation of vesicular-arbuscular mycorrhiza formation and growth of white clover by flavonoid compounds. *New Phytol.* 118: 87-93.

Siqueira, J.O., Sylvia, D.M., Gibson, J., and Hubbell, D.H. (1985) Spores, germination, and germ tubes of vesicular-arbuscular mycorrhizal fungi. *Can. J. Microbiol.* 31: 965-972.

Spanu, P., Boller, T., Ludwig, A., Wiemken, A., Faccio, A., and Bonfante-Fasolo, P. (1989) Chitinase in roots of mycorrhizal *Allium porrum*: regulation and localization. *Planta* 177: 447-455.

Spanu, P. and Bonfante-Fasolo, P. (1988) Cell-wall-bound peroxidase activity in roots of mycorrhizal *Allium porrum*. *New Phytol.* 109: 119-124.

Staples, R.C. and Hoch, H.C. (1982) A possible role for microtubules and microfilaments in the induction of nuclear division in bean rust uredospore germlings. *Exp. Mycol.* 6: 293-302.

Staples, R.C. and Macko, V. (1980) Formation of infection structures as a recognition response in fungi.*Exp. Mycol.* 4: 2-16.

Terhune, B.T., Bojko, R.J., and Hoch,H.C. (1993) Deformation of stomatal guard cell lips and microfabricated artificial topographies during appressorium formation by *Uromyces*. *Exp. Mycol.* 17: 70-78.

Tommerup, I.C. (1984) Development of infection by a vesicular-arbuscular mycorrhizal fungus in *Brassica napus* and *Trifolium subterraneum*. *New Phytol.* 98: 497-502.

72

Tommerup, I.C. (1985) Inhibition of spore germination of vesicular-arbuscular mycorrhizal fungi in soil.*Trans. Brit. Mycol. Soc.* 85: 267-278.

Tsai, S.M. and Phillips, D.A. (1991) Flavonoids released naturally from alfalfa promote development of symbiotic *Glomus* spores in vitro. *Appl. Environm. Microbiol.* 57: 1485-1488.

Tucker, B.E., Hoch, H.C., and Staples, R.C. (1986) The involvement of F-actin in *Uromyces* cell differentiation: the effects of cytochalasin E and phalloidin. *Protoplasma* 135: 88-101.

van Nuffelen, M. and Schenck, N.C. (1984) Spore germination, penetration, and root colonization of six speciesof vesicular-arbuscular mycorrhizal fungi on soybean. *Can. J. Bot.*62: 624-628.

Vierheilig, H. and Ocampo, J.A. (1990) Role of root extract and volatile substances of non-host plants onvesicular-arbuscular mycorrhizal spore germination. *Symbiosis* 9: 199-202.

Impact of Arbuscular Mycorrhizas on
Sustainable Agriculture and Natural Ecosystems
S. Gianinazzi and H. Schüepp (eds.)
© 1994 Birkhäuser Verlag Basel/Switzerland

Ultrastructural analysis reveals the complex interactions between root cells and arbuscular mycorrhizal fungi

P. Bonfante

Dipartimento di Biologia Vegetale dell'Università di Torino e Centro di Studio sulla Micologia del Terreno del CNR, Viale Mattioli, 25 10125 Torino, Italy

Introduction

Mycorrhizas are widespread symbiotic associations established between the roots of land plants and many soil fungi. 2,500 plant species have been estimated to be mycorrhizal, while the symbiotic fungi comprise about 6000 species. This impressive biodiversity gives rise to a number of mycorrhizal types which are usually identified on the basis of morphological, physiological and taxonomical parameters. However, understanding of the mycorrhizal phenomenon in the ecosystems requires both the search for common aspects offering unifying keys, and the development of an adequate theory. A first useful generalization is that mycorrhizas are complex systems formed by cells belonging to two different eukaryotic organisms and communicating to maintain a durable bidirectional nutrient exchange (Harley, 1989). From the seventies onwards, ultrastructure proved a good experimental approach to study the complex organization of mycorrhizas by analysing cell structure of both symbiotic plants and fungi as well as of their contact surfaces (Scannerini and Bonfante, 1983). Ultrastructure has provided a great deal of information on the cell to cell interactions and offered important tools to classify mycorrhizas on the basis of their morphofunctional features, and also offered some of the first keys for the understanding of the mycorrhizal functioning: plant-fungal exchanges - in particular phosphorus transfers - were suggested to occur at the arbuscule level on the basis of ultrastructural observations (Cox and Tinker, 1976). By the middle of the eighties, however, ultrastructure alone seemed to have been overtaken by other more powerful investigation tools, such as those coming from molecular biology. However, the introduction of cryotechniques and affinity techniques (see next paragraphs) produced a conspicuous jump in the field of electron microscopy creating a bridge between morphological, molecular and biochemical observations. These techniques were rapidly applied in the field of mycorrhizas, leading to new substantial results.

The aim of this chapter is therefore to review the contribution made by electron microscopy to the knowledge of the biology of arbuscular mycorrhizas. Since the morphology of plant and fungi in arbuscular mycorrhizas has been analyzed in detail in some recent reviews (Bonfante and Perotto, 1992; Bonfante, 1994) particular attention will be devoted to new technical approaches which have deeply changed our views on plant and fungal interactions, and on the other hand have maintained ultrastructure as a still reliable approach to understand the cellular and molecular bases of functioning endomycorrhizas.

Plant and fungus morphology is improved by the use of protocols based on cryotechniques Our knowledge of the cell structure of the arbuscular mycorrhizas dates back to the seventies (Scannerini and Bonfante, 1983 for a review) and is mostly based on a scenario provided by pictures similar to those shown in Figures 1 and 2. They display light and electron microscope pictures of plant mycorrhizal tissues after chemical fixation with glutaraldehyde and osmium tetroxide (Hall and Hawes, 1991). The colonized cells of the host are filled up by the intracellular hyphae occurring as coils or branched hyphae forming the arbuscule. Intercellular hyphae are present in the intercellular spaces among the host cell walls. Intracellular hyphae are surrounded by the host plasmamembrane. However, preservation of the host membranes and fungal organelles is often *critical* in this type of preparation. In agreement with alterations in membrane structures and extractions in cellular components which have been documented in many experimental systems (Hoch, 1991; Robards, 1991), host membranes and in particular those surrounding the fungal branches are often blebbing and even blurry when osmium is omitted. Observations on fungal morphology are mostly limited to wall organization, while the cytoplasm is described as rich in lipid globules and vacuoles containing electron dense bodies. X-ray microanalysis has shown that these granules are rich in phosphates (Strullu et al.,1981; Turnau et al., 1993). Nuclei, mitochondria, membrane systems are often poorly preserved.

The development of cryotechniques based on cryofixation and on preparation for transmission electron microscopy by freeze-substitution has strongly improved cytological studies, since as Hoch (1991) claims "cellular organization is maintained in a state very similar to that immediately prior to freezing". Cryotechniques are based on the use of a cryogen to freeze the samples, which are then kept frozen (-90°C) in a liquid medium until all cell water has been substituted with another medium, which may contain a chemical fixative (Hoch, 1991), like osmium tetroxide in acetone. The samples are then infiltrated at low or room

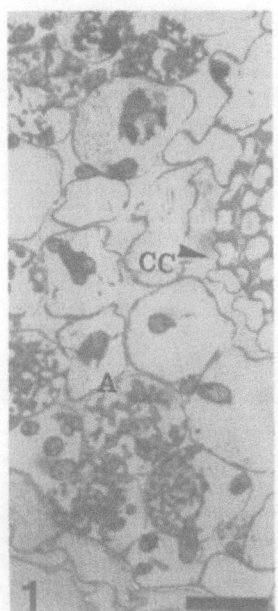 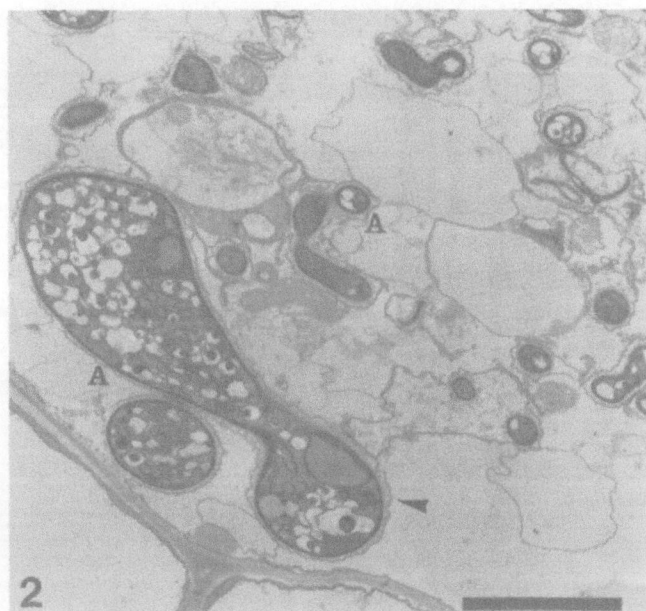

Figure 1. Light micrograph of a tobacco root colonized by *Gigaspora margarita*. All the small parenchyma cells are filled up by hyphae in different steps of their development.A: arbuscule. Central cylinder (CC) is not colonized. Bar corresponds to 50 μm.

Figure 2. Transmission electron micrograph of a *Ginkgo biloba* root colonized by a *Glomus* species. Large and thin arbuscular branches are surrounded by the host membrane (arrow). Bar corresponds to 5 μm.

temperature with resins and polymerised. These techniques have provided new information on fungal and plant cytology (Hoch, 1991; Robards, 1991). A further important improvement is reached when a physical fixation precedes freeze-substitution: samples are placed in an aluminium holder and immediately frozen at high pressure (2000 bar at -180°C) by using apparatus like those described in Muller and Moor (1984) and in Mendgen et al. (1991). Following this procedure membranes are particularly well preserved, leading to the preservation of fragile organelles such as Golgi bodies, not only in plants (Zhang and Staehelin, 1992), but also in fungal cells (Welter et al., 1988).

When this approach was set up for AMF substantial new information was provided during both their sporal and symbiotic phase (Bonfante et al., in preparation), though the samples were not homogenously well preserved. The huge germinating spores of *Gigaspora margarita* displayed a rich cytoplasm, where two areas with different roles are identified: the first is rich in nuclei in

76

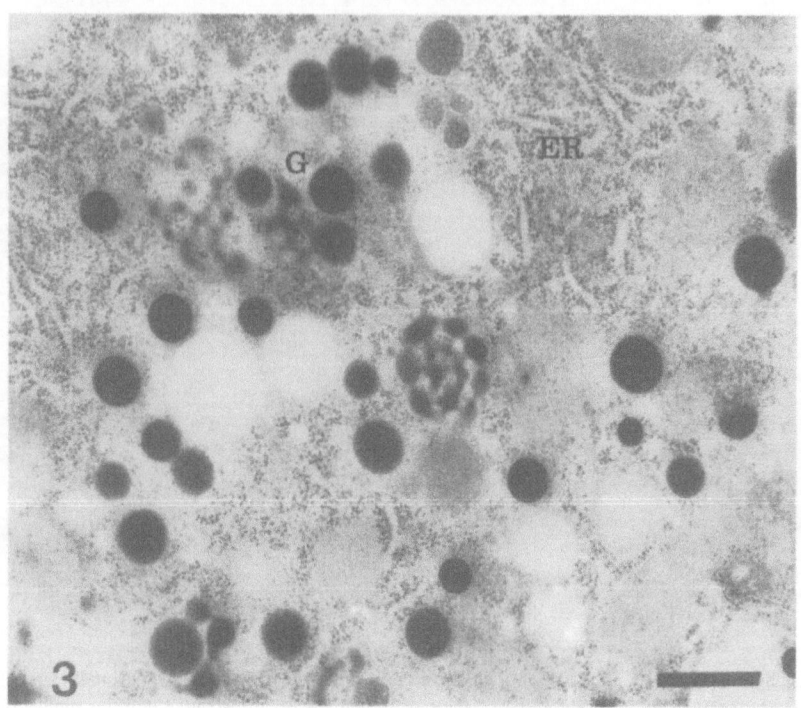

Figure 3. Details of a germinating spore of *Gi margarita* after high pressure-freeze substitution. Ribosomes are associated with membranes of the endoplasmic reticulum (ER), while other membranes with a whirl disposition give rise to electron dense granules (G) (in collaboration with Kurt Mendgen). Bar corresponds to 1 μm.

mitosis (Sward, 1981; Becard and Pfeffer, 1993) as well as in membranous systems. The latter consist of rough endoplasmic reticulum, tubules and balloon-like Golgi equivalents producing vesicles with an electron dense content (Figs.3 and 4). They give rise to an abundant population of granules which are directed towards the wall. The second area can be described as a storage compartment, where i) lipid droplets limited by a semi-membrane, ii) protein-like bodies inside specialized vacuoles and iii) glycogen accumulations are found.

These observations indicate an intense metabolism during the events of spore germination. They strongly support physiological similarities between the spore of an AMF and a germinating seed, and confirm the presence of nuclear mitosis and growth of the germinating mycelium, even in the absence of the host plant (Bianciotto and Bonfante, 1993).

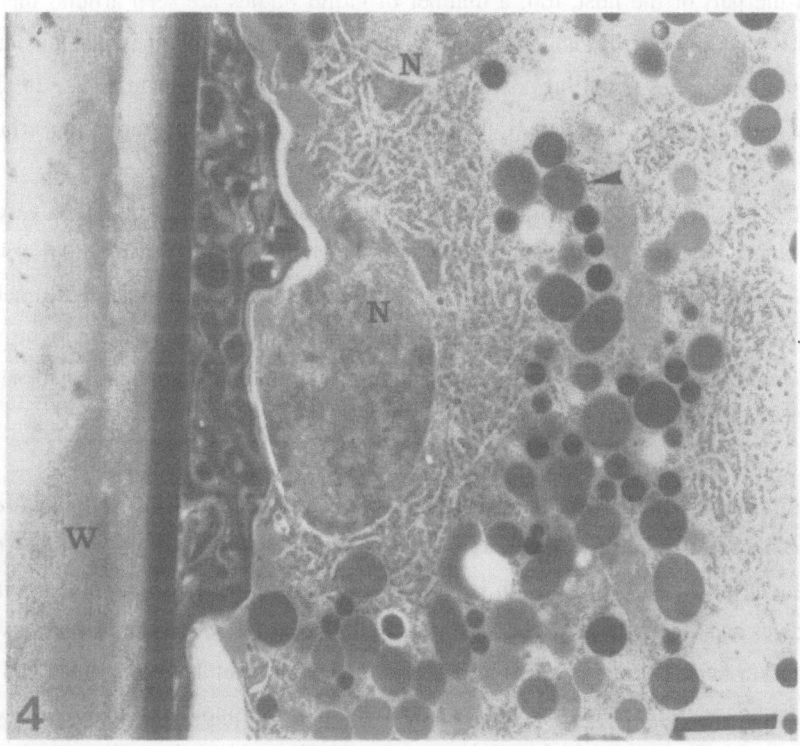

Figure 4. Details of a germinating spore of *Gi. margarita*, after high pressure-freeze substitution. A population of electron dense granules (arrows) is observed close to the thick and laminated wall (W). Nuclei (N) are surrounded by abundant membranes of rough endoplasmic reticulum. Bar corresponds to 1 μm.

The development of the symbiotic structures starting from the germinating mycelium is under the control of fungal and plant factors (Giovannetti et al., 1993). The use of cryotechniques has also revealed new information in this phase: the fungus is rich in organelles including minute mitochondria, a well developed membranous systems, cytoskeletal elements, vacuoles rich in granular content, similar to those already described in *Pisolithus tinctorius* with the same techniques (Orlovich and Ashford, 1993). Interestingly enough, even in this case the electron dense polyphosphate bodies are no longer identified, whereas glycogen particles are abundant. Nuclei are often seen involved in active mitosis, which is a typical endomitosis, as in many other fungi. The membrane systems are usually well preserved. Much new information is offered on the part of the host, too: a number of Golgi bodies are seen around the fungal branches, associated with ribosomes and endoplasmic reticulum membranes. The nucleus with dispersed chromatin has a central position (Berta et al., 1990; Balestrini et al., 1992). The interface membrane (see next paragraph) is smooth and closely surrounds the arbuscular branches.

All these observations shed new light on plant-fungal morphology: the fungus is a very active organism, which - particularly in the early phase of the symbiotic interaction (Garriock et al., 1989) - has a limited number of vacuoles and or lipid globules. By contrast, it is much more involved in the synthesis of glycogen. The host cell reacts to fungus colonisation by moving its nucleus towards the centre of the cell (Balestrini et al., 1992) and by increasing its secretion pathways through endoplasmic reticulum, Golgi bodies and vesicles.

The development of affinity techniques has improved the knowledge of cell to cell contact in arbuscular mycorrhizas The idea that plant cells and the cells of AMF are in a permanent physical contact which leads to the establishment of an apoplastic *interface* structure, dates back many years (Scannerini and Bonfante, 1983). The term *interface* was first used by Bracker and Littlefield (1973) to describe the contact area between plants and fungal pathogens, but now used in a more general sense to indicate morphological contiguity between walled organisms. The role of the interface in mycorrhizal symbiosis is to allow exchanges of signalling and nutritional molecules, thanks to specialised enzymic activities of the plant and fungal membranes (Smith and Smith, 1990; Bonfante and Scannerini, 1992). However, the morphological organization of the interface changes depending on whether the fungus penetrates the host cell or not. In the first case, the fungal cell wall is separated from the host cytoplasm by the invaginated host plasma membrane and by an interfacial material (fig.5), while in the second case the cell walls of both partners are physically in contact. The zone of interface created by the intracellular fungus, particularly during the arbuscular development, s a compartment of high molecular complexity, which therefore proves to be typical of endomycorrhizas, irrespective of their nature (Bonfante and Perotto, 1992; Bonfante 1994). Cryotechniques (see above) have offered convincing evidence of this area as a true space of about 50-80 nm in thickness (fig.6), while the development of affinity techniques, based on the specific non-covalent binding between molecules such as antigen-antibody, sugar-lectin, substrate-enzyme, has provided information on the nature of the interface components. The structural molecules identified in the fungal wall, and in the interfacial material of a number of dicot and monocot hosts, includepolysaccharides, proteins and glycoproteins. They are listed in Table I. Chitin is a constant skeletal component of the fungal wall, even in the thin fungal branches (fig.7), while

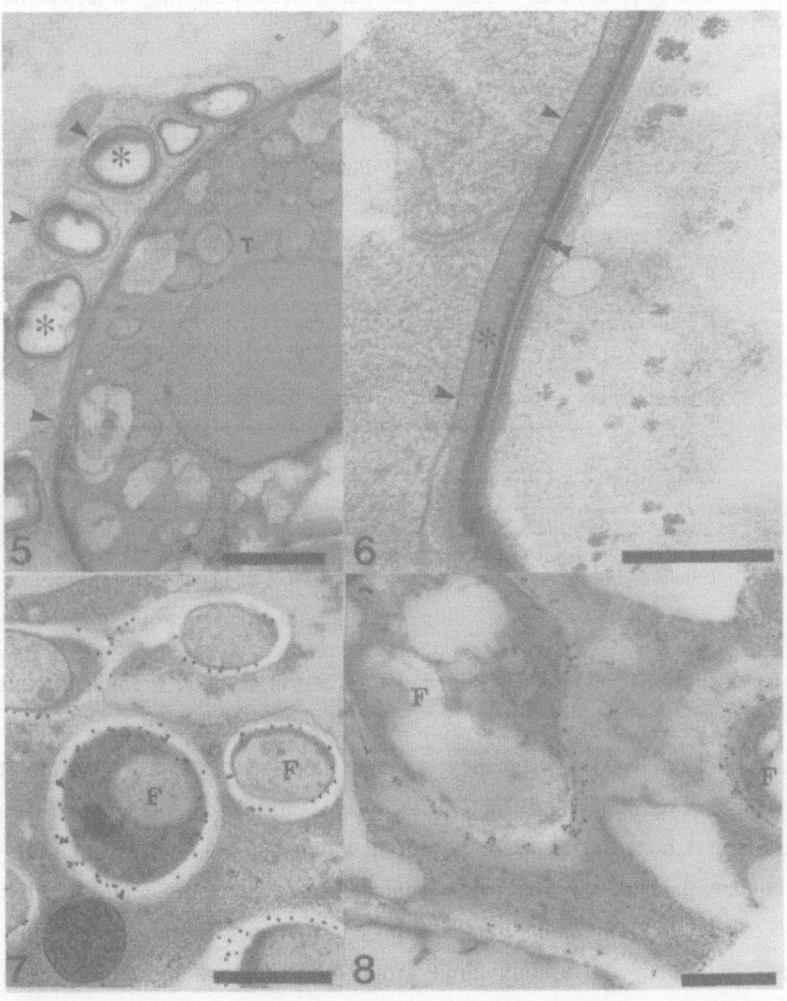

Figure 5. A large trunk (T) of *Gi margarita* surrounded by small arbuscular branches (*) in a root cortical cell of clover. The fungal branches are surrounded by the host plasma membrane limiting the interface area. Bar corresponds to 1 μm. Figure 6. Detail of the interface area between a clover cortical cell and *Gi margarita* after high pressure-freeze substitution (in collaboration with Kurt Mendgen). The host membrane (arrowheads) has a smooth outline and the interfacial material (*) is electron dense. The fungal wall (double arrowhead) is thin and electron dense following the PATAg reaction. Figure 7. High pressure-freeze substituted roots of clover colonized by *Gi margarita*. The interface after treatment with the complex WGA/ gold reveals the presence of gold granules over the fungal wall. Figure 8. Treatment with a McAb which binds to nonesterified pectins (JIM 5) reveals a regular distribution of gold granules around the arbuscular branches (*) of *Glomus versiforme* in a *Pisum* root. Bars corresponds to 0.5 μm.

β1-3 glucans only occur in the AMF such as *Glomus* and *Acaulospora*, but not in the AMF like *Gigaspora* and *Scutellospora* (Gianinazzi-Pearson et al., 1994). On the other hand, pectins,

80

Table I. Specific affinity probes used to analyse cell wall molecules occurring in the interface material established between different plants and arbuscular fungi

Probe	Target molecule	Partner species	Reference
JIM 5 McAb	unsterified pectins	leek, pea, clover, ginkgo	Bonfante et al., 1990b Bonfante and Perotto, 1992
JIM 7 McAb	methylesterified pectins	clover, ginkgo	Bonfante and Perotto, 1992
antiserum HRGP	HRGP molecules	leek, pea, maize	Bonfante et al., 1991 Balestrini et al., in preparation
antiserum β 1-3 glucans	β 1-3 glucans	*Glomus* *G. versiforme*	Gianinazzi-Pearson et al., 1994 Balestrini et al., in preparation
WGA	chitin	*Glomus spp* *Gigaspora margarita*	Bonfante et al., 1990 a Gianinazzi-Pearson et al., 1994
CBH	β 1-3 glucans cellulose	leek, pea, maize	Bonfante et al., 1990 b Balestrini et al., in preparation

cellulose, hemicelluloses and hydroxyproline-rich glycoproteins (HRGP) are constantly present in the interface material (fig.8). All these latter molecules are typical of the plant cell wall (Carpita and Gibeaut, 1993), where they organize themselves into a dynamic extracellular matrix. A great deal of research has led to new concepts on cell wall: it is no more regarded as an inert structural box, but as a compartment which contains surface markers of plant development, components involved in communication, and signal molecules which induce the production of soluble components involved in plant defence. The presence of the interface material as a host cell wall like envelope around the fungus has therefore opened many questions on its function. Different possibilities may exist, among which the most important are: is the deposition of host cell wall material a simple tool to keep the partners separated or is the interfacial material the expression of the defence pathway activation by the part of the host?

Recent observations on mycorrhizal roots of maize by using a specific antibody against HRGP have demonstrated that the protein is present at the interface between the plant and fungus (Balestrini et al., in preparation). It is well known that wounding, pathogenic fungi or fungal elicitors induce genes coding for HRGPs, and cause an accumulation of HRGP transcripts in maize, too (Ludevid et al., 1990; Tagu et al., 1992). HRGP accumulation in the interface area could therefore be regarded as a defence mechanism. Alternatively, due to the high level of compatibility existing between host plants and AMF, HRGP deposition could be regarded as a mechanism limiting the spread of the symbiotic partner, as in bean nodules infected by *Rhizobium* (Benhamou et al., 1991).

Another molecule which is a constitutive cell wall component, but tends to increase in pathogenic and/or in compatible associations, is callose. However, the β1,3-glucans indicative of callose do not show any change in their location following fungal infection of maize roots (fig.9), demonstrating that the callose in the interface compartment is not a component produced by the host (Balestrini et al., in preparation). This suggests that there is no elicitation of the wall modifications usually associated with activation of defence reponses to pathogens. In contrast, a specific location of callose at the contact point beween the extraradical hypha and the wall of the epidermal host cell has been found in a 'locus a' myc⁻ mutant of *Pisum sativum*, (Gollotte et al., 1993). In this experimental system, the infection is not successful, and the penetration attempt of the fungus led to the deposition of callose by the host. These recent experiments support the idea that the deposition of host cell wall molecules at the interface between the plant and the fungus cell is not related to a specific defence response, but to a general mechanism of metabolic activation observed in the infected host cells (see next paragraph) on one hand, and to the necessity to keep the fungal growth under control.

The host membrane represents the other important component of the interface. As shown in Figures 2, 5 and 6 the host plant surrounds its endocellular partner by producing a specialized membrane (sometimes called *periarbuscular membrane*), which is physically continuous with the peripheral cell plasma membrane. Many studies have revealed similarities and differences between the two membranes during the functioning of arbuscular mycorrhizas. Gianinazzi-Pearson et al . (1991) demonstrated that membrane-bound ATPases occur on the periarbuscular membrane, but not on the peripheral plasma membrane. These cytochemical results are consistent with the hypothesis of a two-way nutrient flow occurring in mycorrhizas and mediated by an H⁺-ATPase pump, and contrast with the single-way nutrient flow in plant-pathogen interactions, where no ATPase activity was detected on the perihaustorial membrane (Gay et al., 1987). The different ATPase distribution is mirrored by differences in membrane potential: experiments on mycorrhizal leeks have shown a stable hyperpolarization of the membrane (Fieschi et al., 1991). In pathogenic associations. however, depolarization of the host membrane is commonly observed (Pelissier et al., 1986): this may be indicative of an increased membrane permeability and of an ion efflux toward the fungus. Our understanding of the relationship between the peripheral plasma membrane and the invaginated periarbuscular one has been improved by using monoclonal antibodies (McAb) developed in the Rhizobium/legume association (Perotto et al., 1991). According to the terminology of Martin and Tagu (1994), the use of McAbs represents a non-target approach, since the nature of the molecules with antigenic properties is firstly unknown, but their expression is analyzed and related with defined morphogenetical and developmental events, providing substantial information (Perotto et al., 1990). During the experiments, a panel of McAbs was used to investigate the composition of

82

the interface formed around the AMF in pea plants. Many of the probes reacted over both the peribacteroid and the periarbuscular membranes, revealing the presence of glycoproteins and of arabinogalactan proteins. However, other antibodies suggested some differences between the two types of symbiosis, as specific antigens (for example, those recognized by the monoclonal antibody MAC 266) were more abundantly expressed during the mycorrhizal infection than during the nodule symbiosis (Perotto et al.,1994). Interestingly enough, the component identified by MAC 266 was present in the soluble and in the insoluble interface components, and its expression was considerably increased in cells containing the fungal arbuscules at the end of their cycle.

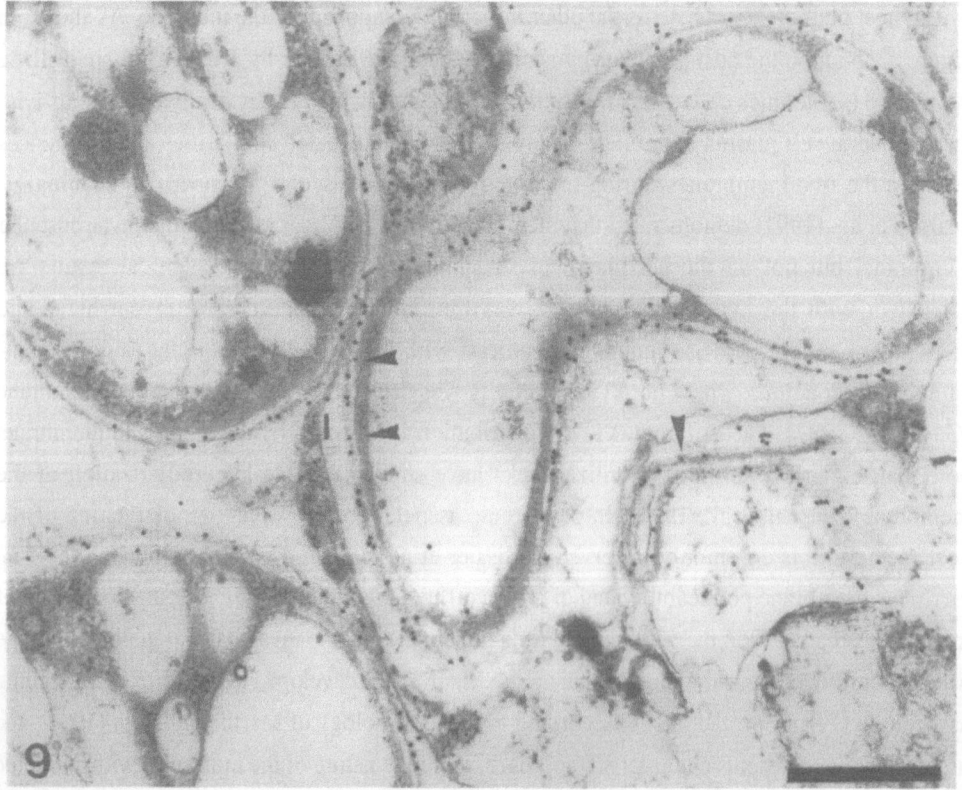

Figure 9. Maize /*Glomus versiforme* interface after treatment with the antiserum against β1-3 glucans. The fungal wall is heavely labelled (arrows), while by contrast no gold granules are found over the interface space (I) Bar corresponds to 0.5 µm.

In conclusion, the molecular dissection of the interface based on the use of *in situ* techniques reveals the complexity of this compartment: it may be regarded as a structural expression of the symbiotic status which does not represent a drawback to nutrient exchanges, but prevents a direct cell to cell contact that might provoke a defence reaction by the plant cell.

Ultrastructure is still a powerful tool to study the novel gene expression during arbuscular mycorrhizal symbioses The striking differences between an uninfected parenchyma cell and an arbuscule-containing cell began to impress researchers in the seventies. Many features were described as a response of the host to the fungal colonization: increase of the organelles, disappearance of starch from the plastids, splitting of the vacuoles, moving of the nucleus towards the centre of the cell (Scannerini and Bonfante, 1983; Toth, 1991; Balestrini et al., 1992). All these features were rightly interpreted as the morphological aspects of physiological and metabolic changes. Nuclei strongly changed their morphology, too. Berta et al. (1990) demonstrated that nuclei were larger, even if their ploidy did not change. The morphological change was intepreted as a change in chromatin organization, suggesting a higher transcriptional activity. Interestingly enough, when nuclei from infected peas were studied by flow cytometry and compared with those extracted from mutant peas, where the fungal colonization was only partly successful, a much lower increase in fluorescence was found in mutant root nuclei (Sgorbati et al., 1993). In addition, nuclei from mycorrhizal roots were very sensitive to DNase digestion, suggesting that AMF-infected cells are transcriptionally much more active than control cells. This hypothesis is supported by biochemical and immunological evidences showing modifications in host/fungus gene expression during mycorrhizal infection (Wyss, 1990, Dumas et al., 1989 Schellenbaum et al., 1992, Lambais and Medhy, 1993).

Ultrastructure coupled to immunogold techniques has been a useful tool to locate the products of genes whose expression changes after infection. Among these, a pathogenesis related protein (PR-b1) was located in the interfacial material surrounding living arbuscular hyphae (Gianinazzi-Pearson et al.,1992). The related antigen was not revealed in Western blot experiments, while Northern analysis of RNAs from mycorrhizal roots hybridized to PR-b1 cDNA. These results suggest that the mycorrhizal fungus leads to a weak expression of a protein usually associated with activation of plant defence systems. Comparable observations have been reached studying the expression of a hydrolytic enzyme, a polygalacturonase (PG) during the infection (Peretto et al., 1994). The mycorrhizal fungus was hypothesized to release a low quantity of PG during the infection process in order to open its way through the host tissues. Quantitative biochemical experiments did not reveal an important increase in PG activities in comparison with the uninfected controls. Western blot experiments did not show any change, either. However, FLC analysis displayed qualitative differences between extracts from mycorrhizal and non-mycorrhizal roots. In addition, by using sensitive immuno-fluorescence and immunogold techniques PG enzyme was constantly detected at the interface between the plant and the intracellular fungus (Peretto et al., 1994). As previously demonstrated (compare fig.8), pectins are present in this compartment, suggesting a close relationship between fungal enzyme and host substrate. These results suggest that AMF probably accomplish their infection process thanks to a low and regulated PG production; the enzyme works by breaking a host molecule which could be useful as nutritional substrate, but its quantity is so small that the plant oligomers produced do not activate the defence mechanisms of the host (Peretto et al., 1994).

Gene activation and the corresponding translation of specific genes are usually studied with *in situ* hybridization in many different experimental systems (Mc Fadden, 1991). This approach is a dawning feature of mycorrhizal research. However, some examples are already available: in addition to the demonstration of tubulin transcripts in ectomycorrhizas (Martin and Tagu, 1994) Harrison and Dixon (1993 and unpublished results) recently demonstrated that activation of some genes involved in the phenylpropanoid and isoflavonoid biosynthesis in *Medicago truncatula* colonized by *Glomus versiforme*. Phenylalanine ammonia lyase and chalcone synthase transcripts, levels of which were elevated in mycorrhizal root, were detected only in the arbuscule-containing cells. These experiments offer the first experimental evidence of a specific gene activation in the infected cells of mycorrhizal roots.

Conclusions

This short review of the extensive literature concerning ultrastructural analysis of arbuscular mycorrhizas from the seventies onwards demonstrates that ultrastructural techniques have been a powerful tool to understand the basic functioning of mycorrhizal symbioses. In experimental systems where the fungal partner is an obligate biotroph, as AMF are, the biochemical approach has seemed almost impracticable, whereas morphological, cytochemical and immunocyto-chemical methods have offered substantial information. Lastly, the development of refined molecular analysis will improve the understanding of arbuscular mycorrhizas. Here too, however, morphological studies will be still important, since they will provide comprehensive cues for the specific spatial and not only temporal gene expression.

Acknowledgements

The author wishes to thank Professor Kurt Mendgen and Dr. Raffaella Balestrini for the collaborative work on high-pressure and freeze-substitution of mycorrhizal samples. Travel expenses were funded by the ECC project COST 810.

The research referenced in this review was funded by CNR, Special project RAISA, by a MURST grant (40%).

References

Balestrini, R., Berta, G. and Bonfante, P. (1992) The plant nucleus in mycorrhizal roots: positional and structural modifications. *Biology of the Cell* 75: 235-243.

Becard, G. and Pfeffer, P.E. (1993) Status of nuclear division in arbuscular mycorrhizal fungi during *in vitro* development. *Protoplasma* 174: 62-68.

Benhamou, N., Lafontaine, P.J., Mazau, D. and Esquerré-Tugayé, M.T. (1991) Differential accumulation of hydroxyproline-rich glycoproteins in bean root nodule cells infected with a wild type strain or a C_4-dicarboxylic acid mutant of *Rhizobium leguminosarum* bv. *phaseoli*. *Planta* 184: 457-467.

Berta, G., Sgorbati, S., Soler, V., Fusconi, A., Trotta, A., Citterio, A., Bottone, MG., Sparvoli, E. and Scannerini, S. (1990) Variations in chromatin structure in host nuclei of a vesicular-arbuscular mycorrhiza. *New Phytol.* 114: 199-205.

Bianciotto, V. and Bonfante, P. (1993) Evidence of DNA replication in an arbuscular mycorrhizal fungus in the absence of the host plant. *Protoplasma* 176: 100-107.

Bonfante, P. (1994) Alteration of host surfaces by mycorrhizal fungi. In: O. Petrini and D. Marois (eds): *Alteration of host walls by fungi* APS Press, in press.

Bonfante, P. and Perotto, S. (1992) Plants and endomycorrhizal fungi: the cellular and molecular basis of their interaction. In: D.P.S.Verma (ed) *Molecular signals in plant-microbe communications*. CRC Press, Boca Raton Ann Arbor, pp 445-470.

Bonfante, P. and Scannerini, S. (1992) The cellular basis of plant-fungus interchanges in mycorrhizal association. In: M.F. Allen (ed) *Mycorrhizal functioning*. Chapman and Hall, New York London, pp. 65-101.

Bonfante, P., Faccio, A., Perotto, S. and Schubert, A. (1990a) Correlation between chitin distribution and cell wall morphology in the mycorrhizal fungus *Glomus versiforme*. *Mycol. Res.* 94: 157-165.

Bonfante, P., Vian, B., Perotto, S., Faccio, A. and Knox, J.P. (1990b) Cellulose and pectin localization in roots of mycorrhizal *Allium porrum:* labelling continuity between host cell wall and interfacial material. *Planta* 180: 537-547.

Bonfante, P., Tamagnone, L., Peretto, R., Esquerre-Tugaye, M.T., Mazau, D., Mosiniak, M. and Vian, B. (1991) Immunocytochemical location of hydroxyproline rich glycoproteins at the interface between a mycorrhizal fungus and its host plant. *Protoplasma* 165: 127-138.

Bracker, C.E. and Littlefield, L.J. (1973) Structural concepts of host-pathogen interfaces. In: R.J.W Byrde and C.V. Cutting (eds) *Fungal pathogenicity and the Plant's response*, Academic Press, London, pp. 159-317.

Carpita, N.C. and Gibeaut, D.M. (1993) Structural models of primary cell walls in flowering plants: consistency of molecular structure with the physical properties of the walls during growth. *Plant Journal* 3:1-30.

Cox, G. and Tinker, P.B. (1976) Translocation and transfer of nutrients in vesicular-arbuscular mycorrhizas. I. The arbuscule and the phosphorus transfer: a quantitative ultrastructural study. *New Phytol.* 77: 371-381.

Dumas, E., Tahiri-Alaoui, A., Gianinazzi, S. and Gianinazzi-Pearson, V. (1989) Observations on modifications in gene expression with VA endomycorrhiza development in tobacco: qualitative and quantitative changes in protein profiles. In: P. Nardon, V. Gianinazzi-Pearson, A.M. Grenier, L. Margulis and D.C. Smith (eds) *Endocytobiology IV*, INRA Press, Paris, 153-155.

Fieschi, M., Alloatti, G., Sacco, S. and Berta, G. (1992) Membrane potential hyperpolarisation in vesicular arbuscular mycorrhizae of *Allium porrum* L.: a non-nutritional long-distance effect of the fungus. *Protoplasma* 168: 136-140.

Garriock, M. L., Peterson, R. L. and Ackerley, C.A. 1989. Early stages in colonization of *Allium porrum* (leek) roots by the VAM fungus *Glomus versiforme*. *New Phytol.* 112: 85-94.

Gay, J.L. and Woods, A.M. (1987). Induced modifications in the plasma membranes of infected cells. In: G.F. Pegg and P.G. Ayres (eds): *Fungal infection of plants*, Cambridge University Press, Cambridge, pp.79- 91.

Gianinazzi-Pearson, V., Smith, S.E., Gianinazzi, S. and Smith, F.A. (1991) Enzymatic studies on the metabolism of vesicular-arbuscular mycorrhizas. V. Is H^+-ATPase a component of ATP-hydrolysing enzyme activities in plant-fungus interfaces? *New Phytol.* 117: 61-74.

Gianinazzi-Pearson, V., Tahiri-Alaoui, A., Antoniw, J.F., Gianinazzi, S. and Dumas, E. (1992) Weak expression of the pathogeneis related Pr-b1 gene and localization of related protein during symbiotic endomycorrhizal interactions in tobacco roots. *Endocytobiosis & Cell Research* 8:177-185.

86

Gianinazzi-Pearson, V., Lemoine, M.C., Arnould, C. and Morton J.B. (1994) Localization of β (1-3) glucans in spore and hyphal walls of fungi in the *Glomales*: taxonomic and phylogenetic implications. *Mycologia* in press.

Giovannetti, M., Avio, L., Sbrana, C. and Citernesi, A.S. (1993) Factors affecting appressorium development in the vesicular-arbuscular mycorrhizal fungus *Glomus mosseae* (Nicol. & Gerd.) Gerd. & Trappe. *New Phytol.* 123: 115-122.

Gollotte, A., Gianinazzi-Pearson, V., Giovannetti, M., Sbrana, C., Avio, L. and Gianinazzi, S. (1993) Cellular localization and cytochemical probing of resistance reactions to arbuscular mycorrhizal fungi in a 'locus a' myc⁻ mutant of *Pisum sativum* L. *Planta*, 191: 112-122.

Hall, J.L. and Hawes, C. (1991) *Electron microscopy of plant cells*. Academic Press, London, New York.

Harley, J. L. 1989. The significance of mycorrhiza. *Mycol.Res.* 92: 92-129.

Harrison, M.J and Dixon, R.A. (1993) Isoflavonoid accumulation and expression of defense gene transcripts during the establishment of vesicular-arbscular mycorrhizal associations in roots of *Medicago truncatula*. *Mol.Plant Micr. Int.* 6: 643-65.

Hoch, H.C. (1991) Preservation of Cell Ultrastructure by Freeze Substitution. In: K Mendgen, D.E. Lesemann (eds) *Electron Microscopy of Plant Pathogens*. Springer Verlag, Berlin, pp. 1-16.

Lambais, M.R. and Medhy, M.C. (1993) Suppression of endochitinase, β-1,3- Endoglucanase, and chalcone isomerase expression in bean vesicular-arbuscular mycorrhizal roots under different soil phosphate conditions. *Mol.Plant Micr. Int.* 6: 75-83.

Ludevid, M.D., Ruiz-Avila, L., Valles, M.P., Stiefel, V., Torrent, M., Tornè, J.M. and Puigdomenech, P. (1990) Expression of genes for cell-wall proteins in dividing and wounded tissues of *Zea mays* L. *Planta* 180: 524-529.

Martin, F. and Tagu, D. (1994) The ectomycorrhiza development: A molecular perspective. In: B. Hoch and A. Varma (eds) *Mycorrhiza: Structure, function, molecular biology and biotechnology*. Berlin: Springer Verlag, in press.

McFadden, G.I. (1991) In situ hybridization techniques: Molecular cytology goes ultrastructural. In: J.L. Hall and C. Hawes (eds) *Electron microscopy of plant cells*. Academic Press, London, pp.219-255.

Mendgen, K., Welter, K., Scheffold, F. and Knauf-Beiter, G. (1991) High pressure freezing of rust infected plant leaves. In: K. Mendgen, D.E. Lesemann (eds) *Electron Microscopy of Plant Pathogens*. Springer Verlag, Berlin, pp 31-42.

Müller, H. and Moor, H. (1984) Cryofixation of thick specimens by high pressure freezing. In: J.P. Revel, T. Barnard and G.H. Hagis (eds) *The science of biological specimen preparation* - SEM. AMF: O'Hare, IL, USA, pp131-138.

Orlovich, D.A. and Ashford, A.E. (1993) Polyphosphate granules are an artefact of specimen preparation in the ectomycorrhizal fungus *Pisolithis tinctorius.. Protoplasma* 173: 91-102.

Pelissier, B., Thibaud, J.B., Grignon, C. and Esquerré-Tugaye, M.T. (1986) Cell surfaces in plant - microorganism interactions. VII.Elicitor preparations from two fungal pathogens depolarize plant membranes. *Plant Science* 46: 103-109.

Peretto, R., Bettini, V., Favaron F., Alghisi, P. Bonfante P. (1994) Polygalacturonase activity and location of arbuscular mycorrhizal roots of *Allium porrum* L. *Mycorrhiza* In press.

Perotto, S., VandenBosch, K.A., Butcher, G.W. and Brewin, N.J. (1991) Molecular composition and development of the plant gycocalix associated with the peribacteroid membrane of pea root nodules. *Development* 112: 763-774.

Perotto, S., Malavasi, F. and Butcher, G.W. (1991) Use of Monoclonal Antibodies to study mycorrhiza: present applications and perspectives. In: J.R. Norris, D.J. Read and K. Varma (eds) *Methods in Microbiology*, Vol.24. Academic Press, London, pp. 221-248.

Perotto, S., Brewin, N.J. and Bonfante P. (1994) Colonisation of pea roots by arbuscular mycorrhizal fungi and rhizobia: an immunological comparison using monoclonal antibodies as probes for plant cell surface components. *Mol.Plant Micr. Int.* 7: 91-98.

Robards, A.W. (1991) Rapid-freezing Methods and their applications. In: J.L. Hall and C. Hawes (eds). *Electron microscopy of plant cells*. Academic Press, London, pp.257-312.

Scannerini, S. and Bonfante, P. (1983) Comparative ultrastructural analysis of mycorrhizal associations. *Canadian Journal of Botany* 61, 917-943.

Schellenbaum, L., Gianinazzi, S. and Gianinazzi-Pearson, V. (1992) Comparison of acid soluble protein synthesis in roots of endomycorrhizal wild type *Pisum sativum* and corresponding isogenic mutants. *Journal of Plant Physiology* 141: 2-6.

Sgorbati, S., Berta, G., Trotta, A., Schellenbaum, L., Citterio, S., De la Pierre, M., Gianinazzi-Pearson, V. and Scannerini, S. (1993) Chromatin structure variation in successful and unsuccessful arbscular mycorrhizas of pea. *Protoplasma* 175: 1-8.

Smith, S.E. and Smith, F.A.(1990) Structure and function of the biotrophic symbioses as they relate to nutrient transport. *New Phytol.* 114: 1-38.

Strullu, D.J., Gourret, J.P., Garrec, J.P. and Fourcy, A. (1981) Ultrastructure and electron-probe microanalysis of the metachromatic vacuolar granules occurring in *Taxus* mycorrhiza. *New Phytol.* 87:537-547.

Sward, RJ. (1981) The structure of the spores of *Gigaspora margarita*. I. The dormant spore. *New Phytol.* 87: 761-768.

Tagu, D., Walter, N., Ruiz-Avila, L., Burgess, S., Martinez-Izquierdo, J.A., Leguay, J.J.and Puigdomenech, P. (1992) Regulation of the maize HRGP gene expression by ethylene and wounding. mRNA accumulation and qualitative expression analysis of the promoter by microprojectile bombardament. *Plant Molecular Biology* 20, 529-538.

Toth , R. (1991) The quantification of arbuscules and related structures using morphometric cytology. In: J.R. Norris, D.J. Read and A.K.Varma (eds) *Methods in Microbiology*, Vol.24. Academic Press, London, pp. 275-299.

Turnau, K., Kottke, I. and Oberwinkler, F. (1993) *Paxillus involutus-Pinus sylvestris* Mycorrhizae from heavily polluted forest. I. Element localization using Energy Loss Spectroscopy and Imaging. *Botanica Acta* 106: 213-219.

Welter, K., Muller, M. and Mendgen, K. (1988) The hyphae of *Uromyces appendiculatus* within the leaf tissue after High Pressure Freezing and Freeze Substitution. *Protoplasma* 147: 91-99.

Wyss, P., Mellor, R.B. and Wiemken, A. (1990) Vesicular-arbscular mycorrhizas of wild type soybean and non-nodulating mutants with *Glomus mosseae* contains symbiosis specific polypetides (mycorhizins), immunogically cross-reactive with nodulins. *Planta* 182: 22-26.

Zhang,G.F. and Staehelin, L.A. (1992) Functional compartmentation of the Golgi apparatus of plant cells. *Plant Physiol.* 99:1070-1083.

Impact of mycorrhizal colonisation on root architecture, root longevity and the formation of growth regulators

D. Atkinson, G. Berta[1] and J. E. Hooker[2]

SAC, West Mains Road, Edinburgh, EH9 3JG UK
[1] *Dipartimento di Biologia Vegetale dell'Università di Torino ,Viale Mattioli, 25 10125 Torino,*
[2] *Soil Biology Unit, Department of Land Resources, SAC, Mill of Craibstone, Aberdeen, AB2 9TQ, UK*

Introduction

Until relatively recently, it was assumed that arbuscular mycorrhizal fungi (AMF) did not influence the morphology of the root system. Harley and Smith, (1983) concluded "little change in root morphology occurs following infection. Root apices continue to grow in an apparently normal manner. "...It is usually not possible to tell if a root system is mycorrhizal..." A number of recent studies, however, reviewed by Berta et al. (1993a) have demonstrated effects of AMF on a range of morphological parameters. While effects of AMF are less extreme than those of ecto-mycorrhizas they have the potential to be of equal importance. The existence of effects raises questions as to how these effects are caused.

Recent studies of the development of the form or architecture of the root system have tended to have been dominated by the application of topological analysis (Fitter 1985). This method uses a mathematical classification to describe branching. In a topological model, first order roots are all of those with an apical meristem while second order roots occur at the junction of two first order roots and third order roots at the junction of a first and second order root. This description can be summed using the terms "magnitude", the number of exterior links within the system and "exterior pathlength", the sum of all pathlengths from all exterior links to the base. The application of this taxonomy to root systems infected by AMF has been discussed by Berta et al. (1993a). The topological model differs from a developmental model which is normally concerned primarily with chronological development, i.e. with the development of different orders of lateral roots. While the two systems both result in a morphological description of the root system, they ultimately describe different properties of the root system and so complement each other rather than duplicating information. The advantages and disadvantages of the two classifications are summarised in Table I.

Table I. The relative strengths and weaknesses of topological and developmental classification of root systems infected with AMF

Attribute	Topological Classification	Developmental Classification
Simplicity of terminology	Complex with elaborate numerical system	Relative simple and logical
Ease of use	Easily applicable to seedling root systems. More difficult on large systems	Easily applicable to young systems. With image analysis can be used on large systems
Use on perennial root systems	Potentially difficult to use on root systems with woody roots and prior death of root members	Can be used on root systems of perennial crops such as trees
Relation to nutrient use	Can be related to soil exploitation potential. Gives an integrative parameter for soil condition effects	Not obviously related to soil factors. May indicate potential activity of individual roots
Temporal value	A static model which requires complex modelling to represent growth	Inherently dynamic and so valuable in describing change. Can be used to reflect developmental processes
Spatial value	Good at summarising integrated architecture	Poor. Developments could be associated with many distributions
Use in relation to mycorrhizal infection	Lack of branch/age relationship makes potential dynamics of infection hard to understand	Allows potential prediction of numbers of infectable roots
Use in relation to mycorrhizal impact	Valuable for summarising over architectural effects of AMF and their impact upon root activity	Valuable for summarising development linked effects such as carbon flow in rhizosphere
Application to minirhizotron images	Not applicable unless a high proportion of tips visible	Depending upon size image relative to scale of branching can be achieved
Application to core samples	Can be applied if sample contains appropriate samples of system	Can be applied especially if image analysis is available

Classification of root systems can be academic. The question "Why does it matter if infection of a root system with AMF changes either its development or its topology?" can be asked. The question is important to both the allocation of resources within the root system and to its functioning. This review therefore discusses the effects of AMF infection on (1) root system architecture and development, (2) the mechanism of AMF effects and (3) the ecological and agricultural significance of these effects.

The effect of AMF on root architecture

Atkinson (1992) divided effects of AMF and other factors on the form of the root system into four principal types of morphological modification i.e. structural, spatial, quantitative and temporal. Root architecture, especially as revealed by topological analysis, relate principally to spatial morphology although it will also tend to reflect quantitative aspects. Berta et al. (1993a) applied topological methods to characterise the effects of *E3* on *Allium porrum*. They found that the pattern of branching was unaffected, i.e. total exterior pathlength (Pe) was unchanged, despite large effects upon the numbers of lateral roots per unit length of root axis. This effect differed from the effects reported for *G. fasciculatum* and either *Vitis vinifera* or *Platanus acerifolia* (Schellenbaum et al., 1991, Tisserant et al., 1991). In *V. vinifera* infection with AMF induced a more random pattern of root branching; in control plants branching was herringbone-like in form. In *P. acerifolia* infection with AMF resulted, initially, in a more herringbone-like system but ultimately in a more dichotomous system than was the case for the control plants. The morphological effect of the fungus, in this study, was related to the degree of AMF infection. These changes have potentially important implications for nutrient capture by the plant, for the energy needed to construct the root system; herringbone structures are more energy intensive, and for effects upon root system dynamics and thus carbon flow into soil organic matter. The direction and magnitude of effects of AMF on root topology clearly are variable. Too few species have been assess to allow the magnitude of effects to be related to taxonomy.

The effect of AMF on root system development

Berta et al.(1993a) identified significant effects of infection with *Glomus E3* on *A. porrum* roots. Infected root systems were more highly branched, i.e. they contained shorter more branched adventitious roots of larger diameter and lower specific root length. In *V. vinifera* Schellenbaum et al. (1991) reported a similar effect. Here the difference between AMF and control roots increased with increasing root order. The number of primary laterals developed per unit length of axes was increasing by 125% by AMF infection. The increases for 2° and in 3° roots were 185% and 230% respectively. These latter studies were carried out on young plants, i.e. 8 weeks post transplanting. Hooker et al. (1992a) assessed the effects of infection with AMF (*Scutellispora calospora*, *Glomus E3* or *Glomus caledonium*) on the development of

92

115 day old *Populus* (var Beupre). As this experiment was conducted using a relatively fertile soil, where the growth of all plants was good and there were no major nutrient deficiencies, it was therefore possible to largely separate the effects of plant size and nutrient supply from those of AMF infection. Here the level of soil fertility was such that neither AMF or the application of nutrients influenced total growth. In a system where all roots were adventitious AMF increased the number of every

root order (2° to 5°). Effects again tended to be greater for the higher order laterals, i.e. *Glomus E3* increased the branching of tertiary roots by 616%, but of secondary roots by only

Table II.   The effect of AMF infection on root branching (% of control)

Parameter	A. porrum Berta et al. (1993)	Populus Hooker et al. (1992a)	V. vinifera Schellenbaum et al. (1991)	P aceryfolia Tisserant et al. (1991)
Plant age (d)	105	115	56	49
Fungus	E3	E3	G. fasciculatum	G. fasciculatum
Length of individual roots				
1°	77	92	95	83
2°	81	200	89	91
3°		219	93	111
4°		107		
No. of branches on				
1°		98	140	
2°	164	181	200	145
3°		717	266	

81%. The addition of nutrients had little effect on branching. As with the studies on *A. porrum,* described above, the magnitude of the branching effect increased with an increase in root colonisation. The 22.5% colonisation of secondary roots by *S. calospora* produced a 40% increase in the number of lateral roots per unit length. A 51.9% colonisation by *G. caledonium* led to 60% increase in laterals. Branching in tertiary roots was affected by a lower intensity of colonisation. A 22.6% colonisation of tertiary roots by *G. caledonium* resulted in a 600% increase in laterals. Effects of AMF infection in the experiments discussed above are summarised in Table II.

Potential and actual effects of AMF are important not only because of their effects on a plant's potential for nutrient uptake, but also because of its relationship to the longer term development of the root system. This is especially important for woody perennials. The normal development of a tree root system has been described by Atkinson (1992). In the tree root system, a proportion of the new roots produced are destined to survive as woody roots although most roots die at an early stage (Atkinson, 1983). Survival is not a random process. The probability of survival is highest for roots of low order ie 1º, 2º laterals. Thus, any factor which increases the allocation of resources to higher order lateral roots should increase the proportion of roots which die and as a consequence the flow of carbon into the soil. It should also decrease the mean age of the root system, and as a consequence influence root activity. Such systems will be more dynamic in nature than uncolonized root systems and a single time point classification, which is common for topological analyses, will be less descriptive of the system as a whole.

The process of root infection by AMF is imperfectly understood. The development of the symbiosis does however involve extensive interaction with the release of key molecules in root exudates and root surface factors both being important (Anderson, 1992). It clearly follows from this that not all roots within a root system will be equally susceptible to colonisation and that the ability of a root to be colonised is likely to decrease with increasing age. The branching pattern of a root system and the related root age factor are thus likely to influence the systems potential to become mycorrhizal. Some implications of this are discussed later.

It is the nature of science that the measurements made by different workers are often too different to allow of easy comparison or that the range of measurements needed to characterise a developmental process are made in such a way that they are difficult to relate. In an attempt to overcome this, in relation to the effects of AMF on root morphology, Berta et al. (1992), Berta et al. (submitted) designed an experiment with *Prunus cerasifera* which was infected by either *G. mosseae* or *G. intraradices*. The aims of this study were to assess effects of AMF on (a) the development of micropropagated plantlets, (b) root system branching, (c) phosphorus nutrition, (d) meristematic activity and (e) gene products.

AMF infection improved both plant survival and growth and most individual growth parameters, i.e. root, stem and leaf weight, root length, leaf area. Data for the effects of AMF on root length and branching are shown in Table III. The AMF treatments did not influence the length of individual roots, but increased the intensity of branching on all root orders and greatly increased the proportion of the root system present as higher order laterals. Other than effects on individual root lengths these effects are consistent with results presented earlier. The uptake of P and its concentration in tissues was increased but the course of their change with time differed markedly from that of the intensity of mycorrhizal colonisation. This suggested a role for non-nutritional factors. Both fungi influenced cell number and size in the root tip, the levels of soluble proteins and changed the patterns of proteins which were expressed. AMF clearly modify root development at a range of scales and as a consequence may result in effects which differ between species.

Table III. The effect of AMF infection on root length (cm) and branching in *Prunus cerasifera* after 75d growth

Treatment	Root Length	Mean Length of adventitious roots	Adventitious root branching
Control	492.5 34.1	21.66 1.53	2.78 0.31
G. mosseae	1244.0 176.0	24.31 3.40	3.81 0.19
G. intraradices	1157.0 132.0	17.45 1.63	3.45 0.36

The mechanism of AMF effects

Because infection of roots with AMF commonly influences plant nutrition, especially P nutrition, effects of AMF on root systems have commonly been ascribed to either nutrition related growth effects or to direct effects of nutrition on development (Drew and Saker, 1978). Many studies, e.g. Trotta et al. (1991a) for E3 and *A. porrum*, have concluded that fungal effects upon the plant could be ascribed to improved nutrition because effects were not obvious at high external P levels. Many of these studies, however, have been complicated by the presence of effects on plant size and physiological age. In the study of Trotta et al. (1991b) clear differences in morphology between control and AMF plants at low P levels were reversed at high P levels while effects on the length of 1º and 2º laterals were decreased by AMF at all P levels. Similarly, Hooker et al. (1992a, b) found that with *Populus* grown under fertile conditions infection with AMF modified the root system in a way which was distinct to that induced by the addition of high levels of phosphate. In this study, because none of the treatments influenced total growth, effects of size, age and nutrition could at least to a degree be eliminated.

The development of a root system depends upon the initiation of root primordia, their out-growth and within the initiated roots the processes of cell division, cell growth and cell death. In assessments of the effects of AMF, it is helpful to characterise effects on these processes. This is summarised in Table IV. Infection with AMF increases primordial formation, cell division and cell death and has a variable effect upon cell growth.

AMF effects involve changes in both the rate and extent of cell growth and in cell division. AMF infected plants are known to contain modified concentrations of both cytokinins and auxins (Berta et al.1993; Dannenberg et al 1992). Clearly therefore effects of AMF on root morphogenesis may be regulated via hormones, either as a result of fungal production or by modification of transport to the root tip. Allen et al. (1980) found increased levels of cytokinins and gibberelins in infected *Bouteloua gracilis*, while Dixon et al. (1988) found increased translocation of cytokinin in infected *Citrus*. Studies by Baas and Kuiper (1989) with *Plantago major* suggested that the modified hormonal balance was not completely P mediated. Measurements of the hyper-polarisation of the membrane potential of cortical cells in AMF infected *A. porrum* indicated effects at all levels of infection and prior to the initiation of nutrition related effects (Fieschi et al., 1992). AMF fungi are known to produce hormones themselves (Barea and Azcon-Aguilar, 1982), but little information exists as to the extent to which they can be translocated to the host. In the studies of Berta et al. (1992) and Berta et al.

Table IV. Effects of AMF infection on component root system development of *Allium porrum L.* (Berta et al., 1993; Fusconi et al., 1994)

Process	Effect of AMF	Consequence
Primordial formation	increased	more numerous and branched roots
Variations in size of root apices	diameter at the base of the meristem, meristem length, cap length increased	large root apices
$3-$ H-Thymidine incorporation in root apical meristems	reduced labelling index	low DNA synthesis
3-H-uridine incorporation in the quiescent centre area	reduced labelling density	low metabolic activity
Mitotic cycles in root apical cells	lengthened	low meristem activity
Cell death in apical meristem	number of abscised apices increased	short roots
Mitotic cycle in root apical cells at different P concentrations	lengthened, but less than in controls	persisting small differences in root system architecture

(in press), the growth of *P. cerasifera* was unaffected by the presence of the *Hymenoschyphus ericae* a non-infecting erecoid mycorrhizal fungus which is known to produce hormones. This suggests that true infection is essential for the development of effects. The same study assessed the effects of AMF on a range of gene products. Soluble proteins differed both qualitatively and quantitatively between control and AMF infected plants. AMF plants both synthesised proteins which were not present in control plants and expressed others to a greater extent. These results thus suggest that the effects of AMF infection are more fundamental than a mere promotion of growth arising from improved nutrition.

The root system of infected *A. porrum* is characterised by a higher number of more branched, larger adventitious roots than in controls, depends on variations in structure and on the activity of root apices. Fusconi et al. (1994) showed that root apices in infected leek are larger than in controls, with a more complex vascular pattern which corresponds to a larger meristem and quiescent centre. In addition in this study there was an inverse relation between the size and activity of the meristem in mycorrhizal and control apices. This is contrary to what generally occurs, i.e. an increase in size corresponding to an increase in growth (Cahn *et al.*, 1989). Fusconi et al. (1994) eliminated the possibility that the lower activity of mycorrhizal apices could be explained by low sucrose concentrations at the apex generated by the absorption of photosynthate upstream by the fungus. As a low sucrose concentration causes a reduction of apex size this would be the opposite of what occurred here in the mycorrhizal root apices. In this study one of the most probable factors is the enhanced phosphorus nutrition occurring in mycorrhizal plants, as the addition of P leads to a lessening of the differences between control and mycorrhizal plants. However data on the duration of the mitotic cycle at different P levels, showed a lengthening of the cycle with increasing P in mycorrhizae which was lower than observed in controls. This might explain the persistence of small differences in root system architecture (Berta et al., 1993b). Variations in hormonal balance could influence cell size, the cell division cycle and possibly vary the amount and activity of protein kinases and their interactions with cyclins, which are key factors in the regulation of the cell cycle in eukaryotes.

The agricultural consequences of AMF induced modification to morphology

Studies of the effects of AMF fungi on root system architecture have only been conducted in recent years. Few studies of effects under field conditions have been carried out. Atkinson (1983) assessed the effects of grass-based and bare-soil management on the development of the root systems of apple trees. These studies indicated that trees under grass management had a higher level of AMF infection, a higher proportion of new root growth present as higher order lateral roots, a lower proportion of new growth surviving as perennial roots and a higher concentration of P in the tissues. These older results, from a study not aimed at understanding AMF effects, are wholly consistent with the effects of AMF described as a result of more recent detailed studies. Studies of the effect of AMF colonisation on root turnover (Black, unpublished) have indicated that this is increased in colonized *Populus*. Extrapolation of effects to the recycling of nitrogen, using data from an agriforestry site in Scotland, indicated probable increases from 77 to 132kg ha^{-1} yr^{-1} in grass. The results of these studies suggest that AMF are likely to have a major effect on nutrient cycling through influences on root system longevity.

The intensive agricultural systems developed in recent years have been characterised by the addition of substantial quantities of nutrients (fertilisers) and crop protection materials (fungicides, herbicides, etc). In such systems, the activities of soil-micro-organisms and their effects on nutrient supply become of relatively little importance. Now, however, questions are being asked about the wisdom of providing these major energy subsidies for agricultural production and there now is interest in the development of systems based on non-renewable inputs i.e. of systems which are more sustainable. A consequence of this (Atkinson and Hooker, 1993) is that such low-input agricultural systems will be more biologically based and that mycorrhizal infection will have a larger impact upon crop performance.

Results from studies on a range of species over many years have suggested that in pot culture infection with AMF consistently increases growth under conditions of restricted nutrient supply. Among the key questions in relation to the design of more sustainable agricultural systems are whether it is possible to inoculate the plants in such systems to increase colonization or whether the level of AMF infection can be increased by soil management and other husbandry methods. Past experience of these strategies (Atkinson and Hooker, 1993) suggest that the latter is the more likely to be effective. Inoculation has been effective in a limited number of field trials perhaps because added AMF inoculum must complete with native species which will normally be present at much higher levels. In contrast to attempts to artificially modify soil microbial composition by the addition of exotic strains or species, soil management merely aims to increase the infection of roots by native AMF species, both via effects on the infection process and through changes to make the root system more infectable. The contribution of these two components are hard to separate but it is important for the development of optimal management systems to be able to define their relative importance. Root system effects may also be increased by the selection of an optimal combination of morphological and physiological characteristics. Studies by Atkinson (1983) of the susceptibility to infection of apple rootstocks showed variation which was at least partly associated with the increased development of higher order laterals. Although root system variation within the genotypes of single species is poorly documented, studies by Atkinson (1989) for barley and lavender, (1993) for birch demonstrate that such variation exists in relation to a range of within root system allocation strategies which seem likely to influence infection. For example in birch, (*Betula pendula*) the proportion of photosynthate allocated to woody roots (which are clearly not infectable) and fine roots (which are infectable) varied between a range of clones.

Further development of such systems need detailed studies of the development of both roots and associated AMF under field conditions. Although always difficult to assess in the field such studies will be aided by the development of the min-rhizotron method (Atkinson et al., 1992; Hooker and Atkinson, 1992). This system which allows root systems and their development to be visualised in the field and effects on infection on the development of populations of roots recorded should allow major AMF effects to be quantified. This will meet a key future need.

The future

Future needs are of two major kinds. The mechanism by which AMF influence root system form needs to be understood at a basic level so that root system design and modification can be used as a management tool, especially within intensive horticultural systems. In addition the use of AMF within sustainable systems needs to be better understood in relation to nutrient cycling.

Acknowledgements

We thank Drs A. Trotta and A. Fusconi and numerous colleagues with Cost 8.10, but especially S. Gianinazzi and H. Schüepp, for useful discussion. Collaboration which made this work possible was funded by Cost 810 DA and JEH thank the Scottish Office Agriculture and Fisheries Department for their funding of this research activity.

References

Allen, M.F., Moore, T.S. and Christensen, M. (1980) Phytohormone changes in *Bouteloua gracilis* infected by Vesicular Arbuscular Mycorrhizae. I Cytokinin increases in the host plant. *Can. J. Bot.* 58: 371-374.

Anderson, A.J. (1992) The influence of the plant root system on mycorrhizal formation. In: *Mycorrhizal Functioning* M.F. Allen (ed) Chapman and Hall New York pp 37-64.

Atkinson, D. (1983) The growth, activity and distribution of the fruit tree root system. *Plant Soil* 71: 23-35.

Atkinson, D. (1989) Root growth and activity: Current performance and future potential. *Aspects Applied Biology* 22: 1-13.

Atkinson, D. (1992) Tree root development: the role of models in understanding the consequences of arbuscular endo-mycorrhizal infection. *Agronomie* 12: 817-820.

Atkinson, D., Hooker, J.E., Pauline, O., Perry, R.L., Blasing, D. and Fogel, R. (1992) The use of mini-rhizotron and micro-rhizotrons to quantify root turnover. In: *Root Ecology and its practical Applications* L. Kutschera et al (eds) Verein für Wurzelforschung, Klagenfurt, Austria pp 291-294.

Atkinson, D. and Hooker, J.E. (1993) Using roots in sustainable agriculture. Chemistry and Industry 4 January pp 14-17.

Baas, R. and Kuiper, D. (1989) Effects of vesicular arbuscular mycorrhizal infection and phosphate on *Platnayo major* ssp. *pleiosperma* in relation to internal cytokinin concentrations. *Physiol.Planta.*76: 211-215.

Barea, J.M. and Azcon-Aguilar, C. (1982). Production of plant growth regulating substances by the vesicular arbuscular mycorrhizal fungus *Glomus mosseae. Appl. Environm. Microbiol.* 43: 810-813.

Berta, G., Trotta, A., Fusconi, A., Cardinale, F., Hooker, J.E., Atkinson, D., Giovannetti, M., Loreti, F., Branzanti, B., Tisserant, B., Gianinazzi-Pearson, V. and Gianinazzi, S. (1992) Root Morphogenis in a micropropagated fruit plant as influenced by endo-mycorrhizal infection. *Giornale Botanico Italiano* 126: 338.

Berta, G., Fusconi, A. and Trotta, A. (1993a) VA mycorrhizal infection and the morphology and function of root systems. *Environm. Exp. Bot.* 33: 159-173.

Berta, G., Fusconi, A., Trotta, A., Brazzaventre, S., Scannerini, S. and Tagliasacchi, A.M. (1993b) Duration of the mitotic cycle in an endomycorrhizal system as related to phosphorus nutrition. 4th International Symposium on "Structure and Function of Roots", June 20-26, 1993, Starà Lesnà, Slovakia.

Berta, G., Trotta, A., Fusconi, A., Hooker, J.E., Munro, M., Atkinson, D., Giovannetti, M., Marini, S., Loreti, S., Tisserant, B., Gianinazzi-Pearson, V. and Gianinazzi, S. (1994) The effects of arbuscular mycorrhizal infection on root system morphology in *Prunus Cerasifera* L. *Tree Physiol.* (in press)

Cahn, M.D., Zobel, R.W. and Bouldin, D.R. 1989 Relationship between root elongation rate and diameter and duration of growth of lateral roots of maize. *Plant Soil* 119: 271-279.

Dannenberg, G., Latus, C., Zimmer, W., Hundeshagen, B., Schneider-Poetsc, H.Jj. and Bothe, H. 1992 Influence of vesicular ararbuscular mycorrhiza on phytohormone balance in maize (*Zea mays* L.) *J. Plant Physiol.* 141: 33-39.

Dixon, R.K., Garret, H.E. and Cox, G.S. (1988) Cytokinins in the root pressure exudate of *Citrus Jambhire* Lush. Colonized by vesicular arbuscular mycorrhiza. *Tree Physiology* 4: 9-18.

Drew, M.C. and Saker, L.R. (1978) Nutrient supply and the growth of the seminal root system in barley III compensatory increases in growth of lateral root system, and in rates of phosphate uptake in response to a localized supply of phosphate. *J. Exp. Bot.* 29: 435-451.

Edriss, M.H., Davis, R.M. and Burger, D.W. 1984 Influence of mycorrhizal fungi on cytokinin production of sour orange. *J Amer. Soc. Hort. Sci.* 109, 4: 587-590.

Fitter, A H (1985) Functional significance of root morphology and root system architecture. In: *Ecological Interactions in Soil* A.H. Fitter, D. Atkinson, D.J. Read and M.B. Usher (eds) Blackwell, Oxford, UK pp 87-106.

Harley, J.L. and Smith, S.E. (1983) *Mycorrhizal Symbiosis.* Academic Press, London, UK pp 15.

Fieschi, M., Alloatti, G., Sacco, S. and Berta, G. (1992) Membrane potential hyperpolarisation in vesicular arbuscular mycorrhizae of *Allium porrum* L.: a non-nutritional long distance effect of the fungus. *Protoplasma* 168: 131-140.

Fusconi, A., Berta, G., Tagliasacchi, A.M., Scannerini, S., Trotta, A,. Gnavi, E. and De Padova, S. (1994) Root apical meristems of arbuscular mycorrhizae of *Allium porrum* L. *Environm. and Expe.Bot.* 34 in press.

Hooker, J.E. and Atkinson, D. (1992) Application of computer-aided image analysis to studies of arbuscular endo-mycorrhizal fungi effects on plant root system morphology and dynamics *Agronomie* 12: 821-824.

Hooker, J.E., Munro, M. and Atkinson, D. (1992a) Vesicular-arbuscular mycorrhizal fungi induced alteration in poplar root system morphology. *Plant Soil* 145: 207-214.

Hooker, J.E., Munro, M., Atkinson, D., Perry, R.L. (1992b) The effects of VAM fungi on the root morphology of poplar. In: *Root Ecology and its Practical Applications* Kutschere et al (eds) Vereim für Wurzelforschung, Klagenfurt, Austria pp 579-582.

Lavender, E.A., Atkinson, D. and Mackie-Dawson, L.A. (1993) Variation in root development in genotypes of *Betula pendula.* Aspects *Applied Biology* 34: 183-192.

Schellenbaum, L., Berta, G., Ravolanirina, F., Tisserant, B., Gianinazzi, S. and Fitter, A.H. (1991). Influence of endo-mycorrhizal infection on root morphology in a micro-propagated woody plant species (*Vitis Vinifera* L) *Ann. Bot.* 68: 135-141.

Tisserant, B., Schellenbaum, L., Gianinazzi-Pearson, V., Gianinazzi, S. and Berta, G. (1991) Influence of infection by an endo-mycorrhizal fungus on root development and architecture in *Platanus acerifolia. Allionia* 30 171-181.

Trotta, A,. Carminati, C., Schellenbaum, L., Scannerini, S., Fusconi, A. and Berta, G. (1991a) Correlation between root morphogenesis, VA mycorrhizal infection and phosphorus nutrition. In: *Plant Roots and their Environment* B.L. McMichael and H. Perrson (eds) Elsevier, Amsterdam pp 333-339.

Trotta, A., Berta, G., Fusconi, A. and Scannerini, S. (1991b) Root development in VA mycorrhiza as related to phosphorus nutrition. Abstracts 3rd ISRR Symposium p 134.

Biogeochemical cycling and arbuscular mycorrhizas in the sustainability of plant-soil systems

P. Jeffries and J. M. Barea[1]

Biological Laboratory, University of Kent, Canterbury, Kent CT2 6NJ, U.K.
[1]Estación Experimental del Zaidín CSIC Prof. Albareda 1, 18008 Granada, Spain

Introduction

The fundamental importance of microbial processes for the efficient cycling of mineral nutrients within biosphere has been realized for many years (fig.1), and the involvement of the arbuscular mycorrhizal fungi (AMF) fully recognized (Mosse, 1986; Bethlenfalvay and Lindermann, 1992; Barea and Jeffries, 1994).

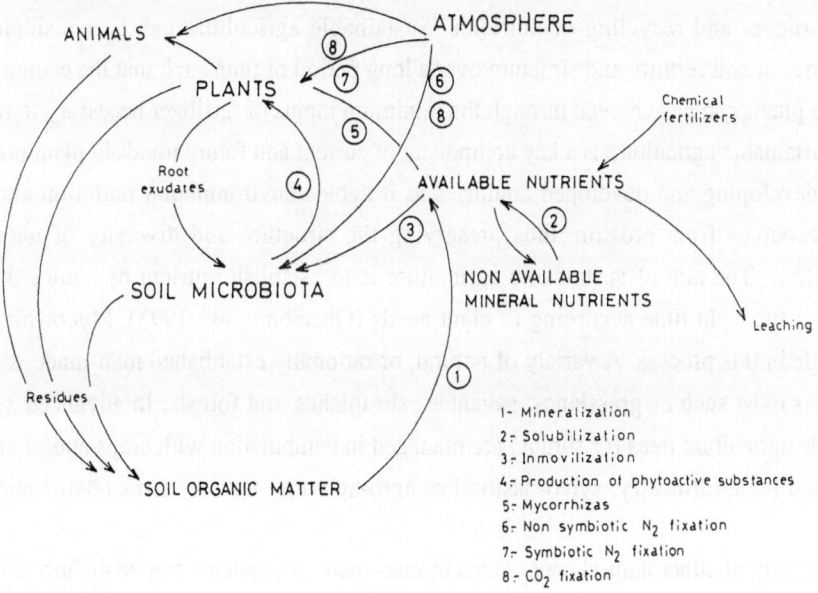

Figure 1. Main microbiologically–mediated processes in nutrient cycling.

Mycorrhizal associations make up a large biomass component in many ecosystems (Allen, 1991), and becuase both the fungi and the associated roots are turned over rapidly, mycorrhizas tend to be the largest throughout component in the ecosystem. By altering plant resource acquisition and plant production, mycorrhizas also dictate nutrient cycling rates and patterns. Nutrients from plants are recycled through six major pathways in ecosystems: (i) grazing, (ii) seed consumption, (iii) feeding on nectar, (iv) loss of soluble exudates, (v) active extraction by parasitic and mutualistic organisms and (vi) decomposition of plant structures (Odum and Biever, 1984). Mycorrhizas are particularly important in capturing nutrients in these latter three categories. Mycorrhizas link the biotic and geochemical portions of the ecosystem (O'Neil et al., 1991) and their essential bridging role in sustainable ecosystems is beyond doubt. What is sometimes in question, however, is the precise meaning of the term 'sustainable soil–plant systems'. In this chapter we have taken sustainability as the rational use of the natural resources rather than their exploitation. It is based on the use of renewable inputs and in the optimization of resource utilization to reach a balanced environmental relationship (Lal, 1989). By definition, most stable (climax) natural ecosystems are sustainable and the importance of mycorrhizas in maintaining an equitable cycling of nutrients cannot be overemphasized. A mature ecosystem is usually characterized by a nutrient conservative system in which nutrients are rapidly cycled between the biotic parts and do not leak out of the system. The mycelium of mycorrhizal fungi permeates the soil and takes up nutrients and channels them rapidly to their plant partners. In practice, however, the concept of sustainability is often applied to agricultural situations when the aim is to conserve the productive capacity of the land, whilst minimizing energy and resource use, and optimizing the rate of turnover and recycling of nutrients. Sustainable agriculture can be considered as the maintenance of soil fertility and structure over a long period of time such that the economic yields from crop plants can be achieved through the minimum inputs of fertilizer necessary to reach such yields. Sustainable agriculture is a key component of current and future trends in plant productivity for both developing and developed countries as it avoids environmental pollution and protects natural resources from erosion, thus preserving the structure and diversity of natural plant communities. The aim of sustainable agriculture is to establish nutrient dynamics that supply nutrients at the right time according to plant needs (Oberson et al., 1993). Mycorrhizas have a critical role in this process. A variety of natural, or rationally established man-made, sustainable ecosystems exist such as grasslands, savannas, shrublands and forests. In advanced systems of sustainable agriculture trees (or shrubs) are managed in combination with crops and/or animals, as exemplified by agroforestry, sylvo-pastoral or agro-sylvo-pastoral systems (Barea and Jeffries, 1994).

The stability of either natural ecosystems or man-made ecosystems can be disturbed by several factors of differing origin (Herrera et al., 1993). For example, the equilibrium of natural ecosystems, can be disturbed by natural agents (climatic, geomorphic, or paleotectonic processes). These can induce an irreversible damage (desertization) of the biological, chemical, and physical status of the soil, concomitant with a degeneration of the structure, morphology,and species

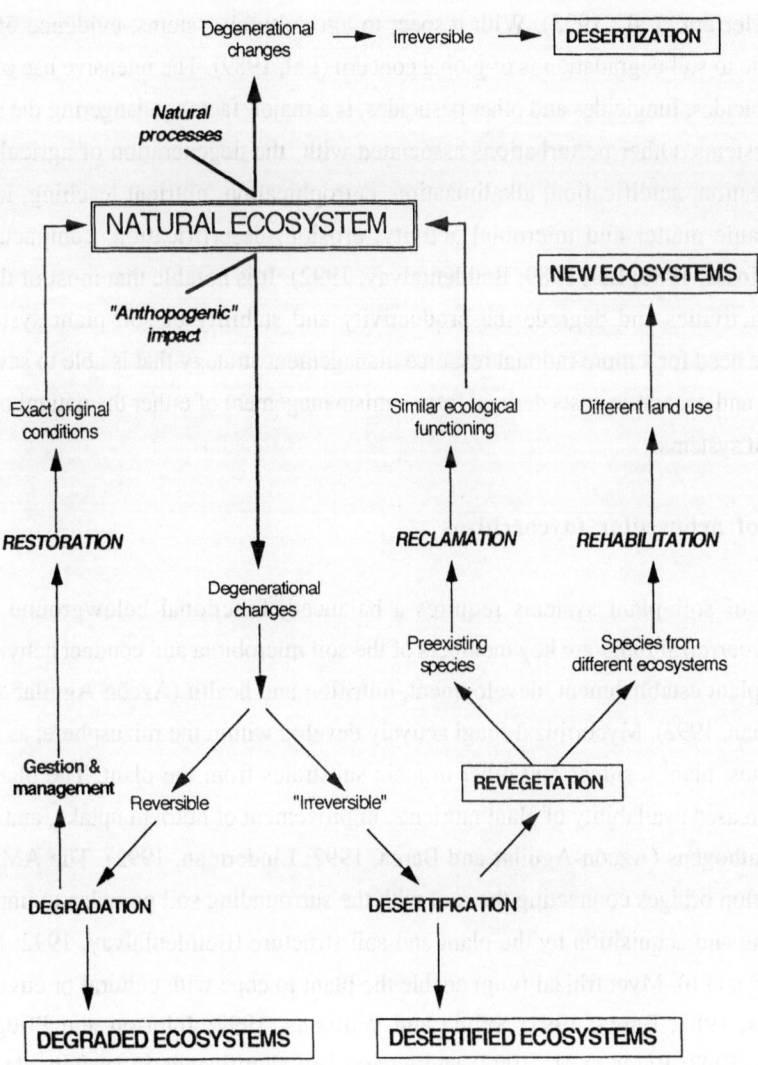

Figure 2. Factors involved in the degradation of natural ecosystems and action/concept approaches to recover them

combination of the vegetation cover (Francis and Thornes, 1990; Morgan et al., 1990; Allen et al., 1992). A natural desertization process can be accelerated, exacerbated, or even initiated by man through overgrazing, deforestation, and poor agricultural and recreational practices, causing desert-like situations to develop. The degradative, man-mediated, process is then called desertification (Fig.2).

A number of factors, characteristic of desertification, act either as causes or effects to produce a decline in both soil quality and plant development. These include loss or disturbance of the vegetation cover; loss of soil structure; increase in soil erosion; loss of available nutrients and organic matter; loss of microbial propagules, and/or diminution in microbiota activity (Skujins and Allen, 1986; Herrera et al., 1993). With respect to agricultural systems, evidence of a loss of arable lands due to soil degradation is of global concern (Lal, 1989). The intensive use of chemical fertilizers, herbicides, fungicides and other pesticides, is a major factor endangering the stability of agricultural systems. Other perturbations associated with the degeneration of agricultural soils include salinization, acidification, alkalinization, eutrophication, nutrient leaching, loss of soil structure, organic matter and microbial activity, erosion, desertification, compactation, and laterization (Mosse, 1986; Lal, 1989; Bethlenfalvay, 1992). It is notable that most of these result from human activities and degrade the productivity and stability of soil-plant systems. This emphasizes the need for a more rational resource management strategy that is able to save both the environmental and economic costs derived from a mismanagement of either the natural or the man-made soil-plant systems.

Importance of arbuscular mycorrhizas

Sustainability of soil–plant systems requires a balanced, functional belowground microbial ecosystem. Mycorrhizal fungi are key members of the soil microbiota and conduct activities which are crucial to plant establishment, development, nutrition and health (Azcón-Aguilar and Barea, 1992; Linderman, 1992). Mycorrhizal fungi actively develop within the rhizosphere, as stimulated by root exudates, plant residues and other organic substrates from the plant. The plant benefits through an increased availability of plant nutrients, improvement of nutrient uptake, and protection against root pathogens (Azcón-Aguilar and Barea, 1992; Linderman, 1992). The AMF are also known to develop bridges connecting the root with the surrounding soil particles to improve both nutrient cycling and acquisition by the plant and soil structure (Bethlenfalvay, 1992; Miller and Jastrow, 1992a and b). Mycorrhizal fungi enable the plant to cope with cultural or environmental stress (Jeffries, 1987; Barea, 1991; Sylvia and Williams, 1992; Johnson and Pfleger, 1992: Bethlenfalvay, 1992; Barea et al., 1993) and have a key significance in sustainable soil-plant systems. They can affect processes important to the sustainability of either disturbed or natural landscapes (Allen et al., 1992). Bethlenfalvay and Linderman (1992) envisage AMF as mediators of nutrient exchange in the soil-plant system. In fact, the AMF mycelium is acting in a close cause-and-effect interchange of mineral nutrients, C compounds, signals between the plant and rhizosphere populations and soil aggregation. As these authors argued, the maintenance of an optimal biological balance in the soil is critical in sustainability. Since soil degradation reduces the organic matter content, and the size and balance of the microbial population, it is obvious that the coordinated manipulation of AMF and suitable microorganisms is the only way to restore the biological component of sustainability.

This concept is closely allied to the use of leguminous plants as essential components of sustainable ecosystems as they are able to form two types of mutualistic symbiotic associations with soil microorganisms: N_2-fixing rhizobial nodules and arbuscular mycorrhizas (Azcón-Aguilar et al., 1979). This has great ecological importance with respect to sustainability. Firstly, N_2 fixation is a key source of N, a crucial nutrient for the biosphere. The N_2-fixation process is, however, dependent on the supply of phosphate and other nutrients. Mycorrhizal associations can satisfy these demands for both the plant and for N_2 fixation (Barea and Azcón-Aguilar, 1983; Barea et al., 1992a; Bethlenfalvay, 1992) such that a synergistic relationship exists between the two symbioses. Nodulated legumes with arbuscular mycorrhizas are well-adapted to cope with nutrient-deficient situations. Legumes are therefore widely used for protecting soil from erosion, for stabilizing soil structure, for fuel and timber, for food and fodder (proteins), for mulch and green manure. They play a key role in some cropping systems (e.g.rotation, intercropping), and in grazing, plantation, agroforestry and silvopastoral systems; and are being used for revegetation purposes (Olivares et al., 1988; LeTacon and Harley, 1990; Francis and Thornes, 1990; Fujita et al., 1992; Danso et al., 1992; Peoples and Craswell, 1992; Sanginga, 1992; Herrera et al., 1993).

Role of AMF in closing nutrient cycles

The hyphal network of AMF within the soil is a vital component of the soil ecosystem. This mycelium is the functional organ for the uptake and translocation of nutrients to and from mycorrhizas. Many reviews (e.g. Harley and Smith, 1983) describe the well-established role of the extraradical mycelium in the uptake of water and mineral nutrients, especially phosphorus, and the mechanisms of transfer of these elements to the plant in exchange for carbon metabolites derived from photosynthesis. The extensiveness, responsiveness and activity of a plant root system will determine its ability to obtain relatively immobile soil resources and these root characteristics are often negatively correlated with mycorrhizal colonization or mycorrhizal dependency (Brundett, 1991). Some mycorrhizal species in natural ecosystems have coarse, relatively inactive roots that would be inefficient at direct nutrient absorption. Mycorrhizal fungi must be important to these plants for the aquisition of nutrients. It is clear that P, C, N and other mineral nutrients can be transported from remote sources in the soil by AMF hyphae (e.g. Ames et al., 1983; Francis and Read, 1984; Cooper and Tinker, 1978). AMF hyphae are believed to contribute to biogeochemical cycling of nutrients by more than just providing a greater surface area for scavenging mineral nutrients that may be relatively immobile in soil or in short supply (Linderman and Pfleger, 1994). For example, they retain nutrients by altering their concentration ratios in vegetation and by decreasing their mobility by retention within the biomass. The combination of these functions means that the soil–plant system relies more on mineralization processes than on degrading parent soil materials.

A well-developed hyphal network can only become established under conditions which favour hyphal growth, and will cease to function under adverse environmental conditions or as a result of soil disturbance. The extent of the network can vary as these conditions fluctuate and this can influence the success or failure of establishment of plants introduced into that soil. It is thus important to distinguish mycorrhizal effects on plants being introduced where a stable hyphal network is in place from those where plants are being established without the benefit of a functional network. The network is essential for the continued cycling of nutrients within the plant community, and once it is lost nutrient sequestration or leaching will occur at a faster rate than in its presence. The hyphal network is thus vital for the maintainence of sustainable plant yields, both in a natural ecosystem and in low-input agricultural situations where nutrients are limiting. In a study of a tropical grazing ecosystem, the amount of AMF mycelium in the soil was found to be negatively correlated with soil fertility, but was positively correlated with the plant nutrient contents (McNaughton and Oesterheld, 1990). These results suggested that mycorrhizal associations act to stabilize ecosystem nutrient fluxes across edaphic gradients, and that they functionally compensate for wide variation in soil fertility since the nutritional status of the plants was not proportionally affected by differences in soil nutritional status. Mycorrhizal plants at sites of low fertility maintained mineral contents similar to, or only slightly lower than, those found at sites of high fertility with low mycorrhizal abundance.

The mycelium of AMF extends out for several centimetres (>9cm) so that it bridges the gap between the zone of nutrient depletion around the root and the bulk soil, where it can absorb immobile nutrient elements. The total amount of mycelium produced by AMF is difficult to quantify but a range of values from $2.6 - 54$ m·g^{-1} soil have been reported (Miller and Jastrow, 1994). The plant thus benefits from being able to exploit a much larger volume of soil in its vicinity. It is rare, however, for a plant to grow in isolation and it may be of greater significance that the mycelium can spread from plant to plant to form an linked nutrient-absorbing network that has access to the soil in which all the roots are growing. Many studies have demonstrated that inter-plant bridges formed by AMF can provide channels for direct nutrient transfer between the mycorrhizas of different plants. This may be sufficient to sustain significant enhancement of both growth and nutrient composition of receiver plants, in some cases within six weeks of commencement of experimentation (Francis et al., 1986). Transfers of C (Hirrel and Gerdemann, 1979; Francis and Read, 1984; Martins, 1992), N (van Kessel et al., 1985; Francis et al., 1986), and P (Heap and Newman, 1980; Chiariello et al., 1982; Whittingham and Read, 1982; Francis et al., 1986) have all been shown using laboratory systems, yet the significance in the field remains uncertain. Newman et al. (1992) have suggested that transfer of P between plant via mycorrhizal links may be too slow to be of much ecological significance, but there is evidence that N can be much more rapidly transferred, and this could be important in transfer from legumes to non–legumes. It may also be true that hyphae from different closely-related AMF mycelia are able to anastomose and thus create further links within the network. In laboratory culture, anastomosis between mycelia is quite frequent (Tommerup, 1988) but the frequency of its occurrence in the soil is not known. Despite this reservation, it is clear that the mycorrhizal mycelium of an undisturbed soil is extensive and can provide the conduit for the uptake of slowly-exchangeable soil ions and sparingly soluble mineral ions that form the basis of sustainable systems. The fungi take up P from the same pool of soluble ions as do roots, although it has been suggested that hyphae are able to absorb P from lower concentrations than can uninfected roots (Howeler et al., 1981). Unlike N_2-fixing bacteria, that function as biological fertilizers, AMF do not add phosphate to the soil but only improve its availibility to the plant. There is evidence that phosphatase activity is higher in the rhizosphere around arbuscular mycorrhizas than in that of non-mycorrhizal roots (Dodd et al., 1987) but there is no clear evidence that this is a fungus-mediated phenomenon that allows alternative P sources to be accessed.

There are several additional factors to be considered in relation to the effectiveness of the soil mycelium. It cannot be assumed, for example, that all AMF produce hyphae that are equally capable of transporting nutrients. In a comparison of hyphal transport of ^{32}P over defined distances, Jakobsen et al., (1992) found that *Scutellospora calospora* transported much less of this element to plants, but accumulated more of it in its hyphae, than either *Acaulospora laevis* or a *Glomus* sp. Neither can it be assumed that once present the hyphal network is long-lived. As much of the soil biomass may consist of mycorrhizal hyphae they provide a major food source for the mycophagic soil fauna (Fitter, 1985). Grazing by the soil fauna can thus influence the effectiveness of nutrient transport by the mycorrhizal network. The effect of soil disturbance on the beneficial growth response conferred on mycorrhizal plants is also well-documented (Johnson and Pfleger, 1992). Tillage or similar agricultural practices which disrupt the mycelial network have serious effects on its capacity to translocate nutrients over any significant distance. For example, the shoot P and N concentrations were much lower in maize plants grown in disturbed soil (Evans and Miller, 1990) indicating that the effects were a result of the destruction of a pre-existing mycelial network. Finally the biomass of the mycorrhizal mycelium must be considered, as the hyphae must represent a considerable nutrient sink. For example, they represent approximately 26% of the labelled extraradical organic carbon pool (Jakobsen and Rosendhal, 1990). Although it is assumed that hyphal turnover is relatively rapid, there is little experimental information available, however, about the rates of such hyphal turnover (Miller and Jastrow, 1994).

There is little doubt that mycorrhizas are important for the cycling of some elements such as P, but in a system where nutrient influx is limited, sustainability can only be achieved if all nutrient cycles are closed, thus minimizing losses. Mycorrhizal fungi might effect such closure by decomposing litter and transporting the mineral nutrients thus released (Janos, 1984). Such direct nutrient cycling, however, requires that mycorrhizal fungi decompose organic material and although there is good evidence that orchidaceous mycorrhizal fungi and some ectomycorrhizal fungi can degrade cellulose and/or lignin in the soil, there is no such evidence for AMF. Production of pectinolytic and cellulolytic enzymes by *Glomus* has been demonstrated *in vitro* (García-Romera et al., 1990) but it would seem unlikely that these are important in nature in the competitive saprotrophic breakdown of dead organic remains.

The closure of nutrient cycles that do not require degradative ability of AMF is much clearer. The intradical mycelium of the mycorrhiza remains in place once the arbuscules have degenerated. It is thus ideally placed to absorb nutrients released as the root senesces or is broken down by extracellular enzymes from other soil microorganisms. By virtue of their position, these hyphae could be extremely effective in competitively scavenging nutrients from dying roots (Heap and Newman, 1980), even though mineralization of the plant remains might have been initiated by the activity of other microorganisms. Thus the mycorrhizal hyphal links promote direct nutrient cycling by avoiding the mineralization process. The intraradical mycelium is continuous with the soil mycelium, thus if significant quantities of mineral nutrients are absorbed from the dying root the resulting fluxes serve to close the nutrient cycles. Dying plants can lose 60% of their nitrogen and 70% of their phosphorus within three weeks and much of this ends up in neighbouring plants (Eason and Newman, 1990). The magnitude of this transfer is substantially greater if cohabiting plants are mycorrhizal, providing evidence that the transfer occurs through the hyphal bridges of the AMF (Newman and Eason, 1989). In sustainable agrosystems such as alleycropping, this mycorrhizal effect could be crucial in maintaining soil fertility. This mycorrhizal function could also be important with respect to effective recycling of C from dead roots, thus restricting its loss to the general soil microbial community. Martins (1992) has shown that AMF provide hyphal channels for direct transfer of C between individual plants.

Mycorrhizal symbioses influence the relative abilities of plants to compete for limiting nutrients. Integration of individual plants into the collective nutrient-gathering capacity of the community assures survival and the role of the soil mycelium is again crucial. Under adverse conditions the stability of both natural and agricultural systems can be threatened, and plants must be able to cope with these stresses. In sustainable agriculture the arbuscular mycorrhizal symbiosis plays a key role in helping the plant not only to survive but to efficiently recycle nutrients and thus remain productive under adverse conditions (Mosse, 1986) such as water stress, salt stress, low nutrient availibility and degraded or eroded soils (Barea and Jeffries, 1994). Mycorrhizal hyphae are also involved in the formation of stable soil aggregates, a process crucial for soil conservation. This effect is critical in the development of sustainable ecosystems especially in eroding soils. Miller and Jastrow (1992 a and b) have raised the interesting question as to whether nutrient–rich aggregates in soil arise because of the proliferative effects of mycorrhizal hyphae. The encapsulation of organic debris in this way within a soil aggregate may be a means for creating a more conservative nutrient cycle (Miller and Jastrow, 1992 a and b). In some soils these effects may be further enhanced by the ablity of AMF to weather mica (Bethlenfalvay et al., 1984), resulting in an increase in the capacity of the soil to retain water and to provide nutrients for associated microsymbionts.

Once the importance of the mycorrhizal symbiosis is recognized, the management of AMF populations becomes a potential tool for improving sustainability. A decision must be made whether the native population of AMF suffices as the starting material from which to develop a sustainable system. If not it will be necessary to supplement the native species with inoculum. Alternatively, if the native population is adequate but of low infectivity, it may be possible to rapidly increase inoculum levels using appropriate management practices (Barea and Jeffries, 1994) such as crop rotation, fallowing, pre-cropping, intercropping or the use of appropriate P and N sources.

Sustainable systems with limited inputs of P and N exist naturally in many ecosystems. In agricultural situations large P and N inputs can drastically affect the mycorrhizal symbiosis and result in reduced or zero colonization. Although high inputs of P may depress AMF colonization, continued fertilization may eventually select for AMF adapted to high plant P content (Thomson et al., 1986), a situation with undesirable consequences for low-input regimes. It is also apparent that modern crop breeding programmes have selected high–yielding cultivars adapted to high inputs of fertilizer. This process has also inadvertently selected genotypes that are unresponsive to mycorrhizal colonization. Reversion to a low–input sustainable system will need a re–examination of the older varieties of crop plants as in these systems there is an opportunity to balance fertilizer inputs with the maximum mycorrhizal benefit such that crop yields are obtained that approach those of some high-input systems. Azcon-Aguilar et al., (1979) have discussed the role of mycorrhizas in this context, and it is clear that the efficiency of the AMF will be closely correlated with the availability of nutrients in the soil (Furlan and Bernier-Cardou, 1989). In the tropics many crops are grown in acid-infertile soils, where their establishment is limited by the low levels of available P (Arias et al., 1991). In such soils AMF can be particularly useful as a low-input technology (Sieverding, 1991), especially in combination with locally-available slow-release sources of P, such as rock phosphate. Recently concern has been raised regarding the potential impact of transgenic plants in agriculture (Miller, 1993). Cultivars that have been modified to constitutively produce transgenic pathogenesis–related proteins such as chitinase may be more resistant to plant pathogens, yet may suffer a concomitant reduction in susceptibility to mycorrhizal infection. This unintentional non–target effect must be closely monitored before such cultivars gain wide acceptance within sustainable agrosystems.

Mycorrhizas and habitat restoration

We have already commented on the importance of a mycorrhizal network within a stable, mature ecosystem. Allen et al. (1992) have suggested that at the ecosystem scale the presence of mycorrhizas represents a key difference between a weedy, unstable site where nutrients are predominantly abiotic in form, and a grass or shrub–dominated site that is relatively stable and in which organic nutrients prevail. Nutrient cycling patterns have been shown to differ between the nonmycorrhizal early successional systems and the mycorrhizal later successional systems as discussed by (Allen et al., 1992) who cite as an example, at sites dominated by nonmycotrophic annuals such as *Salsola kali*, the organic matter and N content of the soil was low and most of the phosphate was found in the bound inorganic fractions. The addition of mycorrhizal fungi to the same soils resulted in the mineralization of P (Knight et al., 1989) and the increased incorporation of P into plant tissue. The main role of mycorrhizas is thus in the later stages of succession where they short–circuit nutrient cycles by reaquiring nutrients in organic form from plant (and fungal) litte. They may then reallocate resources between different plant individuals, preventing loss of resources from the entire ecosystem (Pankow et al., 1991).

Several experiments have been carried out where pioneer plant species were inoculated with AMF to improve their establisment and development. For example, Carpenter and Allen (1988) found that AMF inoculation (topsoil inoculum) of *Hedysarum boreale* increased growth, survival and the production of flowers and seeds by this shrub legume in the field. It was therefore expected that this would contribute to the establishment of a self-sustainable ecosystem. More recently, Herrera et al., (1993) have reported the results of a 4-year field revegetation trial carried out in a semi-arid desertified ecosystem in south–eastern Spain. Degraded ecosystems are common in regions with a Mediterranean climate, particularly in the south-east of Spain, where low precipitation and human disturbance have resulted in erosion and desertification of already fragile ecosystems. These desertified ecosystems are appropriate for testing rehabilitation programmes (Allen, 1988; Francis and Thornes, 1990; Morgan et al., 1990). Herrera et al. (1993) assessed the significance and effectiveness of plant-microbe symbioses as a component of a revegetation strategy in which a number of woody species, common in revegetation programs in Mediterranean regions, were used. These included two native shrubs (*Anthyllis cytisoides* and *Spartium junceum*) and four exotic tree legumes (*Robinia pseudoacacia, Medicago arborea, Acacia caven* and *Prosopis chilensis*). Plant species and microsymbionts were screened for appropriate combinations, and a simple procedure to produce plantlets with an optimized mycorrhizal and nodulated status was developed. During a four-year period after outplanting, the results showed that: (i) only the native shrub legumes were able to establish under the local environmental conditions; (ii) inoculation with rhizobia and AMF improved plant survival, and biomass development. Since the two native shrubs are found in the natural plant community are particularly appropriate for revegetation of these desertified areas. A reclamation strategy has therefore been proposed, using *Anthyllis cytisoides*, a particularly drought-tolerant species. *Anthyllis cytisoides* is known to be mycorrhizally responsive at low P levels (López–Sanchez et al., 1992 and unpublished observations). This technique involving the artifical acceleration of natural revegetation could be accomplished by replanting randomly spaced groups of shrubs according to the natural pattern and structure of the undisturbed ecosystem (Francis and Thornes, 1990; Morgan et al., 1990; Herrera et al., 1993). Management of appropriate microsymbionts can help legumes to promote the stabilization of a self-sustaining ecosystem. The mycorrhizal shrubs, acting as a "fertile islands" (Skujins and Allen, 1986), could serve as sources of inoculum for the surrounding area and to improve N nutrition for the non N-fixing vegetation in these and other similar semi-arid ecosystems.

Conclusions

We have stressed the importance of the arbuscular mycorrhizal network in the preservation of nutrient cycles within sustainable soil–plant ecosystems. The same considerations are also true in ecosystems where other types of mycorrhizas are dominant (Read, 1993). For the future, it is clear that there is no "single mycorrhizal effect" on plant communities (Allen, 1991) and for the detailed study of mycorrhizal behaviour, one may no longer think of "mycorrhizas" (in general) to be merely present or absent, but that mycorrhizal populations differ in their effects on plant and soil. As a consequence, it is necessary to have information about the individual fungi within an ecosystem in order to be able to predict their behaviour. It is also necessary to understand how individual mycelia interact with one another at an interspecific and intraspecific level. This may seem a daunting prospect, but the molecular tools are now available for a challenging dissection of AMF population dynamics. Comparison of isozyme profiles can be used to differentiate AMF at a number of levels, including individual isolates in some cases (Hepper et al., 1988; Rosendhal and

Sen, 1992; Rozycka et al., 1992). They can also be used to identify infection by different AMF within plant roots (Hepper et al., 1988). Immunological approaches, using polyclonal or monoclonal antibodies, also have applications (Aldwell et al., 1985; Wright et al., 1987), but careful screening procedures must be employed in order to determine cross-reactivity across a range of isolates. The development of PCR-based techniques using molecular primers offers great promise for the development of specific probes to differentiate AMF at a species or isolate level. Such approaches have already been used to fingerprint DNA from spores of a range of isolates (Wyss and Bonfante, 1993) and the technique can be taken further to identify mycelia within roots, once appropriate probes are developed that do not cross-react with plant nucleic acid. We hope that collaborative work within the European framework can enable this aim to be achieved in investigations of the role of microsymbionts within the desertified ecosystems around the Mediterranean.

Highly-mycotrophic plants are characteristic of stable, sustainable ecosystems and any attempts to convert non-sustainable ecosystems to sustainable ones must consider the mycorrhizal component.The optimization of plant–microbe symbioses is essential for enhancing nutrient uptake in low–input ecosystems. Mycorrhizal benefits can be increased by choosing plant cultivars responsive to the symbiosis, by ensuring that highly effective strains of symbionts are present and, if necessary, by modifying the habitat to encourage mycorrhization. This type of approach has been exemplified in a misture of traditional and modern agricultural technologies used in Costa Rica for the cultivation of beans (Rosemeyer and Fliessman, 1992), but further studies of this nature are urgently required. Bethlenfalvay and Linderman (1992) have suggested that within sustainable ecosystems AMF are "the universal compensators needed to accomplish the mission of sustainable agriculture". We fully agree and hope that this review has stressed their importance in non-agricultural ecosystems also.

Acknowledgements

We are grateful to the EC COST 810 programme, the British Council (Acciones Integradas) and the AFRC Wain Fund for financial assistance to facilitate exhange visits between our laboratories. Some experimental work referred to in this Chapter was sponsored by the CICYT Project NAT–91–1127.

References

Aldwell, F.E.D., Hall, I.R., and Smith, J.M.B. (1985) Enzyme-linked immunosorbent assay as an aid to taxonomy of the Endogonaceae. *Trans. Brit. Mycol. Soc.* 84: 399-412.
Allen, M.F. (1991) *The Ecology of Mycorrhizae*. Cambridge University Press, Cambridge.

Allen, M.F., Clouse, S. D., Weinbaum, B.S., Jeakins, S., Friese, C.F., and Allen, E.B. (1992) Mycorrhizae and the integration of scales: From molecules to ecosystems. In: M.F. Allen (ed) *Mycorrhizal Functioning: an Integrative Plant-Fungal Process*. Chapman & Hall, New York, London, pp. 488-516.

Ames, R.N., Reid, C.P.P., Porter, L.K., and Cambardella, C. (1983) Hyphal uptake and transport of nitrogen from two ^{15}N labelled sources by *Glomus mosseae*, a vesicular-arbuscular mycorrhizal fungus. *New Phytol.* 95: 381-396.

Arias, I., Koomen, I., Dodd, J.C., White, R.P., and Hayman, D.S. (1991) Growth responses of mycorrhizal and non-mycorrhizal tropical forage species to different levels of soil phosphate. *Plant and Soil* 132: 253-260.

Azcón-Aguilar, C., Azcón, R., and Barea, J.M. (1979) Endomycorrhizal fungi and *Rhizobium* as biological fertilizers for *Medicago sativa* in normal cultivation. *Nature* 279: 325-327.

Azcón-Aguilar, C. and Barea, J.M. (1992) Interactions between mycorrhizal fungi and other rhizosphere microorganisms. In: M.F. Allen (ed) *Mycorrhizal Functioning: an Integrative Plant-Fungal Process*. Chapman & Hall, New York, pp. 163-198.

Barea, J.M. (1991) Vesicular-arbuscular mycorrhizae as modifiers of soil fertility. *Adv. Soil Sci.* 15: 1-40.

Barea, J.M., Azcón, R. and Azcón-Aguilar, C. (1993) Mycorrhiza and crops. *Adv. Pl. Path.* 9: 167-189.

Barea, J.M., and Jeffries, P. (1994). Arbuscular mycorrhiza in sustainable soil-plant systems. In: B. Hock, and A. Varma (eds) *Mycorrhiza: Structure, Function, Molecular Biology and Biotechnology*. Springer, Heidelberg (in press).

Bethlenfalvay, G.J. (1992) Mycorrhizae and crop productivity. In: G.J. Bethlenfalvay and R.G. Linderman (eds) *Mycorrhizae in Sustainable Agriculture*. ASA Special Publication, Madison, pp. 1-28.

Bethlenfalvay, G.J., Dakessian, S., and Pacovsky, R.S. (1984) Mycorrhizae in a southern California desert: ecological implications. *Can. J. Bot.* 62: 519-524.

Bethlenfalvay, G.J., and Linderman, R.G. (1992) *Mycorrhizae in Sustainable Agriculture*. ASA Special Publication, Madison, WI.

Brundrett, M. (1991) Mycorrhizas in natural ecosystems. *Adv. Ecol. Res.* 21: 171-313.

Carpenter, A.T., and Allen, M.F. (1988) Responses of *Hedysarum borelae* Nutt to mycorrhizas and *Rhizobium*: plant and soil nutrient changes in a disturbed shrub-steppe. *New Phytol.* 109:125-132.

Chiariello, N., Hickman, J.C., and Mooney, H.A. (1982) Endomycorrhizal role for interspecific transfer of phosphorus in a community of annual plants. *Science* 217: 941-943.

Cooper, K.M., and Tinker, P.B. (1978) Translocation and transfer of nutrients in vesicular-arbuscular mycorrhizas II. Uptake and translocation of phosphorus, zinc and sulphur. *New Phytol.* 81: 43-52.

Danso, S.K.A., Bowen, G.D., and Sanginga, N. (1992) Biological nitrogen fixation in trees in agro-ecosystems. *Plant and Soil* 141:177-196.

Dodd, J.C., Burton, C.C., Burns, R.G., and Jeffries, P. (1987) Phosphatase activity associated with the roots and rhizosphere of plants infected with vesicular-arbuscular mycorrhizal fungi. *New Phytol.* 107: 163-172.

Eason, W.R., Newman, E.I. (1990) Rapid cycling of nitrogen and phosphorus from dying roots of *Lolium perenee*. *Oecologia* 82: 424-436.

Evans, D.G., and Miller, M.H. (1988) Vesicular-arbuscular mycorrhiza and the soil-diturbance-induced reduction of nutrient absorption in maize. *New Phytol.* 110: 75-84.

Fitter, A.H. (1985) Functioning of vesicular-arbuscular mycorrhizas under field conditions. *New Phytol.* 99: 257-265.

Francis, R., Finlay, R.D., and Read, D.J. (1986) Vesicular-arbuscular mycorrhiza in natural vegetation systems IV. Transfer of nutrients in inter- and intra-specific combinations of host plants. *New Phytol.* 102: 103-111.

Francis, R., and Read, D.J. (1984) Direct transfer of carbon between plants connected by vesicular-arbuscular mycorrhizal mycelium. *Nature* 307: 53-56.

Francis, C.F., Thornes, J.B. (1990) In: J. Albaladejo, M.A.Stocking and E. Díaz (eds) *Soil Degradation and Rehabilitation in Mediterranean Environmental Conditions*. CSIC, Murcia, pp. 87-115.

Fujita, K., Ofosu-Budu, K.G., and Ogata, S. (1992) Biological nitrogen fixation in mixed legume-cereal cropping systems. *Plant and Soil* 141: 155-175.

Furlan, V., and Bernier-Cardou, M. (1989) Effects of N, P, and K on formation of vesicular-arbuscular mycorrhizae, growth and mineral content of onion. *Plant and Soil* 113: 167-174.

García-Romera, I., Garcia-Garrido, J.M., Martinez-Molina, E., and Ocampo, J.A. (1990) Possible influence of hydrolytic enzymes on vesicular arbuscular mycorrhizal infection of alfalfa. *Soil Biol.Biochem.* 22: 149-152.

Harley, J.L., and Smith, S.E. (1983) *Mycorrhizal Symbiosis*. Academic Press, New York.

Heap, A.J., and Newman, E.I. (1980) The influence of vesicular-arbuscular mycorrhizas on phosphorus transfer between plants. *New Phytol.* 85: 173-179.

Hepper, C.M., Azcón-Aguilar, C., Rosendhal, S., and Sen, R. (1988) Competition between three species of *Glomus* used as spacially separated introduced and indigenous mycorrhizal inocula for leek (*Allium porrum* L.). *New Phytol.* 110: 207-211.

Herrera, M.A., Salamanca, C.P., and Barea, J.M. (1993) Inoculation of wooody legumes with selected arbuscular mycorrhizal fungi and rhizobia to recover desertified mediterranean ecosystems. *Appli. Environm. Microbiol.* 59: 129-133.

Hirrell, M.C., and Gerdemann, J.W. (1979) Enhanced carbon transfer between onion infected with a vesicular-arbuscular mycorrhizal fungus. *New Phytol.* 83: 731-738.

Howeler, R.H., Edwards, D.G., and Asher, C.J. (1981) Application of the flowing solution culture techniques to studies involving mycorrhizas. *Plant and Soil* 59: 179-183.

Jakobsen, I., Abbott, L.K., and Robson, A.D. (1992) External hyphae of vesicular-arbuscular mycorrhizal fungi associated with *Trifolium subterraneum* L. 2. Hyphal transport of ^{32}P over defined distances. *New Phytol.* 120: 509-516.

Jakobsen, I., Rosenthal, L. (1990) Carbon flow into soil and external hyphae from roots of mycorrhizal cucumber plants. *New Phytol.* 115: 77-83.

Janos, D.P. (1984) Tropical mycorrhizas, nutrient cycles and plant growth. In: E. Medina, H.A. Mooney and C. Vazquez-Yanes (eds) *Physiological Ecology of Plants in the Wet Tropics*. Junk Publ., The Hague, pp. 327-345.

Jeffries, P. (1987) Use of mycorrhizae in agriculture. CRC *Critical Review Biotechnology* 5: 319-358.

Johnson, N.C., and Pfleger, F.L. (1992) Vesicular-arbuscular mycorrhizae and cultural stresses. In: G.J. Bethlenfalvay and R.G. Linderman (eds) *Mycorrhizae in Sustainable Agriculture*. ASA Special Publication, Madison, WI. pp 71-99.

Knight, W.G., Allen, M.F., Jurianak, J.J., and Dudley, L.M. (1989) Elevated carbon dioxide and solution phosphorus in soil with vesicular-arbuscular mycorrhizal western wheatgrass. *Soil Sci.Soc.Am.J.* 53: 1075-1082.

Lal, R. (1989) Conservation tillage for sustainable agriculture: tropics versus temperate environments. *Adv.Agronomy* 42: 85-185.

Le Tacon, F., and Harley, J.L. (1990) Deforestation in the tropics and proposals to arrest it. *Ambio* 19: 372-378.

Linderman, R.G. (1992) Vescicular-arbuscular mycorrhizae and soil microbial interactions. In: G.J. Bethlenfalvay and R.G. Linderman (eds) *Mycorrhizae in Sustainable Agriculture*. ASA Special Publication, Madison, WI., pp. 45-70.

Linderman, R.G., Pfleger, F.L. (1994) General summery. In: Pfleger, F. L., Linderman, R. G., (eds.): *Mycorrhizae and Plant Health*. APS Press, St. Paul Minnesota, pp. 337-344.

López-Sánchez, M.E., Díaz, G. and Honrubia, M. (1992) Influence of vesicular arbuscular mycorrhizal infection and P addition on growth and P nutrition of *Anthyllis cytisoides* L. and *Brachypodium retusum* (Pers.) Beauv. *Mycorrhiza* 2: 41-45.

Martins, M.A. (1992) The role of the external mycelial network of vesicular-arbuscular mycorrhizal fungi: a study of carbon transfer between plants interconnected by a common mycelium. *Mycorrhiza* 2: 69-73.

McNaughton, S.J., and Oesterheld M. (1990) Extramatrical mycorrhizal abundance and grass nutrition in a tropical grazing ecosystem, the Serengeti National Park, Tanzania. *Oikos* 59: 92-96.

Miller, R. M. (1993) Nontarget and ecological effects of transgenically altered disease resistance in crops Possible effects on the mycorrhizal symbiosis. *Mol. Ecol.* 2: 327-335.

Miller, R.M., and Jastrow J.D. (1992a) The application of VA mycorrhizae to ecosystem restoration and reclamation. In: M.F. Allen (ed) *Mycorrhizal Functioning an Integrative Plant Fungal Process*. Chapman & Hall, New York, pp 438-467.

Miller, R.M., and Jastrow J.D. (1992b) The role of mycorrhizal fungi in soil conservation. In: G.J. Bethlenfalvay and R.G. Linderman (eds) *Mycorrhizae in Sustainable Agriculture*. ASA Special Publication, Madison, Wisconsin, pp. 29-44.

Miller, R.M., and Jastrow J.D. (1994) Vesicular-arbuscular mycorrhizae and biogeochemical cycling. In: F.L. Pfleger, and R.G. Linderman (eds) *Mycorrhizae and Plant Health*. APS Press, St. Paul Minnesota, pp. 189-212.

Morgan, R.P.C., Rickson, R. J., and W. Wright (1990) In: J. Albaladejo, M.A. Stocking and E. Díaz (eds) *Soil degradation and rehabilitation in Mediterranean environmental conditions*. CSIC Murcia, pp. 69-85.

Mosse, B. (1986) Mycorrhiza in a sustainable agriculture. *Biological Agricultural Horticulture* 3: 191-209.

Newman, E.I., and Eason,W.R. (1989) Cycling of nutrients from dying roots to living plants including the role of mycorrhizas. In: M. Charholm and L. Berfstrom (eds) *Ecology of arable Lands*. Kluwer Academic Publ. pp. 133-137.

Newman, E.I., Eason,W.R., Eissenstat, D.M., and Ramos, M.I.R.F. (1992) Interactions between plants: the role of mycorrhizae. *Mycorrhiza* 1: 47-53.

Oberson, A., Fardeau, J.C., Besson, J.M., and Sticher, H. (1993) Soil phosphorus dynamics in cropping systems managed according to conventional and biological agricultural methods. *Biology and Fertility of Soils* 16: 111-189.

Odum, E.P., and Biever, L.J. (1984) Resource quality, mutualism and energy partitioning in food chains. *American Naturalist* 124: 360-376

Olivares, J., Herrera, M.A., and Bedmar E.J. (1988) In: D.P. Beck and L.A. Materon (eds) *Nitrogen fixation legumes in mediterranean agriculture.* ICARDA, Martinus Nijhoff, Dordrecht, The Netherlands, pp. 65-72.

O'Neil, E.G., O'Neil, R.V., Norby, R.J. (1991) Hierachy theory as a guide to mycorrhizal research on large-scale problems. *Environmental Pollution* 73: 271-284.

Pankow, W., Boller, T., and Wiemken, A. (1991) The significance of mycorrhizas in protective ecosystems. *Experientia* 47: 391-394.

Peoples, M.B., and Craswell, E.T. (1992) Biological nitrogen fixation: Investments, expectations and actual contributions to agriculture. *Plant and Soil* 141: 13-39.

Read, D.J. (1991) Mycorrhizas in ecosystems. *Experientia* 47: 376-391.

Read, D.J. (1993) Mycorrhiza in plants communities. *Adv. in Plant Path.* 9: 1-31.

Rosemeyer, M.E., Gliessman, S.T. (1992) Modifying traditional and high-input agroecosystems for optimization of microbial symbioses: a case study of dry beans in Costa Rica. *Agric. Ecosyst. Environment* 40: 61-70.

Rosendhal, S., Sen, R. (1992) Isozyme analysis of mycorrhizal fungi and their mycorrhiza. *Methods in Microbiology* 24: 169-194.

Rozycka, M., Jeffries, P., and Dodd, J.C. (1992) Immunological identification and characterization of arbuscular mycorrhizal fungi (AMF). *International Symposium on Management of Mycorrhizas in Agriculture Horticulture and Forestry.* Perth, p. 159.

Sanginga, N. (1992) In: IFS (ed.):*Interactions Between Plants and Microorganisms.* Dakar, Senegal, pp. 14-32.

Sieverding, E. (1991) *Vesicular-Arbuscular Mycorrhiza Management in Tropical Agrosystems.* Deutsche GTZ GmbH. Eschborn. Germany.

Skujins, J. and Allen, M. F. (1986) Use of mycorrhizae for land rehabilitation. *MIRCEN Journal* 2:161-176.

Sylvia, D.M., and Williams, S.E. (1992) Vesicular-arbuscular mycorrhizae and environmental stress. In: G.J. Bethlenfalvay and R.G. Linderman (eds) *Mycorrhizae in Sustainable Agriculture.* ASA Special Publication, Madison, pp. 101-124.

Thomson, B.D., Robson, A.D., and Abbott, L.K. (1986) Effects of phosphorus on the formation of mycorrhizas by Gigaspora calospora and *Glomus fasiculatum* in relation to root carbohydrates. *New Phytol.* 103: 751-765.

Tommerup, I.C. (1988) The vesicular arbuscular mycorrhizas. *Adv.Plant Path.* 6: 81-92.

Van Kessel, C., Singleton, P.W., and Hoben, H.J. (1985) Enhanced nitrogen-transfer from a soybean to maize by vesicular arbuscular mycorrhizal (VAM) fungi. *Plant Phys.* 79: 562-563.

Whittingham, J., and Read, D.J. (1982) Vesicular-arbuscular mycorrhizae in natural vegetation systems III. Nutrient transfer between plants with mycorrhizal interconnections. *New Phytol.* 97: 413-426.

Wright, S.F., Morton, J.B., and Sworobuk, J.E. (1987) Identification of a vesicular-arbuscular mycorrhizal fungus by using monoclonal antibodies in an enzyme-linked immunosorbent assay. *Appl. Environm. Microbiology* 53: 2222-2225.

Wyss, P., Bonfante, P. (1993) Amplification of genomic DNA of arbuscular-mycorrhizal (AM) fungi by PCR using short arbitrary primers. *Mycol. Res.* 97: 1351-1357.

Impact of Arbuscular Mycorrhizas on
Sustainable Agriculture and Natural Ecosystems
S. Gianinazzi and H. Schüepp (eds.)
© 1994 Birkhäuser Verlag Basel/Switzerland

Arbuscular mycorrhizas and agrosystem stability

G.J. Bethlenfalvay and H. Schüepp[1]

US Dept. of Agriculture, Agric. Research Service, Horticultural Crops Research Lab, Corvallis, OR 97330, USA
[1]Swiss Federal Research Station, CH–8820 Wädenswil, Switzerland

Introduction

An agrosystem consists of plant roots, the soil microflora, the soil fauna and the abiotic geochemical soil matrix. Plant shoots, as its source of energy, also form an integral part of the system. In keeping with the role of plants as providers of food and fiber, functions of fundamental societal importance, the agrosystem has been traditionally used, treated and evaluated from a phytocentric and edaphic point of view in agriculture. On a scale of priorities, the soil has always taken second place, with its function of support subordinate to that of the plant, the primary producer. This perception is undergoing radical change in our times. We now recognize the importance of soil not only as an agricultural 'resource base' (Stewart et al., 1991), but as a complex, living, and fragile system that must be protected (Reganold et al., 1990) and managed for its own sake (Pierce and Lal, 1991) to guarantee its long–term stability and productivity. Articulated scientifically in the early 1980s (Bezdicek and Power, 1984; Jackson, 1980; Rodale, 1983), sustainability has a long history in agriculture (Harwood, 1990), and has deep roots in societal consciousness also.

In agricultural research, the goals of sustainability may be summarized in their briefest form as 'maximum plant production with a minimum of soil loss'. Within this context of balanced agrosystem inputs and outputs (Hornick and Parr, 1987), the role of the arbuscular–mycorrhizal fungi (AMF) have been described as that of a fundamental link between plant and soil (Bethlenfalvay and Linderman, 1992; Miller and Jastrow, 1994; O'Neil et al., 1991). In keeping with its importance, the symbiotic association between AMF and their host plants and host soils, and its impact on agrosystem stability, is currently subject to intensive investigation. This work has been reviewed with focus on: (1) plant growth and plant ecology (Barea and Jeffries, 1994; Bethlenfalvay, 1992; Sieverding, 1991), (2) plant health and biocontrol (Linderman, 1992, 1994;

Schönbeck and Dehne, 1989), and (3) cultural (Johnson and Pfleger, 1992; Kurle and Pfleger, 1994; Miller and Jastrow, 1992a) and environmental (Sylvia and Williams 1992; Wright and Millner 1994) plant stress. Recently, perhaps inspired by the impact of sustainability, the AMF have also come to life in the review literature as soil symbionts and agrosystem stabilizers (Bethlenfalvay and Svejcar, 1991; Finlay and Sönderström, 1992; Miller and Jastrow, 1992b; Tisdall, 1991). But no attempt had been made so far to integrate the information available on the interactions between specific AMF isolates and specific groups of soil biota with agrosystem stability.

The purpose of this contribution is to discuss the current concept of mycorrhizal effectiveness and to refocus it by making it applicable not only to the host plant but to the entire agrosystem, within the context of sustainability in agriculture.

Mycorrhizal efficiency and benefits to the agrosystem: changing views

The contributions of AMF in natural or disturbed ecosystems and in experimentation under controlled conditions have been traditionally measured by plant response (Asai, 1943; Gerdeman, 1968; Jeffries, 1987; Pfleger and Linderman, 1994; Schlicht, 1889; Stahl, 1900). The more an AMF was able to improve plant growth relative to other isolates under a given set of conditions, the more 'effective' it was said to be. In turn, the better an arbuscular mycorrhizal plant could approximate growth by an optimally fertilized non-arbuscular-mycorrhizal plant (Abbot and Robson, 1991), the more 'benefits' it was said to derive from its symbiotic status. These benefits (mycotrophy, or mycorrhizal dependence) were evaluated in terms of gains derived from mycotrophic P import against the price paid in reduced carbon by the plant to support its obligatorily biotrophic endophyte (Fitter, 1991; Koide, 1991).

Such an analysis of cost–benefit relationships from the plant's point of view continues to be of interest to agriculture and above–ground ecology, and is particularly applicable to a demonstration of host–endophyte relationships under the conditions of two–component agrosystems that consist only of the mycorrhiza and the sterilized soil used in most mycorrhiza experiments. In the field, however, the relationship between host plants and AMF is altered by the other biotic components of the agrosystem (Fitter, 1985; Hetrick et al., 1988; Safir, 1994), which permit measurable benefits to accrue to the plant only under particular conditions of growth (Fitter, 1986).

The extent to which the concept of mycorrhizal benefit is influenced by phytocentric thinking is illustrated by two important recent studies. Hetrick et al. (1992) found that a decline in dry weight was related to a loss of mycorrhizal dependence in modern wheat (*Triticum aestivum* L.) cultivars. These authors suggested that mycorrhizal dependency should be considered in breeding programs (Hetrick et al., 1993). In doing so, they equated mycorrhizal benefit with mycorrhizal dependence of the host plant, as if mycorrhizal contributions to the soil in which that plant grows were irrelevant. In the second study, Johnson (1993) suggested that an understanding of the mechanisms that let AMF enhance or inhibit plant growth is necessary for managing ecosystems. This view was based on findings that cultural practices select AMF that are inferior mutualists (Johnson et al., 1992). Johnson (1993) recommended a manipulation of AMF communities favouring proliferation of the most beneficial isolates with regard to plant yield instead of the inferior ones that contribute to yield decline. Again, the suggestion does not consider that yield, in the long run, depends on the quality of the soil.

We fully agree with these authors that plant growth response to AMF is important in agriculture, but wish to emphasize that there is more to mycorrhizal benefit than just the yield increase derived from mycotrophy. Stribley's verdict (1989): "mycorrhizal inoculants have failed to fulfil their promise because currently there is little promise to fulfil" was apparently based on the assumption that plant growth enhancement is all that AMF affect in the agrosystem. However, the full range of the promise is far from being elucidated, for many of the benefits are hidden below–ground. It is our hope that our discussion here will contribute to a better understanding of both the promise and the benefits, since the latter clearly fill a current conceptual as well as a practical niche in sustainable agriculture. These extra benefits may be summed up as 'agrosystem stability'. They accrue to plant and soil alike and cannot be weighed at harvest on a scale of dry weights. They result from the inseparable, complex processes that unite all components of the agrosystem, and represent a new agenda for agriculture (Board on Agriculture, 1993).

AMF and soil structure

Shifting the focus from the plant to the agrosystem A message to agriculture fifty years ago advised that "the presence of an effective mycorrhizal symbiosis is essential to plant health" (Howard, 1943). Now, looking back on hundreds of research reports and an extensive collection of mycorrhiza books, it seems that the time to fully appreciate this message has come, if only because the challenge posed by the complexity of interacting, interdependent factors that have a bearing on rhizosphere research is now more clearly delineated (Linderman and Paulitz, 1990; Schroth and Weinhold, 1986). Plant health and productivity are rooted in the soil, and the quality of soil depends on the diversity and viability of its biota (Doran and Linn, 1994; Visser, 1985) which shape the structures that support a stable and healthy agrosystem.

The interest shown by increasing numbers of mycorrhiza workers in the interactions of AMF with the soil and its biota is therefore not so much a sign of masochist's delight in grappling with unmanageably complex systems (Schroth and Weinhold, 1986), but stems from the needs and priorities of the agriculture of our times. In this sense, that the inclusion of soil structure into mycorrhiza research is a necessity whose time has finally arrived. Since Tisdall and Oades (1979) first reported that aggregate stability and AMF status are related in agricultural soils, work has advanced at three levels: (1) collection of evidence for the relationship, (2) elucidation of its mechanisms, and (3) the integration of its process into agricultural concepts. What is largely lacking is a realization of theory in practice.

Impact of mycorrhizas on soil aggregation The roster of reports of AMF effects on agricultural soils is still minuscule compared to the wealth of information available on plant responses. It starts with the pioneering work of Tisdall and Oades (1979, 1980a) on the relationships between crop rotation, fallowing and soil stability, showing a connection between the extent of the soil mycelium and macroaggregate formation and stabilization by arbuscular mycorrhizal roots (Tisdall and Oades 1980b). It moves on to Miller's extensive field studies (1984, 1987) on prairie reconstruction. Miller, in collaboration with Jastrow, reported in a series of important papers how the affinity between mycorrhizas and soil aggregates varies with root characteristics, with the intensity of root colonization, and with the amount of soil mycelium associated with the roots. They further elucidated, or contributed to the understanding of, the mechanisms of the formation of water–stable aggregates (Jastrow, 1987; Miller and Jastrow, 1990; Miller and Jastrow, 1992a and b; Miller and Jastrow, 1994). Thomas et al. (1986, 1993) experimented on effects of AMF on soil in pot cultures. They found that root and fungus effects are difficult to separate, but that the soil mycelium alone is capable to bring about soil effects equivalent to those of roots, while roots and fungi together affect soil aggregation synergistically.

Mycorrhizas and the mechanism of aggregate formation Articles discussing the concept of mycorrhiza contributions to soil structure are as numerous as those offering empirical data. The major recent reviews (Finlay and Sönderstöm, 1992; Miller and Jastrow, 1992a and b, 1994; Tisdall, 1991) agree that all of the biotic components of the agrosystem interact in the forming of its abiotic matrix from the parent materials (Robert and Berthelin, 1986; Emerson et al., 1986), but there are few findings, if any, that show interactions between specific groups of soil organisms and AMF isolates in the aggradative process (Jastrow and Miller, 1991). While each soil organism may have a necessary function in soil structure formation, fungi and filamentous actinomycetes had been shown to be most effective in binding soil particles into crumbs (Harris et al., 1966), even before Tisdall and Oades (1982) developed their concept of aggregate organization with its important niche for AMF.

The contribution of AMF hyphae to soil aggregation was summarized by Miller and Jastrow (1994) as consisting of three related steps. First, hyphal growth into the soil matrix creates the skeletal structure that holds the primary soil particles together through physical entanglement. Second, roots and hyphae together create the physical and chemical conditions and produce organic and amorphous materials (Gupta and Germida, 1988; Tisdall, 1991) for the binding of particles. Third, hyphae and roots enmesh microaggregates into macroaggregate structures. Once formed, the aggregates enhance carbon and nutrient storage (Elliott, 1986; Gupta and Germida, 1988; Cambardella and Elliott, 1993, 1994) and provide microhabitats for soil microorganisms (Foster, 1994; Tisdall, 1991). The quality and size distribution of the aggregates affects pore size distribution (Elliott and Coleman, 1988) and the pores offer improved access to the hyphae for grazing by soil invertebrates (Ingham, 1992). Although there is considerable evidence that interactions between the soil biota and AMF may have negative effects on plants (Hetrick et al., 1988; Ingham, 1988; Rabatin and Stinner, 1991; Ross, 1980), complementary effects of these interactions on the soil and its stability are little–known. One may speculate that these soil responses are largely positive, since they involve enhanced carbon input (Finlay and Sönderström, 1992; Hepper, 1975; Lynch and Whipps, 1991; Wright and Millner, 1994) from plant to soil. Ultimately, however, this loss of carbon by the plant not only improves soil quality, but also benefits plant growth (Burns and Davies, 1986).

Mycorrhizal effectiveness measured by soil responses Because of its importance to both plant and soil, aggregate stability has been suggested as a measure of AMF effectiveness in agroecology (Bethlenfalvay et al., 1988; Bethlenfalvay and Newton, 1991). This idea was expanded by Tisdall (1991), who described the characteristics of effective AM fungal soil stabilizers as (1) the production of greater quantities of more persistent and sticky mucilage, (2) bonding by hyrophobic bonds and polyvalent cations to clay platelets, (3) preferential interactions with plants, microorganisms and animals, (4) effective orientation of clay particles, (5) vigorous soil penetration, and (6) the production of a profuse soil mycelium. Tisdall's list supplements the selection guide for AMF effective in promoting plant growth (Abbott and Robson, 1984a, 1991) and refocuses priorities of mycorrhiza research in sustainable agriculture.

AMF and the soil biota

Different effects on plant and soil? Much effort has gone during the past ten years into the integration of the combined effects of specific plant–fungus combinations and specific groups of soil biota with basic and applied aspects of plant science (Azcón–Aguilar and Barea, 1992; Bagyaraj, 1984; Hendrix et al., 1990; Ingham, 1992; Miller, 1990; Paulitz and Linderman, 1991; Reid, 1990; Tinker, 1984). At the same time, soil biota effects on mycorrhiza formation have also received attention (Azcón et al., 1990; Linderman, 1992; Rabatin and Stinner, 1991). Much less is known, however, about the interactions between specific AMF isolates and distinct groups of soil organisms as it relates to soil structure, even though these interactions had been conceptualized thoroughly in general terms (Burns and Davies, 1986; Newman, 1985; Oades, 1984; Tisdall, 1991). We feel therefore confident in predicting that many experimental designs of future mycorrhiza research will include evaluations of soil responses as routinely as they now report determinations of root colonization. If stimulation of the rhizosphere biota and the resulting improvement of soil structure is in fact an evolutionary mechanism that imparts competitive advantages to plants (Burns and Davies, 1986), then a holistic approach to the joint study of arbuscular mycorrhizal plant and soil responses is indeed an absolute necessity.

The goal of improving or restoring disturbed agrosystems may be approached by studying the conditions that provide stability in natural systems and use it as a model to reconstruct (Linderman, 1986) the disturbed system. Alternatively, a manipulation of the disturbed system may be tried to achieve specific, limited ends. These ends have been production–oriented in the past (Cooke, 1982; Hatfield and Karlen, 1994). Soil microbes have been tried, with or without AMF, to suppress plant pathogens, to control plant pests, to enhance plant nutrition, to promote plant growth, or to relieve environmental stress to plants. Now, regardless of its promise for enhancing plant production, a new agenda for agriculture advises and prescribes that each new biotechnique to be employed in agriculture be also scrutinized as to its effects on agrosystem stability (Board on Agriculture, 1993). This will take some rethinking of the premises, for the use of beneficial soil microorganisms as tools for the control of deleterious ones is evaluated by plant effects (Schippers et al., 1987). Let us speculate here, if only to stimulate controversy, that favorable plant responses to rhizosphere manipulation may not always be accompanied by beneficial soil responses, and that conversely, conditions may exist where processes favorable to soil stability may be unfavorable to plant growth, at least initially.

What are some of these processes in agrosystem biology that may affect plant production and soil stability differently? Among many, we will single out three examples: (1) microbe–mediated nutrient uptake and soil pH, (2) stimulatory or antagonistic relationships between AMF and soil microbes, and (3) soil fauna effects on the mycorrhiza and its microbial associates.

Nutrient uptake and soil pH Long–known as enhancers of symbiotic N_2 fixation in P–deficient soils (Asai, 1948; Mosse et al., 1976; Barea and Azcón–Aguilar, 1982), AMF have also been shown to affect N uptake from soil (Ames et al., 1983; Azcón et al., 1991; Johansen et al., 1992), although a preference for the form of N has not yet been conclusively demonstrated (Barea et al., 1987; Azcón et al., 1992; Vaast and Zasoski, 1992). Frey and Schüepp (1993a) showed, in a cuvette system with root-free soil compartments, separated from the confined rhizosphere of maize plants, that ^{15}N was taken up by the soil mycelium of AMF after addition of $(^{15}NH_4)_2SO_4$ to the soil to be transported in considerable amounts to plant roots being several centimetres apart from the site of application. Similarly N also can be transported via mycorrhizal hyphae from plant to plant (Frey and Schüepp, 1993b). The soil mycelium of AMF also provide channels for transfer of fixed N from legume to non-legume plants (Frey and Schüepp, 1992). This could be demonstrated in a cuvette system separating the nodulated roots of *Trifolium alexandrinum* from the rhizosphere of maize. Further studies are needed to elucidate the impact of AMF in N cycling in relation to N fixation. The function of AMF concerning the N cycle should not be reduced to the N nutrition of the plant by hyphal N uptake or N transport. AMF must be regarded in the dynamic processes resulting in the temporary immobilization of N within its biomass and the N mineralisation at phases of decomposition of arbuscular mycorrhizal mycelium. N losses from the root system and from the soil to the ground water may be reduced or enhanced by mycorrhizal activity.

While soil biologists are said to be more preoccupied with the tripartite legume association than with any other biological process (Lynch, 1987), an increasing number of bacteria with N_2–fixing capabilities are also being discovered. This provides the challenge of supplying, perhaps by means of AMF hyphae (Barea et al., 1992), nonlegumes with biologically–derived N (Döbereiner, 1989; Zuberer, 1990). In addition, an increase in associative diazotroph populations in the presence of AMF (Bagyaraj, 1984) may also improve soil quality, since aggregate stability can be proportional to the biomass of cells present (Lynch, 1987).

Biologically–fixed N always improves productivity, since N availability is one of the major limiting factors in agriculture. However, when N input is in the form of NH_4–N, as is the case with N_2 fixation, extrusion of H^+ and of organic acids is prevalent and results in an acidification of the growth medium not only in the rhizosphere (Marschner and Römheld, 1983; Marschner et al., 1987) but also in the entire mycorrhizosphere (Li et al., 1991). Soil pH effects on AMF have long been known (Wang et al., 1993), but it is little known to what extent mycorrhizas and their associated microflora may create, control, and maintain the pH of their environment through exudation (Schwab et al., 1991) and CO_2 levels (Knight et al., 1989) in the absence of soil disturbance. Elevated soil pH, however, affects the stability of aggregates (Oades, 1984; Reid and Goss, 1981) as well as the composition of the soil microflora (Harris et al., 1966). In fact, negative effects on soil aggregation by legume cropping have been documented (Alberts et al., 1985; Laflen and Moldenhauer, 1979). Soil loss, however is determined by many aggregating and disaggregating forces (Gisch and Browning, 1948; Strickling, 1950), is influenced by climatic, edaphic, cropping and tillage factors, and each of these affect the processes of soil biology. It is therefore not surprising that the connection between soil loss and legume cropping is unresolved (Alberts and Wendt, 1985), but the phenomenon serves as an example how mycorrhiza–microbe relationships may affect cost–benefit ratios in production and conservation.

Mycorrhiza–microbe interactions and their effects on plant and soil An important function of the arbuscular mycorrhizal soil mycelium is the transport of carbon to microbial communities (Jakobsen and Rosendahl, 1990). This is especially significant when root density is low (Abbott and Robson, 1984b), since the hyphae can penetrate several centimeters of soil (Camel et al., 1991) and reach the microfauna of the bulk soil outside the influence of the rhizosphere (Finlay and Sönderström, 1992). In view of their role as mediators of carbon flow

(Whipps, 1990), one would expect the influence of AMF on soil microbes to be positive. This is not always the case, however. Many studies have shown that AMF may alter the soil microflora (Ames, et al., 1984; Bagyaraj and Menge, 1978; Christensen and Jakobsen, 1993; Meyer and Linderman, 1986; Secilia and Bagyaraj, 1987) by stimulating as well as inhibiting total bacterial counts or selected bacterial groups. Soil microbes, in turn, may promote (Ames, 1989; Azcón, 1989, Azcón et al., 1990; Azcón–Aguilar et al., 1986; Staley et al., 1992; Vejsadová et al., 1993) or antagonize (Azcón et al., 1990; Bethlenfalvay et al., 1985; Dhillon, 1992; Krishna et al., 1982) mycorrhiza development.

How do these complex interactions affect plant production and soil stability? A stimulation of plant growth may be achieved by manipulating specific groups of organisms, such as phosphate solubilizing (Azcón–Aguilar et al., 1986) or diazotrophic (Paula et al., 1992) bacteria, or rhizobacteria that promote plant growth by various mechanisms (Burr and Caesar, 1984). However, when the arbuscular mycorrhizal plant is grown in the field, subject to many influences at the same time, growth stimulation by AMF becomes elusive (Fitter, 1991; Hetrick et al., 1988, Ross, 1980). This led some workers to conclude that it is the soil microflora that regulates mycorrhiza formation and plant growth response, regardless of the AMF isolates present (Hetrick and Wilson, 1991). One must keep in mind, however, that the 'growth response' is only one of the ways to evaluate the AMF effect on plants (Koide and Schreiner, 1992), let alone the agrosystem, of which the plant is but one component.

The absence of a plant growth response to AMF, or a negative one, was interpreted as a loss of carbon by the plant, which outweighs the mutualistic advantage of enhanced P uptake by the endophyte (Fitter, 1991). This form of parasitism has been viewed traditionally as a lack of arbuscular mycorrhizal efficiency, and in a wider sense, as a lack of application potential for AMF in agriculture (Stribley, 1989). From the point of view of agrosystem stability, on the other hand, the gain of carbon by the soil represents an increase in substrate availability, resulting in greater microbial activity (Kirchner et al., 1993) and increased organic matter content and soil stability.

Seen in this context, one may even ascribe useful (agrosystem–stabilizing) functions to mycoparasites: although the parasites may limit AMF populations and thereby reduce plant growth (Paulitz and Menge, 1986; Ross and Ruttencutter, 1977), they may also stimulate hyphal regrowth, thus further increasing carbon flux and microbial activity in the soil. Seen from this angle, the utility of chemical control of mycoparasites (Sylvia and Schenck, 1983), may be revised, using soil aggregation measurements as an alternate tool for the evaluation of mycoparasite effects. Microbial biomass and activity (Dinel et al., 1992) play an important role in the formation and stability of soil aggregates, and promise a wide range of applications for AMF and their associated microflora.

Mycorrhizas and the soil fauna Invertebrates and AMF are ubiquitous and abundant coinhabitants of the soil environment. Together they fill important functions in processes which regulate nutrient availability and mineralization (Ingham, 1992). Interactions between the major groups of fungivorous soil invertebrates (nematodes, springtails, mites and microarthropods have been reviewed (Fitter and Sanders, 1992; Paulitz and Linderman, 1991; Rabatin and Stinner, 1991) and discussed in ecological (McGonigle and Fitter, 1988) and agricultural (Sylvia and Williams, 1992) settings. Grazing by invertebrates on the soil mycelium of AMF may limit its development or disconnect it from the root mycelium, but it may also stimulate its growth (Fitter and Sanders, 1992). Damage to the hyphal network would result in an impairment of nutrient uptake, a preponderance of root– over soil–mycelial biomass. Stimulation of hyphae and spore production, on the other hand, would be beneficial to the agrosystem and to plant growth. Such positive effects have been reported with springtails (Harris and Boerner, 1990) and nematodes (Ingham, 1988), and related to grazer density and grazing intensity (Moore, 1988). Plant growth may have been affected by the increased mineralization or mobilization of nutrients by the grazers (Finlay, 1985; Ingham et al., 1985; Harris and Boermer, 1990). Alternatively, the removal of senescing soil mycelia by the grazers may have resulted in the elimination of growth–inhibitory secondary metabolites (Moore, 1988).

The consequences of such trophic interactions between the soil fauna and AMF on soil aggregation specifically, are little–known. Generally, however, all biota found within the agrosystem were shown to contribute to the development of soil structure (Jastrow and Miller, 1991). In mycorrhiza research, it remains to be seen how the soil fauna affects the cost–benefit ratio of plant or soil development as they relate to agrosystem stability.

Summary and conclusions

An agrosystem is that part of the larger (natural or agricultural) agroecosystem that may be subject to experimental control, and where roots, the soil microflora and the soil fauna interact to support plant growth and to form a stable soil matrix. An agroecosystem is sustainable when the biotic components of the agrosystem are in balance. In disturbed ecosystems, this balance depends on the goals of land management: production or conservation. The two goals may be combined if the agricultural manager understands the biological complexity of the land under his stewardship.

Among the multitude of organisms that make up the agrosystem, AMF stands out because of its ability to form a bridge between plant and soil. These fungi penetrate and colonize the cells of host–plant roots, while their soil hyphae are in intimate contact with the microbiota that inhabit soil aggregates and contribute to soil structure formation. By mediating nutrient fluxes between plant and soil, the fungi influence both plant growth and health and the development of communities of soil organisms. In the course of experimental manipulation of the agrosystem, complex relationships between organisms manifest themselves that can be stimulatory, antagonistic, or both, depending on the circumstances. Such relationships may be real or artefacts of artificial environments, and their effects may be beneficial or deleterious to plant and soil in divergent ways, at least initially and in passing. It is one of the challenges of agricultural and ecological research to draw valid inferences from such transient effects achieved under controlled conditions to the reality of the field.

In the field of agriculture, sustainability has become the paradigm of our time, and in biological research sustainability means plant production without soil loss. For mycorrhiza research, this means a rethinking of the concept of mycorrhizal benefit. Synonymous with plant growth enhancement in the past, in the context of sustainability it may be redefined in terms of agrosystem stability, resting on soil biotic communities in harmony with roots and balance with each other within a strong, resilient, life–supporting soil matrix.

Thus, we see a closed chain of cause–effect relationships as the ultimate benefit of mycorrhizal fungi in the agrosystem. The fungi improve plant growth, health, and stress resistance; the plant so strengthened is a more abundant source of energy to the soil, encouraging the development of its biota; the organisms enhance soil aggregate formation; and the life–supporting soil structure so formed permits better plant growth, closing the chain.

References

Abbott, L.K. and Robson, A.D. (1984a) Selection of 'efficient' VA mycorrhizal fungi. Proc. 6th North American Conference on Mycorrhizae, College of Forestry, Oregon State University, Corvallis, pp. 89–90.

Abbott, L.K. and Robson, A.D. (1984b) The effect of root density, inoculum placement and infectivity of inoculum on the development of vesicular–arbuscular mycorrhizas. New Phytol. 97: 285–299.

Abbott, L.K. and Robson, A.D. (1991) Field management of VA mycorrhizal fungi. In: D.L. Keister and P.B. Cregan (eds) The Rhizosphere and Plant Growth, Kluwer Academic Publishers, Dordrecht, pp. 355–362.

Alberts, E.E., and Wendt, R.C. (1985) Influence of soybean and corn cropping on soil aggregate size and stability. Soil Sci. Soc. Am. J. 49: 1534–1537.

Alberts, E.E., Wendt, R.C., and Burwell, R.E. (1985) Corn and soybean cropping effects on soil losses and C factors. Soil Sci. Soc. Am. J. 49: 721–728.

Ames, R.N. (1989) Mycorrhiza development in onion in response to inoculation with chitin–decomposing actino-mycetes. New Phytol. 112: 423–427.

Ames, R.N., Reid, C.P.P., and Ingham, E.R. (1984) Rhizosphere bacterial populations responses to root colonization by a vesicular–arbuscular mycorrhizal fungus. New Phytol. 96: 555–563.

Ames, R.N., Reid, C.P.P., Porter, L.K., and Cambardella, C. (1983) Hyphal uptake and transport of nitrogen from two ^{15}N–labelled sources by Glomus mosseae, a vesicular–arbuscular mycorrhizal fungus. New Phytol. 95: 381–396.

Asai, T. (1943) Die Bedeutung der Mykorrhiza für das Pflanzenleben (The significance of mycorrhizae for plants). Japn. J. Bot. 12: 359–436.

Asai, T. (1948) Über die Mykorrhizenbildung der leguminösen Pflanzen (On mycorrhiza formation in leguminous plants). Japn. J. Bot. 13: 463–485.

Azcón, R. (1989) Selective interaction between free–living rhizosphere bacteria and vesicular–arbuscular mycorrhizal fungi. Soil Biol. Biochem. 21: 639–644.

Azcón, R., Gomez, M., and Tobar R. (1992) Effects of nitrogen source on growth, nutrition, photosynthetic rate and nitrogen metabolism of mycorrhizal and phosphorus–fertilized plants of Lactuca sativa L. New Phytol. 121: 227–234.

Azcón, R., Rubio, R., and Barea, J.M. (1991) Selective interactions between different species of mycorrhizal fungi and Rhizobium meliloti strains, and their effects on growth, N_2 fixation (^{15}N) and nutrition of Medicago sativa L. New Phytol. 117: 399–404.

Azcón, R., Rubio, R., Morales, C., and Tobar R. (1990) Interactions between rhizosphere free–living microorganisms and VAM fungi. *Agric. Ecosyst. Environ.* 29: 11–15.

Azcón–Aguilar, C. and Barea, J.M. (1992) Interactions between mycorrhizal fungi and other rhizosphere microorganisms. In: M.F. Allen (ed) *Mycorrhizal Functioning*, Chapman & Hall, New York, pp. 163–198.

Azcón–Aguilar, C., Gianinazzi–Pearson, V., Fardeau, J.C., and Gianinazzi, S. (1986) Effect of vesicular–arbuscular mycorrhizal fungi and phosphate solubilizing bacteria on growth of soybean in neutral–calcareous soil amended with ^{32}P–^{45}Ca–tricalcium phosphate. *Plant Soil* 96: 3–15.

Bagyaraj, D.J. (1984) Biological interactions with VA mycorrhizal fungi. In: C.L.L. Powell and D.J. Bagyaraj (eds) *VA Mycorrhiza*, CRC Press, Boca Raton, pp. 131–153.

Bagyaraj, D.J. and Menge, J.A. (1978) Interactions between a VA mycorrhiza and *Azotobacter,* and their effects on rhizosphere microflora and plant growth. *New Phytol.* 80: 567–573.

Barea, J.M. and Azcón–Aguilar, C. (1982) Mycorrhizas and their significance in nodulating, nitrogen–fixing plants. *Adv. Agron.* 36: 1–54.

Barea, J.M., Azcón, R., and Azcón–Aguilar, C. (1992) Vesicular–arbuscular mycorrhizal fungi in nitrogen–fixing systems. Methods Microbiol. 24: 391–416.

Barea, J.M., Azcón–Aguilar, C., and Azcón, R. (1987) Vesicular–arbuscular mycorrhiza improve both symbiotic N_2 fixation and N uptake from soil as assessed with a ^{15}N technique under field conditions. *New Phytol.* 106: 717–725.

Barea, J.M. and Jeffries, P. (1994) Arbuscular mycorrhizas in sustainable plant–soil systems. In: A. Varma and B. Hock (eds) *Mycorrhiza: Function, Molecular Biology and Biotechnology*, Springer–Verlag, Heidelberg, in press.

Bethlenfalvay, G.J. (1992) Mycorrhizae and crop productivity. In: G.J. Bethlenfalvay and R.G. Linderman (eds) *Mycorrhizae in Sustainable Agriculture*, Am. Soc. Agron. Special Publication 54, American Society of Agronomy, Madison, pp. 1–25.

Bethlenfalvay, G.J., Brown, M.S., and Stafford, A.E. (1985) *Glycine–Glomus–Rhizobium* symbiosis II. Antagonistic effects between mycorrhizal colonization and nodulation. *Plant Physiol.* 79: 1054–1058.

Bethlenfalvay, G.J. and Linderman, R.G. (1992) *Mycorrhizae in Sustainable Agriculture*, Am. Soc. Agron. Special Publication 54, American Society of Agronomy, Madison.

Bethlenfalvay, G.J. and Newton, W.E. (1991) Agro–ecological aspects of the mycorrhizal, nitrogen–fixing legume symbiosis. In: D.L. Keister and P.B. Cregan (eds) *The Rhizosphere and Plant Growth*: Proceedings of a Symposium held May 8–11, 1989, at the Beltsville Agric. Research Center, Maryland, USA, Kluwer Academic Publishers, Dordrecht, pp. 349–354.

Bethlenfalvay, G.J. and Svejcar, A.J. (1991) *Mycorrhizae in plant productivity and soil conservation.* Actes du Quatrième Congrès International des Terres de Parcours, Association Française de Pastoralisme, Montpellier, pp. 251–254.

Bethlenfalvay, G.J., Thomas, R.S., Dakessian, S., Brown, M.S., and Ames, R.N. (1988) *Mycorrhizae in stressed environments: Effects on plant growth, endophyte development, soil stability, and soil water.* In: Whitehead October 20–25, 1985, Tucson, Arizona, Westview Press, Boulder, pp. 1015–1029.

Bezdicek, D.F. and Power, J.P. (1984) *Organic Farming: Current Technology and its Role in Sustainable Agriculture*, Am. Soc. Agron. Special Publication 46, American society of Agronomy, Madison.

Board on Agriculture, Committee of Long–Range Soil and Water Conservation, National Research Council. (1993) *Soil and Water Quality: an Agenda for Agriculture*, National Academy Press, Washington.

Burns, R.G., and Davies, J.A. (1986) The microbiology of soil structure. *Biol. Agric. Hortic.* 3: 95–113.

Burr, T.J. and Caesar, A. (1984) Beneficial plant bacteria. *Crit. Rev. Plant Sci.* 2: 1–20.

Cambardella, C.A. and Elliott, E.T. (1993) Carbon and nitrogen distribution in aggregates from cultivated and native grassland soils. *Soil Sci. Soc. Am. J.* 57: 1071–1076.

Cambardella, C.A. and Elliott, E.T. (1994) Carbon and nitrogen dynamics of soil organic matter fractions from cultivated grassland soils. *Soil Sci. Soc. Am. J.* 58: 123–130.

Camel, S.B., Reyes–Solis, M.G., Ferrera–Cerrato, R., Franson, R.L., Brown, M.S., and Bethlenfalvay, G.J. (1991) Growth of vesicular–arbuscular mycorrhizal mycelium through bulk soil. *Soil Sci Soc. Am. J.* 55: 389–393.

Christensen, H. and Jakobsen, I. (1993) Reduction of bacterial growth by a vesicular–arbuscular mycorrhizal fungus in the rhizosphere of cucumber (*Cucumis sativus* L.) *Biol. Fert. Soils* 15: 253–258.

Cooke, G.W. (1982) *Fertilizing for Maximum Yield*, Third Edition, Macmillan Publishing, New York.

Dhillon, S.S. (1992) Dual inoculation of pretransplant stage *Oryza sativa* L. plants with indigenous vesicular–arbuscular mycorrhizal fungi and fluorescent *Pseudomonas* spp. *Biol. Fert. Soils* 13: 147–151.

Dinel, H., Lévesque, P.E.M., Jambu, P., and Righi, D. (1992) Microbial activity and long–chain aliphatics in the formation of stable soil aggregates. *Soil Sci. Soc. Am. J.* 56: 1455–1463.

Döbereiner, J. (1989) Recent advances in associations of diazotrophs with plant roots. In: V. Vancura and F. Kunc (eds) *Interrelationships between Microorganisms and Plants in Soil*, Academia, Praha, pp. 229–242.

Doran, J.W. and Linn, D.M. (1994) Microbial ecology of conservation management systems. In: J.L. Hatfield and B.A. Stewart (eds) *Soil biology: Effects on Soil Quality*, Lewis Publishers, Boca Raton, pp. 1–57.

Elliott, E.T. (1986) Aggregate structure and carbon, nitrogen and phosphorus in native and cultivated soils. *Soil Sci. Soc. Am. J.* 50: 627–633.

Elliott, E.T. and Coleman, D.C. (1988) Let the soil work for us. *Ecol. Bull.* 39: 23–32.

Emerson, W.W., Foster, R.C., and Oades, J.M. (1986) Organo–mineral complexes in relation to soil aggregation. In: P.M. Huang and M. Schnitzer (eds) *Interaction of Soil Minerals with Natural Organics and Microbes*, Soil Sci. Soc. Am. Special Publication 17, Soil Science Society of America, Madison, pp. 521–548.

Finlay, R.D. (1985) Interactions between soil micro–arthropods and endomycorrhizal associations of higher plants. In: A.H. Fitter, D. Atkinson, D.J. Read, and M.B. Usher (eds) *Ecological Interactions in Soil, Plants, Microbes, and Animals*, Blackwell Scientific Publications, Boston, pp. 319–331.

Finlay, R.D. and Sönderström, B. (1992) Mycorrhiza and carbon flow to the soil. In: M.F. Allen (ed) *Mycorrhizal Functioning*, Chapman & Hall, New York, pp. 134–160.

Fitter, A.H. (1985) Functioning of vesicular–arbuscular mycorrhizas under field conditions. *New Phytol.* 99: 257–265.

Fitter, A.H. (1986) Effect of benomyl on leaf phosphorus concentration in alpine grasslands: a test of mycorrhizal benefit. *New Phytol.* 103: 767–776.

Fitter, A.H. (1991) Costs and benefits of mycorrhizas: Implications and functioning under natural conditions. *Experientia* 47: 350–355.

Fitter, A.H., and Sanders, I.R. (1992) Interactions with the soil fauna. In: M.F. Allen (ed) *Mycorrhizal Functioning*, Chapman & Hall, New York, pp. 333–354.

Foster, R.E. (1994) Microorganisms and soil aggregates. In: C.E. Pankhurst, B.M. Double, V.V.S.R. Gupta and P.R. Grace (eds) *Soil Biota Management in Sustainable Farming Systems*. CSIRO Information Services, Melbourne. pp. 144-155.

Gerdeman, J.W. (1968) Vesicular–arbuscular mycorrhiza and plant growth. *Annu. Rev. Phytopathol.* 6: 397–418.

Frey, B. and Schüepp, H. (1993a) Acquisition of nitrogen by external hyphae of arbuscular mycorrhizal fungi associated with *Zea mays* L. *New Phytol.* 124: 221-230.

Frey, B. and Schüepp, H. (1993b) A role of vesuçukar-arbuscular (VA) mycorrhizal fungi in facilitating interplant nitrogen transfer. *Soil Biol. Biochem.* 25: 651-658.

Frey, B. and Schüepp, H. (1992) Transfer of symbiotically fixed nitrogen from berseem (*Trifolium alexandrinum* L.) to maize via vesicular-arbuscular mycorrhizal hyphae. *New Phytol.* 122:447-454.

Gish, R.E. and Browning, G.M. (1948) Factors affecting the stability of soil aggregates. *Soil Sci.Soc. Am. Proc.* 12: 51–55.

Gupta, V.V.S.R. and Germida, J.J. (1988) Distribution of microbial biomass and its activity in different soil aggregate size classes as affected by cultivation. *Soil Biol. Biochem.* 20: 777–786.

Harris, K.K. and Boerner, R.E.J. (1990) Effects of belowground grazing by Collembola on growth, mycorrhizal infection, and P uptake of *Geranium robertianuM*. *Plant and Soil* 129: 103-210.

Harris, R.F., Chester, G. and Allen, O.N. (1966) Dynamics of soil aggregation. *Adv. Agron.* 18: 107–169.

Harwood, R.R. (1991) A history of sustainable agriculture. In: R. Lal and F.J. Pierce (eds) *Soil Management for Sustainability*, Soil and Water Conservation Society of America, Ankeny, pp. 3–19.

Hatfield, J.H. and Karlen, D.L. (1994). *Sustainable Agriculture Systems*, Lewis Publishers, Boca Raton.

Hendrix, P.F., Crossley, D.A., Blair, J.M. and Coleman, D.C. (1990) Soil biota as components of sustainable agroecosystems. In: C.A. Edwards, R. Lal, P. Madden, R.H. Miller and G. House (eds) *Sustainable Agricultural Systems*, St. Lucie Press, Delray Beach, pp. 637–654.

Hepper, C.M. (1975) Extracellular polysaccharides of soil bacteria. In: N. Walker (ed) *Soil Microbiology*, Butterworths, London, pp. 93–110.

Hetrick, B.A.D. and Wilson, G.W.T. (1991) Effects of mycorrhizal fungus species and metalaxyl application on microbial suppression of mycorrhizal symbiosis. *Mycologia* 83:97–102.

Hetrick, B.A.D., Wilson, G.W.T. and Cox, T.S. (1992) Mycorrhizal dependence of modern wheat varieties, landraces, and ancestors. *Can. J. Bot.* 70: 2032–2040.

Hetrick, B.A.D., Wilson, G.W.T. and Cox, T.S. (1993) Mycorrhizal dependence of modern wheat cultivars and ancestors: a synthesis. *Can. J. Bot.* 71: 512–518.

Hetrick, B.A.D., Wilson, G.W.T., Kitt, D.G. and Schwab, A.P. (1988) Effects of soil microorganisms on mycorrhizal contribution to growth of big bluestem grass in non–sterile soil. *Soil Biol. Biochem.* 20: 501–507.

Hornick, S.B. and Parr, J.F. (1987) Restoring the productivity of marginal soils with organic amendments. *Am. J. Alternative Agric.* 2: 64–68.

Howard, A. (1943) *An Agricultural Testament*. Oxford University Press, London.

Ingham, R.E. (1988) Interactions between nematodes and vesicular–arbuscular mycorrhizae. *Agric. Ecosyst. Environ.* 24: 169–182.

Ingham, R.E. (1992) Interactions between invertebrates and fungi: Effects on nutrient availability. In: G.C. Carrol and D.T. Wicklow (eds) *The Fungal Community*, Marcel Dekker, New York, pp. 669–690.

Ingham, R.E., Trofymow, J.A., Ingham, E.R. and Coleman, D.C. (1985) Interactions of bacteria, fungi, and their nematode grazers: Effects on nutrient cycling and plant growth. *Ecol. Monogr.* 55: 119–141.

Jackson, W. (1980) *New Roots for Agriculture*, Friends of the Earth, San Francisco.

Jakobsen, I. and Rosendahl, L. (1990) Carbon flow into soil and external hyphae from roots of mycorrhizal cucumber plants. *New Phytol.* 115: 77–83.

Jastrow, J.D. (1987) Changes in soil aggregation associated with tallgrass prairie restoration. *Am. J. Bot.* 74: 1656–1664.

Jastrow, J.D. and Miller, R.M. (1991) Methods for assessing the effects of biota on soil structure. *Agric. Ecosyst. Environ.* 34: 279–303.

Jeffries, P. (1987) Use of mycorrhizae in agriculture. *CRC Crit. Rev. Biotechnol.* 5: 319–357.

Johansen, A., Jakobsen, I. and Jensen, E.S. (1992) Hyphal transport of ^{15}N–labelled nitrogen by a vesicular–arbuscular mycorrhizal fungus and its effect on depletion of inorganic soil N. *New Phytol.* 122, 281–288.

Johnson, N.C. (1993) Can fertilization of soil select less mutualistic mycorrhizae? *Ecol. Applica.* 3: 749–757.

Johnson, N.C. and Pfleger, F.L. (1992) Vesicular–arbuscular mycorrhizae and cultural stress. In: G.J. Bethlenfalvay and R.G. Linderman (eds) *Mycorrhizae in Sustainable Agriculture*, Am. Soc. Agron. Special Publication 54, American Society of Agronomy, Madison, pp. 71–99.

Johnson, N.C., Copeland, P.J., Crookston, R.K. and Pfleger, F.L. (1992) Mycorrhizae: Possible explanation for yield decline with continuous corn and soybean. *Agron. J.* 84: 387–390.

Kirchner, M.J., Wollum II, A.G. and King, L.D. (1993) Soil microbial populations and activities in reduced chemical input agroecosystems. *Soil Sci. Soc. Am J.* 57: 1289–1295.

Knight, W.G., Allen, M.F., Jurinak, J.J. and Dudley, L.M. (1989) Elevated carbon dioxide and solution phosphorus in soil with vesicular–arbuscular mycorrhizal western wheatgrass. *Soil Sci. Soc. Am. J.* 53: 1075–1082.

Koide, R.T. (1991) Nutrient supply, nutrient demand and plant response to mycorrhizal infection. *New Phytol.* 117: 365–386.

Koide, R.T. and Schreiner, R.P. (1992) Regulation of the vesicular–arbuscular mycorrhizal symbiosis. *Annu. Rev. Plant Physiol. Plant Mol. Biol.* 43: 557–581.

Krishna, K.R., Balakrishna, A.N. and Bagyaraj, D.J. (1982) Interactions between a vesicular–arbuscular mycorrhizal fungus and *Streptomycetes cinnamomeous* and their effect on finger millet. *New Phytol.* 92: 401–405.

Kurle, J.E., and Pfleger, F.L. (1994) The effects of cultural practices and pesticides on VAM fungi. In: F.L. Pfleger, and R.G. Linderman (eds) *Mycorrhizae and Plant Health*, APS Press, St. Paul, pp. 101–131.

Laflen, J.M. and Moldenhauer, W.C. (1979) Soil and water losses from corn–soybean rotations. *Soil Sci. Soc. Am. J.* 43: 1213–1215.

Li, X.-L., George, E. and Marschner, H. (1991) Phosphorus depletion and pH decrease at the root–soil and hyphae–soil interfaces of VA mycorrhizal white clover fertilized with ammonium. *New Phytol.* 119: 397–404.

Linderman, R.G. (1986) Managing rhizosphere microorganisms in the production of horticultural crops. *Hort.Science* 21: 1299–1302.

Linderman, R.G. (1992) Vesicular–arbuscular mycorrhizae and soil microbial interactions. In: G.J. Bethlenfalvay and R.G. Linderman (eds) *Mycorrhizae in Sustainable Agriculture*, Am. Soc. Agron. Special Publication 54, American Society of Agronomy, Madison, pp. 45–70.

Linderman, R.G. (1994) Role of VAM fungi in biocontrol. In: F.L. Pfleger, and R.G. Linderman (eds) *Mycorrhizae and Plant Health*, APS Press, St. Paul, pp. 1–26.

Linderman, R.G. and Paulitz, T.C. (1990) Mycorrhizal–rhizobacterial interactions. In: D. Hornby (ed) *Biological Control of Soil–Borne Plant Pathogens*, CAB International, Wallingford.

Lynch, J.M. (1987) Soil biology: accomplishments and potential. *Soil Sci Soc. Am. J.* 51: 1409–1412.

Lynch, J.M. and Whipps, J.M. (1991) Substrate flow in the rhizosphere. In: D.L. Keister and P.B. Cregan (eds) *The Rhizosphere and Plant Growth*, Kluwer Academic Publishers, Dordrecht, pp. 15–24.

Marschner, H., and Römheld, V. (1983) In–vivo measurement of root–induced pH changes at the soil–root interface: effect of plant species and nitrogen source. *Z. Pflanzenphysiol.* 111: 241– 251.

Marschner, H., Römheld, V. and Cakmak, I. (1987) Root–induced changes of nutrient availability in the rhizosphere. *J. Plant Nutr.* 10: 1175–1184.

McGonigle, T.P., and Fitter, A.H. (1988) *Ecological consequences of arthropod grazing on VA mycorrhizal fungi.* Proc. Roy. Soc. Edinborough 94B: 25–32.

Meyer, J.R., and Linderman, R.G. (1986) Selective influence on populations of rhizosphere bacteria and actinomycetes by mycorrhizas formed by *Glomus fasciculatum*. *Soil Biol. Biochem.* 18: 191–196.

Miller, R.M. (1984) Microbial ecology and nutrient cycling in disturbed arid ecosystems. In: A.J. Dvorak (ed) *Ecological Studies of Disturbed Landscapes*, Office of Scientific and Technical Information, U.S. Department of Energy, Argonne, pp. 3–1 to 3–29.

Miller, R.M. (1987) The ecology of vesicular–arbuscular mycorrhizae in grass– and shrublands. In: G.R. Safir (ed) *Ecophysiology of VA mycorrhizal plants*. CRC Press, Boca Raton, pp. 135–170.

Miller, R.H. (1990) Soil microbiological inputs for sustainable agricultural systems. In: C.A. Edwards, R. Lal, P. Madden, R.H. Miller and G. House (eds) *Sustainable Agricultural Systems*, St. Lucie Press, Delray Beach, pp. 614–623.

Miller, R.M., and Jastrow, J.D. (1990) Hierarchy of root and mycorrhizal fungal interactions with soil aggregation. *Soil Biol. Biochem.* 22: 579–584.

Miller, R.M. and Jastrow, J.D. (1992a) The application of VA mycorrhizae to ecosystem restoration and reclamation. In: M.F. Allen (ed) *Mycorrhizal Functioning*, Chapman & Hall, New York, pp. 438–467.

Miller, R.M. and Jastrow, J.D. (1992b) The role of mycorrhizal fungi in soil conservation. In: G.J. Bethlenfalvay and R.G. Linderman (eds) *Mycorrhizae in Sustainable Agriculture*, Am. Soc. Agron. Special Publication 54, American Society of Agronomy, Madison, pp. 29–44.

Miller, R.M. and Jastrow, J.D. (1994) Vesicular–arbuscular mycorrhizae and biogeochemical cycling. In: F.L. Pfleger, and R.G. Linderman (eds) *Mycorrhizae and Plant Health*, APS Press, St. Paul, pp. 189–212.

Moore, J.C. (1988) The influence of microarthropods on symbiotic and non–symbiotic mutualisms in detrital–based below–ground food webs. *Agric. Ecosyst. Environ.* 24: 147–159.

Mosse, B., Powell, C.Ll. and Hayman, D.S. (1976) Plant growth responses to vesicular–arbuscular mycorrhiza. IX. Interactions between VA mycorrhiza, rock phosphate and symbiotic nitrogen fixation. *New Phytol.* 76: 331–342.

Newman, E.I. (1985) The rhizosphere: carbon sources and microbial populations. In: A.H. Fitter, D. Atkinson, D.J. Read and M.B. Usher (eds) *Ecological Interactions in Soil, Plants, Microbes, and Animals*, Blackwell Scientific Publications, Boston, pp. 107–121.

Oades, J.M. (1994) Soil organic matter and structural stability: mechanisms and implications for management. In: J. Tinsley and J.F. Darbyshire (eds) *Biological Processes and Soil Fertility*, Kluwer Academic Publishers, The Hague, pp. 319–337.

O'Neill, R.V. and Waide, J.B. (1991) Hierarchy theory as a guide to mycorrhizal research on large–scale problems. *Environ. Pollu.* 73: 271–284.

Paula, M.A., Urquiaga, S., Siqueira, J.O. and Döbereiner, J. (1992) Synergistic effects of vesicular–arbuscular mycorrhizal fungi and diazotrophic bacteria on nutrition and growth of sweet potato (*Ipomoea batatas*). *Biol. Fert. Soils* 14: 61–66.

Paulitz, T.C. and Linderman, R.G. (1991) Mycorrhizal interactions with soil organisms. In: D.K. Aurora, B. Rai, K.G. Mukerji and G. Knudsen (eds) *Handbook of Applied Mycology*, Vol. 1: *Soils and Plants*, Marcel Dekker, New York, pp. 77–129.

Paulitz T.C., and Menge, J.A., (1986) the effects of a mycoparasite on the mycorrhizal fungus, *Glomus deserticola*. *Phytopathology* 76: 351:354.

Pierce, F.J., and Lal, R. (1991) Soil management in the 21st century. In: R. Lal and F.J. Pierce (eds) *Soil Management for Sustainability*, Soil and Water Conservation Society of America, Ankeny, pp. 175–179.

Pfleger, F.L. and Linderman, R.G. (1994) *Mycorrhizae and Plant Health*. APS Press, St. Paul.

Rabatin, S.A. and Stinner, B.R. (1991) Vesicular–arbuscular mycorrhizae, plant, and invertebrate interactions. In: P. Barbosa, V.A. Krishik and C.G. Jones (eds) *Microbial Mediation of Plant–Herbivore Interactions*, John Wiley & Sons, New York, pp.142–168.

Reganold, J.P., Papendick, R.I. and Parr, J.F. (1990) Sustainable agriculture. *Sci. Am.* 262: 112–120.

Reid, C.P.P. (1990) Mycorrhizas. In: J.M. Lynch (ed) *The Rhizosphere*, John Wiley & Sons, Chichester, pp. 281–315.

Reid, J.B. and Goss, M.J. (1981) Effect of living roots of different plant species on the aggregate stability of two arable soils. *J. Soil Sci.* 33: 47–53.

Robert, M. and Berthelin, J. (1986) Role of biological and biochemical factors in soil mineral weathering. In: P.M. Huang and M. Schnitzer (eds) *Interaction of Soil Minerals with Natural Organics and Microbes*, Soil Sci. Soc. Am. Special Publication 17, Soil Science Society of America, Madison, pp. 453–496

Rodale, J.I. (1983) Breaking new ground: The search for a sustainable agriculture. *The Futurist* 1: 15–20.

Ross, J.P., and Ruttencutter, R. (1977) Population dynamics of two vesicular–arbuscular endomycorrhizal fungi and the role of hyperparasitic fungi. *Phytopathology* 67:490–496.

Safir, G.R. (1994) Involvement of cropping systems, plant–produced compounds, and inoculum production in the functioning of VAM fungi. In: F.L. Pfleger, and R.G. Linderman (eds) *Mycorrhizae and Plant Health*, APS Press, St. Paul, pp. 239–259.

Schippers, B., Bakker, A.W. and Bakker, A.H.M. (1987) Interactions of deleterious and beneficial rhizosphere microorganisms and the effect of cropping practices. *Annu. Rev. Phytopathol.* 25: 339–358.

Schlicht, A. (1889) Beitrag zur Kenntniss der Verbreitung und der Bedeutung der Mykorhizen (Contribution to knowledge on the distribution and significance of mycorrhizae). *Landw. Jahrb.* 18: 477–508.

Schönbeck, F. and Dehne, H.–W. (1989) VA mycorrhiza and plant health. In: V. Vancura and F. Kunc (eds) *Interrelationships between Microorganisms and Plants in Soil*, Academia, Praha, pp. 83–91.

Schroth, M.N. and Weinhold, A.R. (1986) Root–colonizing bacteria and plant health. *Hort.Science* 21: 1295–1298.

Schwab, S.M., Menge, J.A. and Tinker, P.B. (1991) Regulation of nutrient transfer between host and fungus in vesicular–arbuscular mycorrhizas. *New Phytol.* 117: 387–398.

Secilia, J. and Bagyaraj, D.J. (1987) Bacteria and actinomycetes associated with pot cultures of vesicular–arbuscular mycorrhizas. *Can. J. Microbiol.* 33: 1069–1073.

Sieverding, E. (1991) *Vesicular–Arbuscular Mycorrhiza Management in Tropical Agrosystems*, Hartmut Bremer Verlag, Friedland.

Stahl, E. (1900) Der Sinn der Mycorhizenbildung (The meaning of mycorrhiza formation). *Jahrb. Wiss. Bot.* 34: 540–668.

Staley, T.E., Lawrence, E.G., Nance, E.L. (1992) Influence of a plant growth–promoting pseudomonad and vesicular–arbuscular mycorrhizal fungus on alfalfa and birdsfoot trefoil growth and nodulation. *Biol. Fert. Soils* 14: 175–180.

Stewart, B.A., Lal, R., and El–Swaify, S.A. (1991) Sustaining the resource base on an expanding world agriculture. In: R. Lal and F.J. Pierce (eds) *Soil Management for Sustainability*, Soil and Water Conservation Society of America, Ankeny, pp. 125–144.

Stribley, D.P. (1989) Present and future value of mycorrhizal inoculants. In: R. Campbell and R.M. Macdonald (eds) *Microbial Inoculation of Crop Plants*, IRL Press, Oxford, pp. 49–65.

Strickling, E. (1950) The effect of soybeans on volume, weight and and water stability of soil aggregates, soil organic matter content, and crop yield. *Soil Sci. Soc. Am. Proc.* 14: 30–34.

Sylvia, D.M., and Schenck, N.C. (1983) soil fungicides for controlling chytridaceous mycoparasites of *Gigaspora margarita* and *Glomus fasciculatum*. *Appl. Environ. Microbiol.* 45: 1306–1309.

Sylvia, D.M., and Williams, S.E. (1992) Vesicular–arbuscular mycorrhizae and environmental stress. In: G.J. Bethlenfalvay and R.G. Linderman (eds) *Mycorrhizae in Sustainable Agriculture*, Am. Soc. Agron. Special Publication 54, American Society of Agronomy, Madison, pp. 101–124.

Thomas, R.S., Dakessian, S., Ames, R.N., Brown, M.S., and Bethlenfalvay, G.J. (1986) Aggregation of a silty loam soil by mycorrhizal onion roots. *Soil Sci Soc. Am. J.* 50: 1494–1499.

Thomas, R.S., Franson, R.L., and Bethlenfalvay, G.J. (1993) Separation of vesicular–arbuscular mycorrhizal fungus and root effects on soil aggregation. *Soil Sci. Soc Am. J.* 57: 77–81.

Tinker, P.B. (1984) The role of microorganisms in mediating and facilitating the uptake of plant nutrients from soil. In: J. Tinsley and J.F. Darbyshire (eds) *Biological Processes and Soil Fertility*, Kluwer Academic Publishers, The Hague, pp. 77–91.

Tisdall, J.M. (1991) Fungal hyphae and structural stability of soil. *Aust. J. Soil Res.* 29: 729–743.

Tisdall, J.M., and Oades, J.M. (1979) Stabilization of soil aggregates by the root systems of ryegrass. *Aust. J. Soil Res.* 17: 429–441.

Tisdall, J.M., and Oades, J.M. (1980a) The effect of crop rotation on aggregation in a red–brown earth. *Aust. J. Soil Res.* 18: 423–433.

Tisdall, J.M., and Oades, J.M. (1980b) The management of ryegrass to stabilize aggregates of a red–brown earth. *Aust. J. Soil Res.* 18: 415–422.

Tisdall, J.M., and Oades, J.M. (1982) Organic matter and water–stable aggregates in soils. *J. Soil Sci.* 33: 141–163.

Vaast, Ph., and Zasoski, R.J. (1992) Effects of VA–mycorrhizae and nitrogen sources on rhizosphere soil characteristics, growth and nutrient acquisition of coffee seedlings (*Coffea arabica* L.). *Plant Soil* 147: 31–39.

Vejsadová, H., Catská, V., Hrselová, H., and Gryndler, M. (1993) Influence of bacteria on growth and phosphorus nutrition of mycorrhizal corn. *J. Plant Nutr.* 16: 1857–1866.

Visser, S. (1985) Role of the soil invertebrates in determining the composition of soil microbial communities. In: A.H. Fitter, D. Atkinson, D.J. Read and M.B. Usher (eds) *Ecological Interactions in Soil, Plants, Microbes, and Animals*, Blackwell Scientific Publications, Boston, pp. 297–317.

Wang, G.M., Stribley, D.P., Tinker, P.B., and Walker, C. (1993) Effects of pH on arbuscular mycorrhiza I Field observations on the long–term liming experiments at Rothamsted and Woburn. *New Phytol.* 124: 465–472.

Whipps, J.M. (1990) Carbon economy. In: J.M. Lynch (ed) *The Rhizosphere*, John Wiley & Sons, Chichester, pp. 59–97.

Wright, S.F., and Millner, P.D. (1994) Dynamic processes of vesicular–arbuscular mycorrhizae: A mycorrhizo-system within the Agroecosystem. In: J.L. Hatfield and B.A. Stewart (eds) *Soil Biology: Effects on Soil Quality*, Lewis Publishers, Boca Raton, pp. 29–59.

Zuberer, D.A. (1990) Soil and rhizosphere aspects of N_2–fixing plant–microbe associations. In: J.M. Lynch (ed) *The Rhizosphere*, John Wiley & Sons, Chichester, pp. 317–353.

Impact of Arbuscular Mycorrhizas on
Sustainable Agriculture and Natural Ecosystems
S. Gianinazzi and H. Schüepp (eds.)
© 1994 Birkhäuser Verlag Basel/Switzerland

Hyphal phosphorus transport, a keystone to mycorrhizal enhancement of plant growth

I. Jakobsen, E. J. Joner[1] and J. Larsen

Plant Biology Section, Environmental Science and Technology Department, Risø National Laboratory, DK-4000 Roskilde, Denmark
[1]Department of Biotechnological Sciences, Agricultural University of Norway, N-1432 Aas, Norway

Introduction

The beneficial effects of arbuscular mycorrhizas on plant growth are in most cases caused by the transfer of mineral nutrients from the fungus to its host plant. Nutrients involved are those which are transported to the plant root primarily by diffusion and phosphorus (P) is of particular interest. This is because P is not only required by the plant in relatively large amounts, but it is also strongly adsorbed to the surfaces of soil particles and therefore present in very low concentrations in the soil solution. The mycorrhizal enhancement of plant-P uptake is related to the ability of the external hyphae of the mycobiont to cross the P-depletion zone around the roots and thereby get direct access to P which would otherwise be available to the plant only via slow diffusion processes.

The role of mycorrhizas in plant-P uptake has been widely studied in pot experiments in particular, but also in the field. Responses to mycorrhizas are visualized by comparing growth of colonized plants with their uncolonized counterparts grown in semi-sterile soil with no viable mycorrhizal propagules. The magnitude of the mycorrhizal response will depend on the soil-P level and the P requirement of the host plant and should therefore be measured over the full P response curve of the host (Abbott and Robson, 1984). An example of the influence of mycorrhizal fungi on the P response of field-grown flax and oil-seed rape, a mycorrhizal and a non-mycorrhizal crop, respectively, is given in Fig. 1. Growth of flax responded much more to P fertilizer in fumigated than in untreated soil with viable mycorrhizal propagules so that the

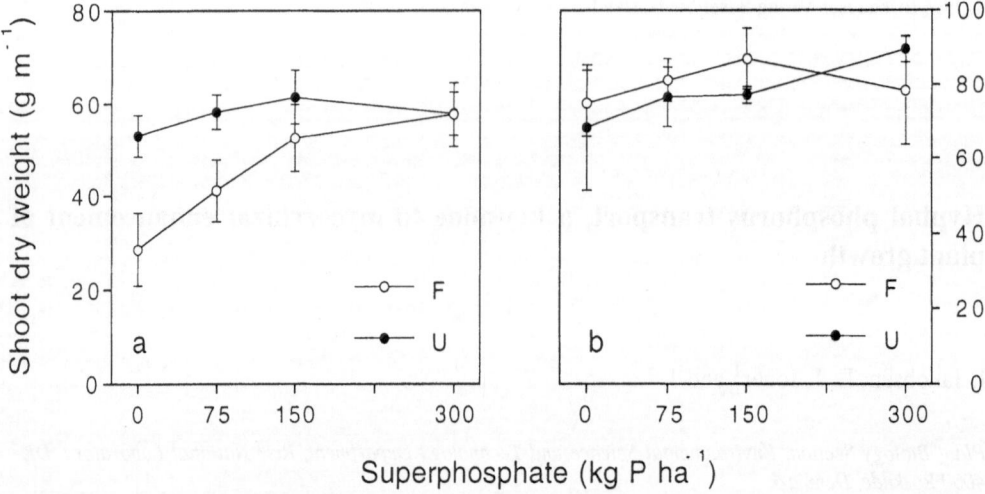

Figure 1. Growth responses to P fertilizer of flax (a) and oil-seed rape (b) grown in untreated (U) and in dazomet fumigated (F) field plots. The four fertilizer levels corresponded to soil-P levels of 20, 28, 44 and 54 mg P kg[-1] (Olsen et al., 1954) (I. Jakobsen, unpublished)

presence of mycorrhizas substituted the application of *ca.* 150 kg P ha[-1]. Oil-seed rape served as a control and responded neither to P fertilizer nor to the presence of mycorrhizal propagules. The shape of the P response curve will depend not only on the presence of arbuscular mycorrhizal fungi (AMF) but also on the effectiveness of the symbiosis. Indigenous populations of fungi may differ markedly in their ability to improve plant growth within a range of soil P levels (Stribley et al., 1980; Dodd et al., 1983) and there is a need to understand the basis for this variation in order to develop strategies for the management of mycorrhizas towards maximum effectiveness. The aim of this paper is to identify some limitations to mycorrhizal functioning with particular reference to the performance of the external hyphae in soil.

Growth of mycorrhizal hyphae in soil

The quantity and distribution of external hyphae in the soil is expected to have a higher predictive value for mycorrhizal effectiveness than the colonized root length, and the missing documentation on hyphal performance in mycorrhizal research papers until the last 4-6 years was caused more by the lack of adequate methods than by ignorance. Methods for the quantification of mycorrhizal hyphae were described by Sylvia (1992). Indirect methods are based on measurements of the colonization of 'receiver plants' (Schüepp et al., 1987), soil aggregation (Graham et al., 1982) and chitin content in the soil (Pacovsky and Bethlenfalvay, 1982). Fatty acids, specific to the mycorrhizal fungi, may represent a realistic alternative for

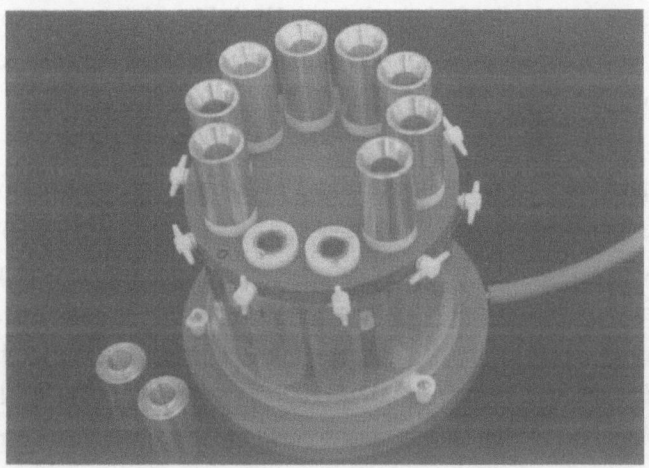

Figure 2. Filtering manifold for the simultaneous processing of ten membrane filters. Each filter is held by the weight of the steel towers (Water Quality Institute, Hørsholm, Denmark).

quantifying viable hyphae in soil (P. A. Olsson et al., unpubl.). However, direct microscopic measurements of hyphal length have provided most information so far. Aqueous extraction of hyphae combined with a membrane filter technique (Hanssen et al., 1974) was first used to quantify arbuscular mycorrhizal hyphae by Abbott et al. (1984). A time-consuming modification, involving dispergence, sonication, centrifugation and wet sieving through a 20 μm mesh may facilitate the extraction and quantification of hyphae in heavy soils (Miller and Jastrow, 1992). The time needed for preparation of filters with hyphae was considerably reduced by using a filtering manifold (Fig. 2) holding ten 25 mm filters (Jakobsen et al., 1992a). The manifold facilitated the preparation of 20 filters per hour, corresponding to five original soil samples with duplicate subsamples and duplicate filters prepared from each subsample. An appropriate vital stain (Schubert et al., 1987; Sylvia, 1988; Saito et al., 1993) is useful at this stage to quantify the proportion of total hyphal length, which can be expected to be active in P transport.

It is difficult to distinguish hyphae of AMF from hyphae of other fungi. Morphological features were used to quantify mycorrhizal hyphae in field soils (Miller and Jastrow, 1992), whereas Abbott et al. (1984) concluded that 'it was not possible to distinguish hyphae of AMF from those of other fungal hyphae on the basis of their general morphology'. This controversy may partly result from differences in the degree of breaking up of hyphae: The method used by

Miller and Jastrow probably leaves longer pieces of hyphae intact than the method of Abbott et al. (1984) where a blending step is included. In experiments involving non-mycorrhizal controls, the most reliable measurements are obtained by correcting hyphal length in mycorrhizal treatments for hyphal length in the controls, although it cannot be excluded that AMF may influence the growth of other soil fungi. Hyphal length in non-mycorrhizal controls as a percentage of the length in mycorrhizal treatments has differed considerably between experiments. This probably depend on the origin and pretreatment (storage, sieving) of the soil.

Published hyphal length densities range from 1 to 50 m·g^{-1} soil with most values in the 5-15 m range (Jakobsen et al., 1992a; Sylvia, 1992). The average hyphal length density is thus about 100 times the typical root length density, which explains why hyphae may efficiently acquire significant amounts of soil P beyond the depletion zone around roots. The spread of hyphae in soil is most easily measured in a growth system including a fine mesh (20-40 μm) to separate a root-free hyphal compartment from the root compartment. The hyphae may spread over considerable distances, but each fungus seems to have its own characteristic pattern of spread with rates of spread in the range 1-3 mm d^{-1} (Fig. 3). The spread of root colonization from a localized source of inoculum appears to simply reflect the spread of hyphae (Scheltema et al., 1985, 1987; Jakobsen et al., 1992a). The length of hyphae in a root-free compartment is usually only slightly less than in a corresponding compartment with roots (Fig. 3).

AMF compete for colonization of roots, so that one fungus may dramatically reduce colonization by other fungi (Hepper et al., 1988; Pearson et al., 1993). The possible mechanisms that determine the outcome of competition would include not only an induced 'defense' response by the plant, but also antagonism between hyphae in the rhizosphere. The external hyphae may also be affected by a range of environmental factors. Growth of hyphae might be limited at poor nutrient availability, although P may not be limiting in spite of the low levels found in soil (Beever and Burns, 1980). High soil-P levels decreased the length of external hyphae relatively more than the colonized root length (Abbott et al., 1984), but this could be an indirect effect, as high soil-P levels in the hyphal environment had no influence on hyphal length density (Li et al., 1991a). Hyphae were observed to proliferate in soil organic matter particles (Mosse, 1959; St. John et al., 1983; Warner, 1984) and hyphal length densities in root-free compartments were higher when soil had been incubated with wheat straw than when the soil received no straw (Joner and Jakobsen, 1992). This experiment was carried out with partially sterilized soil with subsequent reintroduction of the general microflora and the results were later confirmed for another AMF (E. Joner unpubl.). The mechanisms behind such effects of organic matter are unknown so far, but may well be related to a changed activity of the soil biota. Hyphal growth *in vitro* was stimulated markedly by high CO_2 concentrations

(Bécard and Piché, 1989), which would also be associated with the decomposition of organic matter. Although the presence of microorganisms may influence spore germination and mycorrhiza formation (Hetrick et al., 1986, 1988; Wilson et al., 1989; Calvet et al., 1992) direct interactions between growth of mycorrhizal hyphae and other microorganisms have not

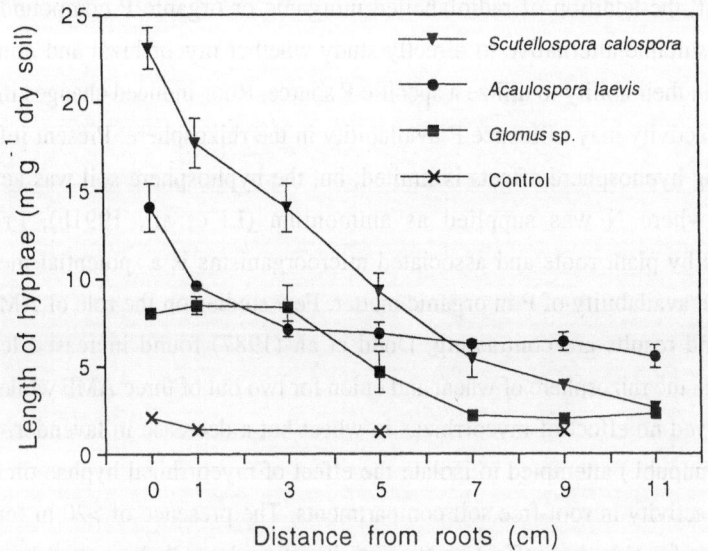

Figure 3. Spread of external hyphae from *Trifolium subterraneum* roots into hyphal compartments during 47 days. Soil cores were taken inside the root compartment (0 cm) and at increasing distances form the root compartment (0-11 cm) (Jakobsen et al., 1992a).

been described. In contrast, the hyphae undoubtedly represent a food source for a range of soil animals including fungivorous collembolans and nematodes. The presence of soil animals might explain why Joner and Jakobsen (1992) found no hyphal growth response to straw amendment when the partial sterilization treatment was omitted.

Phosphorus sources for mycorrhizal hyphae

The primary advantage of mycorrhizal hyphae in P uptake as compared to roots is the ability of the hyphae to extend beyond the P-depletion zone of the root. It is also possible, however, that hyphae may utilize P sources of low plant availability. The question has been studied by means of the isotopic dilution principle, where a tracer isotope (e.g. ^{32}P) is allowed to exchange with the native soil P (Gianinazzi-Pearson et al. 1981; Morel and Plenchette, 1994). Subsequently grown mycorrhizal and non-mycorrhizal plants had similar specific activities in a number of experiments. This observation was used as a proof that mycorrhizal and non-mycorrhizal plants obtain their P from the same 'sources', 'fractions', 'pools' or 'forms' of P from the soil (Bolan, 1991). However, isotopic exchange is a continuing process and the different P fractions in soil will tend to have the same specific activity. Accordingly, the specific activity of P was similar in soil extracts obtained with chemicals of different extraction strength although the total P extracted differed as much as 100 fold (Bolan et al., 1984).

In contrast, the addition of radiolabelled inorganic or organic P compounds to the soil represents a suitable alternative to directly study whether mycorrhizal and nonmycorrhizal plants differ in their ability to utilize a specific P source. Root-induced changes in soil-pH and phosphatase activity may influence P availability in the rhizosphere. Present information on corresponding hyphosphere effects is limited, but the hyphosphere soil was acidified in an experiment where N was supplied as ammonium (Li et al., 1991b). Production of phosphatases by plant roots and associated microorganisms is a potential mechanism for increasing the availability of P in organic matter. Few studies on the role of AMF have been conducted and results are contrasting. Dodd et al. (1987) found increased levels of acid phosphatase in the rhizosphere of wheat and onion for two out of three AMF while Azcon et al. (1982) observed no effect of mycorrhizas in wheat but a decrease in lavender. Recently, E. Joner et al. (unpubl.) attempted to isolate the effect of mycorrhizal hyphae on extracellular phosphatase activity in root-free soil compartments. The presence of >20 m $(cm^3 soil)^{-1}$ of hyphae of two fungi had no effect on the activity of acid or alkaline phosphatase. Another experiment by E. Joner showed that alkaline phosphatase activity was stimulated by mycorrhizal hyphae, when fresh clover leaves had been mixed into the soil. This effect was not observed in soil which had been incubated with wheat straw for 12 months. Interactions between hyphae and other microorganisms are obviously important in this context. However, two recent experiments provide evidence that mycorrhizal plants can utilize organic P sources more effectively than nonmycorrhizal plants. Jayachandran et al. (1992) found that the uptake over four weeks of ^{32}P from labelled cytidine 3'- and 5'-diphosphate was 500 times larger by mycorrhizal *Andropogon gerardii* Vit. than by nonmycorrhizal P-fertilized plants. The relationship to length of roots or hyphae growing into the radiolabelled soil and the specific activity of P were not presented. Joner and Jakobsen (1994) found that hyphae of two AMF increased net release of P from organic matter in root-free compartments by 33 and 48%, respectively. The mechanism behind this benefit was attributed to interception of soil P fixation after mineralization by a reintroduced saprophytic microflora.

Dying roots of neighbouring plants in a mixed community may also serve as a P source for mycorrhizal hyphae (Eason et al., 1991). This was recently confirmed with two plant species, which were connected by a mycorrhizal fungus growing from one root compartment, through a 20 mm root-free soil layer and into the root compartment of the other plant (A. Johansen, unpubl.). A split-root technique was used to label a donor plant with ^{32}P, whereafter its shoot was removed. During the following 50 days, 7% of the ^{32}P in the donor was transferred to the other plant. It is unclear whether the interplant transfer occurred directly via hyphal connection with the donor root or indirectly via ^{32}P leaked from the dying roots.

P transport by mycorrhizal hyphae in relation to host-fungus characteristics

Differences between AMF in their ability to improve P uptake and plant growth is often unrelated to the extent of mycorrhiza formation. The development of novel methods now enable us to study these differences in relation to the hyphal P transport, which is a keystone in mycorrhizal functioning. The root-external hyphae are allowed to grow through a mesh into a root-free soil compartment, and hyphal P transport may be measured either directly by means of tracer isotopes (Jakobsen et al., 1992b) or indirectly from hyphal depletion of soil P (Li et al., 1991ab). While the capacity for P transport by a mycorrhizal fungus is generally unrelated to its hyphal length density, the rate and distance of hyphal spread could explain differences in P transport between two fungi, especially in the case of transport over more than 1 cm (Jakobsen et al., 1992b). Some fungi may spread at least 10-12 cm from the roots (Li et al., 1991a; Jakobsen et al., 1992a). This would be an important characteristic in soils with a low density or uneven distribution of mycorrhizal propagules.

However, the hyphal P transport may be rather limited even in the presence of a well-developed mycelium in the soil. This has been most clearly demonstrated for an Australian isolate of *Scutellospora calospora* in a number of experiments (Jakobsen et al. 1992a and b; Pearson and Jakobsen, 1993). The P uptake systems of this fungus appears to be efficient, as its external hyphae accumulated much more ^{32}P than the hyphae of two other fungi which transported large amount of ^{32}P to the plant (Jakobsen et al., 1992b). These observations suggest that the rate-limiting step in P transport by this fungus was located at the host-fungus interface. In consequence, while *S. calospora* is compatible with both subterranean clover and cucumber with respect to mycorrhiza formation and therefore also host-fungal C transport, these symbioses can be regarded as functionally incompatible with respect to P transport. It is likely that *S. calospora* is functionally compatible with one or several plant species growing in its natural habitat and variation in functional compatibility as determined by the host genome was recently demonstrated with two fungi, which both formed well-developed mycorrhizas with three plant species (S. Ravnskov, unpubl.). Hyphae of a *Glomus* sp. isolate transported large amounts of P to flax, but not to wheat or cucumber, whereas an isolate of *Glomus caledonium* was functionally compatible with all three plant species. Evidently, the functional compatibility of a given plant-fungus combination can be predicted only if we fully understand the mechanisms involved in the nutrient transfer across the complex interface between the membranes of the plant and the fungus. The transfer of P is assumed to result from passive conditioned efflux from the fungus and active absorption across the plant plasmalemma, and the opposite directed transfer of C is assumed to occur in a similar way (see Smith et al., 1994 for a detailed discussion).

The contribution of the fungus to total P uptake by mycorrhizas can hardly be estimated from comparisons between mycorrhizal and non-mycorrhizal plants as P uptake by root tissues is probably influenced by the presence of the fungus. Root P uptake was thus influenced both by their P status via feedback regulation (Jungk et al., 1990) and by the presence of pathogens (Paul, 1989). A dual labelling method has been developed as an attempt to directly assess the relative contribution from the fungus. The main compartment for the roots is equipped with not only a root-free hyphal compartment, but also with an identical compartment containing both roots and hyphae (Pearson and Jakobsen, 1993). The P transport from each compartment is studied by means of ^{32}P supplied to the hyphal compartment and ^{33}P supplied to the compartment containing both roots and hyphae. The total and relative uptake of each isotope varied with the fungal isolate, but most surprisingly, one fungus transported just as much ^{32}P from the hyphal compartment as the combined transport of ^{33}P by roots and fungus (Pearson and Jakobsen, 1993). This is further demonstrated by the data in Table I, where the transport by hyphae of *G. caledonium* was similar to the combined uptake of roots and hyphae.

Table I. Dry weight-specific content of ^{32}P from hyphal compartments (HC) and ^{33}P from compartments with roots and hyphae (RHC) in cucumber shoots. Plants were grown in dazomet-fumigated or unfumigated soil either uninoculated or inoculated with two different mycorrhizal fungi (SE in brackets).

Inoculum	Fumigated soil		Untreated soil	
	^{32}P (HC)	^{33}P (RHC)	^{32}P (HC)	^{33}P (RHC)
		(kBq g^{-1} shoot dry wt.)		
None	7.7 (0.8)	14.9 (1.8)	11.2 (1.3)	19.2 (2.3)
Glomus sp.	16.5 (1.0)	21.2 (1.3)	11.0 (1.1)	15.4 (1.4)
G. caledonium	22.4 (2.2)	22.7 (1.9)	18.7 (2.3)	22.8 (2.0)

The soil was sampled from fumigated or untreated field plots, where similar results were obtained with the same fungus in combination with pea (Jakobsen, unpubl.). Two explanations seem possible: A high P input from the hyphae may reduce the direct root uptake by feed-back regulation, or the presence of roots may impair the optimal functioning of the hyphae by competition for P or via interactions with rhizosphere microorganisms. The exact mechanism is unknown so far, but the hyphal P transport by a given fungus from a root-free compartment with semi-sterile soil appears to represent its maximum possible transport. It is important to notice that interfungal differences in P transport from a root-free compartment are not necessarily reflected in the combined P transport by roots and hyphae. This is shown in Table I, where inoculation with *G. caledonium* in unfumigated soil increased P transport from HC but not from RHC. The hyphal P transport by a field population of AMF may also be studied in undisturbed soil by means of a mesh bag containing soil with a tracer isotope (Jakobsen, 1992).

Hyphal P transport in relation to environmental factors

The functioning of mycorrhizas can be modified by a range of soil biological and physio-chemical environmental factors. Reported interactions between soil microbes, bacteria in particular, and AMF range from suppression (Hetrick et al., 1988) to amplification (Meyer and Linderman, 1986) of the growth response to mycorrhizas. Microbial suppression of hyphal P transport may be partly caused by competition for P between bacteria and mycorrhizal fungi (Hetrick, 1989). Fitter and Garbaye (1994) have further suggested that some of the plant growth promoting rhizobacteria (PGPR) may actually function via a close association with mycorrhizal fungi, similar to the effect of the mycorrhization helper bacteria (MHB) in ectomycorrhizas.

The study of P transport from a root-free hyphal compartment in relation to the presence of single isolates or populations of microorganisms appears to represent an appropriate approach for identifying mechanisms involved in the functional interactions between AMF and other microorganisms. A similar approach is useful for the study of soil fauna - mycorrhiza interactions which were reviewed by Fitter and Sanders (1992). Special attention has been paid to fungivorous collembolans and fragments of mycorrhizal hyphae have been identified in the guts of collembolans (Warnock et al., 1982) which were also directly observed to feed on the hyphae of two AMF (Moore et al., 1985). Selective biocides were used to demonstrate simultaneous changes in mycorrhizal growth responses and in numbers of collembolans (Finlay, 1985; McGonigle and Fitter, 1988). Those results suggest that collembolans may influence plant growth via grazing of mycorrhizal hyphae. This is supported by an observed non-linear impact of collembolans, both on growth and P content of mycorrhizal plants (Finlay, 1985; Harris and Boerner, 1990) and on metabolic activity of saprophytic fungi

(Hedlund et al., 1991). Direct evidence for effects of collembolans on hyphal P transport has now been obtained by means of a model system where 0, 20 or 200 individuals of *Folsomia candida* were added to a root-free hyphal compartment containing 200 g soil. The hyphal P transport was unaffected by 20 collembolans, but reduced by the presence of 200 animals (Fig. 4). The P transport at the low grazing intensity was lower without, than in the presence of yeast, which represents an additional food source for the collembolans. This suggests that the animals preferred yeast to the mycorrhizal hyphae, but a direct interaction between the mycor-

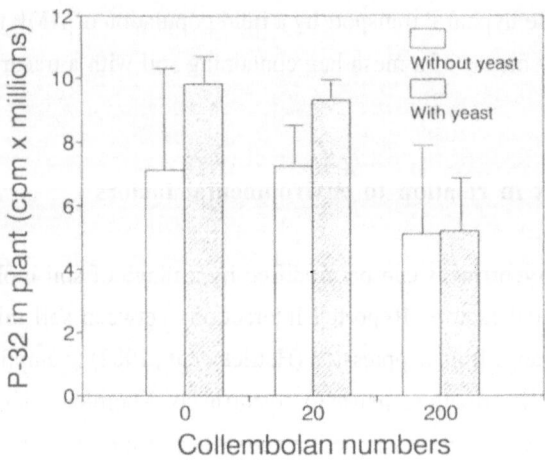

Figure 4. The effect of *Folsomia candida* on the hyphal transport of ^{32}P to subterranean clover in symbiosis with *Glomus* sp. The soil of the hyphal compartments contained dry yeast in half of the treatments.

rhizal hyphae and the yeast cannot be excluded. Direct evidence for grazing may be obtained with this model systems by measuring the presence of ^{32}P in collembolans contained in soil with radiolabelled hyphae.

Through a better understanding of effects of the soil biota on P transport by AMF, we can ultimately hope to be able to include the different biotic components in future management strategies towards a maximum utilization of AMF in crop production. However, it is already well documented that agricultural practices like soil tillage and pesticide use may have adverse effects on hyphal P transport. Mycorrhiza formation and P uptake in the field was lower during early growth stages of maize under conventional than under reduced tillage (McGonigle and Miller, 1993). This effect of tillage on P uptake may functionally be caused by disruption of the external mycelium from the previous crop, as soil disturbance reduced P uptake without affecting mycorrhiza formation (McGonigle et al., 1990). The functional longevity of the undisturbed mycelium is crucial in this context, but has so far only been studied in terms of infectivity (Warner and Mosse, 1980; Jasper et al., 1989).

A range of fungicides reduces mycorrhiza formation, even at field application rates (Dodd and Jeffries, 1989), and can have dramatic effects on P inflow into roots of mycotrophic host plants (Hale and Sanders, 1982). Accordingly, benomyl applied to a root-free hyphal compartment was recently found to immediately inhibit hyphal P uptake (J. Larsen et al., unpubl.) and future studies of pesticide effects on mycorrhizas should focus more on hyphal nutrient transport and less on mycorrhiza formation.

Conclusions

Positive effects of mycorrhizas are more difficult to demonstrate in the field, than in pots with partially sterilized soil, where a mycorrhizal fungus has been reintroduced. Consequently, the reported evidence for a marked interfungal variation in P transport by mycorrhizal hyphae has little practical relevance unless we become able to use the information at the level of field populations. It is crucial that effects related to the characteristics of the population of AMF are separated from environmental effects. The first step will be to identify field sites where the AMF function at a suboptimal level. Those fungi, which dominate in root colonization, must then be identified with subsequent measurement, under controlled conditions, of the P transport efficiency of each fungus and of the fungi in combination. If the dominating fungi are found to be functionally suboptimal, a change in composition of the fungal population will be the primary management target. Alternatively, if the actively colonizing fungi turns out to be efficient, attention must be paid to their interaction with the biotic and abiotic environment.

References

Abbott, L.K. and Robson, A.D. (1984) The effect of VA mycorrhizae on plant growth. In: C.L. Powell and D.J. Bagyaraj (eds) *VA Mycorrhiza*. CRC, Boca Raton, Florida, pp. 113-130.

Abbott, L.K., Robson, A.D. and De Boer, G. (1984) The effect of phosphorus on the formation of hyphae in soil by the vesicular-arbuscular mycorrhizal fungus, *Glomus fasciculatum*. *New Phytol.* 97: 437-446.

Azcon, R., Borie, F. and Barea, J.M. (1982) Exocellular phosphatase activity of lavender and wheat roots as affected by phytate and mycorrhizal inoculation. In: S. Gianinazzi, V. Gianinazzi-Pearson and A. Trouvelot (eds) *Les Mycorhizes: Biologie et utilisation*. INRA, Dijon, pp. 83-85.

Bécard, G. and Piché, Y. (1989) Fungal growth stimulation by CO_2 and root exudates in vesicular-arbuscular mycorrhizal symbiosis. *Appl. Environ. Microbiol.* 55: 2320-2325.

Beever, R.E. and Burns, D.J.W. (1980) Phosphorus uptake, storage and utilization by fungi. In: H.W. Woolhouse (ed) *Advances in Botanical Research*, vol 8. Academic Press, London, pp. 127-192.

Bolan, N.S. (1991) A critical review on the role of mycorrhizal fungi in the uptake of phosphorus by plants. *Plant Soil* 134: 189-207.

Bolan, N.S., Robson, A.D., Barrow, N.J. and Aylmore, L.A.G. (1984) Specific activity of phosphorus in mycorrhizal and non-mycorrhizal plants in relation to the availability of phosphorus to plants. *Soil. Biol. Biochem.* 16: 299-304.

Calvet, C., Estaun, V. and Camprubi, A. (1992) Germination, early mycelial growth and infectivity of a vesicular-arbuscular mycorrhizal fungus in organic substrates. *Symbiosis* 14: 405-411.

144

Dodd, J.C., Burton, C.C., Burns, R.G. and Jeffries, P. (1987) Phosphatase activity associated with the roots and the rhizosphere of plants infected with vesicular-arbuscular mycorrhizal fungi. *New Phytol.* 107: 163-172.

Dodd, J.C. and Jeffries, P. (1989) Effect of fungicides on three vesicular-arbuscular mycorrhizal fungi associated with winter wheat (*Triticum aestivum* L.). *Biol. Fertil. Soils.* 7: 120-128.

Dodd, J., Krikun, J. and Haas, J. (1983) Relative effectiveness of indigenous populations of vesicular-arbuscular mycorrhizal fungi from four sites in the Negev. *Israel J. Bot.* 32: 10-21.

Eason, W.R., Newman, E.I. and Chuba, P.N. (1991) Specificity of interplant cycling of phosphorus: The role of mycorrhizas. *Plant Soil* 137: 267-274.

Finlay, R.D. (1985) Interactions between soil micro-arthropods and endomycorrhizal associations of higher plants. In: A.H. Fitter (ed.) *Ecological Interaction in Soil*, Blackwell Scientific Publications, Oxford, pp. 319-331.

Fitter, A.H. and Garbaye, J. (1994) Interactions between mycorrhizal fungi and other soil organisms. *Plant Soil* 159: 123-132.

Fitter, A.H. and Sanders, I.R. (1992) Interactions with the soil fauna. In: M.F. Allen (ed) *Mycorrhizal Functioning*, Chapman and Hall, New York, pp. 333-354.

Gianinazzi-Pearson, V., Fardeau J.C., Asimi, S. and Gianinazzi, S. (1981) Source of additional phosphorus absorbed fro, soil by vesicular-arbuscular mycorrhizal soybean. *Physiologie végétale* 19: 33-43.

Graham, J.H., Linderman, R.G. and Menge, J.A. (1982) Development of external hyphae by different isolates of mycorrhizal *Glomus* ssp. in relation to root colonization and growth of troyer citrange. *New Phytol.* 91: 183-189.

Hale, K.A. and Sanders, F.E. (1982) Effects of benlate on vesicular-arbuscular mycorrhizal infection of red clover (*Trifolium pratense* L) and consequences for phosphorus inflow. *J.Plant Nutr.* 5: 1355-1357.

Hanssen, J.F., Thingstad, T.F. and Goksøyr, J. (1974) Evaluation of hyphal lengths and fungal biomass in soil by a membrane filter technique. *Oikos* 25: 102-107.

Harris, K.K. and Boerner, R.E.J. (1990) Effects of belowground grazing by collembola on growth, mycorrhizal infection and P uptake of *Geranium robertianum*. *Plant Soil* 129: 203-210.

Hedlund, K., Boddy, L. and Preston, C.M. (1991) Mycelial responses of the soil fungus, *Mortierella isabellina*, to grazing by *Onychiurus armatus* (Collembola). *Soil. Biol. Biochem.* 23: 361-366.

Hepper, C.M., Azcon-Aguilar, C., Rosendahl, S. and Sen, R. (1988) Competition between three species of *Glomus* used as spatially separated introduced and indigenous mycorrhizal inocula for leek (*Allium porrum* L.). *New Phytol.* 110: 207-215.

Hetrick, B.A.D. (1989) Acquisition of phosphorus by VA mycorrhizal fungi and the growth responses of their host plants. In: L. Boddy, R. Marchant and D.J. Read (eds) *Nitrogen, Phosphorus and Sulphur Utilization by Fungi.* Cambridge University Press, Cambridge, pp. 205-226.

Hetrick, B.A.D., Kitt, D.G. and Wilson, G.T. (1986) The influence of phosphorus fertilization, drought, fungal species, and nonsterile soil on mycorrhizal growth response in tall grass prairie plants. *Can. J. Bot.* 64: 1199-1203.

Hetrick, B.A.D., Wilson, G.T., Kitt, D.G. and Schwab, A.P. (1988) Effects of soil microorganisms on mycorrhizal contribution to growth of big bluestem grass in non-sterile soil. *Soil. Biol. Biochem.* 20: 501-507.

Jakobsen, I. (1992) Phosphorus transport by external hyphae of vesicular-arbuscular mycorrhizas. In: D.J. Read, D.H. Lewis, A.H. Fitter and I.J. Alexander (eds) *Mycorrhizas in Ecosystems*. CAB International, Wallingford, pp. 48-58.

Jakobsen, I., Abbott, L.K. and Robson, A.D. (1992a) External hyphae of vesicular-arbuscular mycorrhizal fungi associated with *Trifolium subterraneum* L. 1. Spread of hyphae and phosphorus inflow into roots. *New Phytol.* 120: 371-380.

Jakobsen, I., Abbott, L.K. and Robson, A.D. (1992b) External hyphae of vesicular-arbuscular mycorrhizal fungi associated with *Trifolium subterraneum* L. 2. Hyphal transport of ^{32}P over defined distances. *New Phytol.* 120: 509-516.

Jasper, D.A., Abbott, L.K. and Robson, A.D. (1989) Hyphae of a vesicular-arbuscular mycorrhizal fungus maintain infectivity in dry soil, except where the soil is disturbed. *New Phytol.* 112: 101-107.

Jayachandran, K., Schwab, A.P. and Hetrick, B.A.D. (1992) Mineralization of organic phosphorus by vesicular-arbuscular mycorrhizal fungi. *Soil. Biol. Biochem.* 24: 897-903.

Joner, E. and Jakobsen, I. (1994) Contribution by two arbuscular mycorrhizal fungi to P uptake by cucumber (*Cucumis sativus* L.) from ^{32}P-labelled organic matter during mineralization in soil. *Plant Soil* (in press).

Joner, E. and Jakobsen, I. (1992) Enhanced growth of external VA mycorrhizal hyphae in soil amended with straw. In: D.J. Read, D.H. Lewis, A.H. Fitter and I.J. Alexander (eds) *Mycorrhizas in Ecosystems*. CAB International, Wallingford, pp. 387.

Jungk, A., Asher, C.J., Edwards, D.G. and Meyer, D. (1990) Influence of phosphate status on phosphate uptake kinetics of maize (*Zea mays*) and soybean (*Glycine max*). In: van Beusichem, M.L. (ed.) *Plant Nutrition - Physiology and Applications*, Kluwer Academic Publishers, pp. 135-142.

Li, X.-L., George, E. and Marschner, H. (1991a) Extension of the phosphorus depletion zone in VA-mycorrhizal white clover in a calcareous soil. *Plant Soil* 136: 41-48.

Li, X.-L., George, E. and Marschner, H. (1991b) Phosphorus depletion and pH decrease at the root-soil and hyphae-soil interfaces of VA mycorrhizal white clover fertilized with ammonium. *New Phytol.* 119: 397-404.

McGonigle, T.P.,and Fitter, A.H. (1988) Ecological consequences of arthropod grazing on VA mycorrhizal fungi. *Proc. Royal Soc. Edinburgh* 94b: 25-32.

McGonigle, T.P. and Miller, M.H. (1993) Mycorrhizal development and phosphorus absorption in maize under conventional and reduced tillage. *Soil. Sci. Soc. Am. J.* 57: 1002-1006.

McGonigle, T.P., Evans, D.G. and Miller, M.H. (1990) Effect of degree of soil disturbance on mycorrhizal colonization and phosphorus absorption by maize in growth chamber and field experiments. *New Phytol.* 116: 629-636.

Meyer, J.R. and Linderman, R.G. (1986) Response of subterranean clover to dual inoculation with vesicular-arbuscular mycorrhizal fungi and a plant growth-promoting bacterium, *Pseudomonas putida*. *Soil Biol. Biochem.* 18: 185-190.

Miller, R.M. and Jastrow, J.D. (1992) Extraradical hyphal development of vesicular-arbuscular mycorrhizal fungi in a chronosequence of prairie restorations. In: D.J. Read, D.H. Lewis, A.H. Fitter and I.J. Alexander (eds) *Mycorrhizas in Ecosystems*. CAB International, Wallingford, pp. 171-176.

Moore, J.C., St. John, T.V. and Coleman, D.C. (1985) Ingestion of vesicular-arbuscular mycorrhizal hyphae and spores by soil microarthropods. *Ecology* 66: 1979-1981.

Morel, C. and Plenchette, C. (1994) Is the isotopically exchangeable phosphate of a loamy soil the plant-available P? *Plant and Soil* 158: 287-297.

Mosse, B. (1959) Observations on the extra-matrical mycelium of a vesicular-arbuscular endophyte. *Trans. Br. mycol. Soc.* 42(4): 439-448.

Olsen, R., Cole, C.V., Watanabe, F.S. and Dean, L.A. (1954) Estimation of available phosphorus in soils by extraction with $NaHCO_3$. *Circ 939*, United States Department of Agriculture, Washington, DC.

Pacovsky, R.S. and Bethlenfalvay, G.J. (1982) Measurement of the extraradical mycelium of a vesicular-arbuscular mycorrhizal fungus in soil by chitin determination. *Plant Soil* 68: 143.

Paul, N.D. (1989) Effects of fungal pathogens on nitrogen, phosphorus and sulphur relations of individual plants and populations. In: L. Boddy, R. Marchant and D.J. Read (eds) *Nitrogen, Phosphorus and Sulphur Utilization by Fungi*. Cambridge University Press, Cambridge, pp. 155-180.

Pearson, J.N. and Jakobsen, I. (1993) The relative contribution of hyphae and roots to phosphorus uptake by arbuscular mycorrhizal plants, measured by dual labelling with ^{32}P and ^{33}P. *New Phytol.* 124: 489-494.

Pearson, J.N., Abbott, L.K. and Jasper, D.A. (1993) Mediation of competition between two colonizing VA mycorrhizal fungi by the host plant. *New Phytol.* 123: 93-98.

Saito, M., Stribley, D.P. and Hepper, C.M. (1993) Succinate dehydrogenase activity of external and internal hyphae of a vesicular-arbuscular mycorrhizal fungus, *Glomus mosseae* (Nicol. & Gerd.) Gerdmann and Trappe, during mycorrhizal colonization of roots of leek (*Allium porrum* L.), as revealed by in situ histochemical staining. *Mycorrhiza* 4: 59-62.

Scheltema, M.A., Abbott, L.K., Robson, A.D. and De'Ath, G. (1985) The spread of *Glomus fasciculatum* through roots of *Trifolium subterraneum* and *Lolium rigidum*. *New Phytol.* 100: 105-114.

Scheltema, M.A., Abbott, L.K., Robson, A.D. and De'Ath, G. (1987) The spread of mycorrhizal infection by *Gigaspora calospora* from a localized inoculum. *New Phytol.* 106: 727-734.

Schubert, A., Marzachi, C., Mazzitelli, M., Cravero, M.C. and Bonfante-Fasolo, P. (1987) Development of total and viable extraradical mycelium in the vesicular-arbuscular mycorrhizal fungus *Glomus clarum* Nicol. & Schenck. *New Phytol.* 107: 183-190.

146

Schüepp, H., Miller, D.D. and Bodmer, M. (1987) A new technique for monitoring hyphal growth of vesicular-arbuscular mycorrhizal fungi through soil. *Trans. Br. mycol Soc.* 89: 429-435.

Smith, S.E., Gianinazzi-Pearson, V., Koide, R. and Cairney, J.W.G. (1994) Nutrient transport in mycorrhizas: structure, physiology and consequences for efficiency of the symbiosis. In: A.D. Robson, L.K. Abbott and N. Malajczuk (eds) Proc Int Symp on Management of Mycorrhizas in Agriculture, Horticulture and Forestry. Perth, Western Australia. *Plant Soil, Special Issue* 159: 103-114.

St. John, T.V., Coleman, D.C. and Reid, C.P.P. (1983) Association of vesicular-arbuscular mycorrhizal hyphae with soil organic particles. *Ecology* 64: 957-959.

Stribley, D.P., Tinker, P.B. and Snellgrove, R.C. (1980) Effect of vesicular-arbuscular mycorrhizal fungi on the relations of plant growth, internal phosphorus concentration and soil phosphate analyses. *J. Soil Sci.* 31: 655-672.

Sylvia, D.M. (1988) Activity of external hyphae of vesicular-arbuscular mycorrhizal fungi. *Soil. Biol. Biochem.* 20: 39-43.

Sylvia, D.M. (1992) Quantification of external hyphae of vesicular-arbuscular mycorrhizal fungi. In: J.R. Norris, D.J. Read and A.K. Varma (eds) *Methods in Microbiology*, Vol. 24, pp. 53-65.

Warner, A. (1984) Colonization of organic matter by vesicular-arbuscular mycorrhizal fungi. *Trans. Br. mycol. Soc.* 82(2): 352-354.

Warner, A. and Mosse, B. (1980) Independent spread of vesicular-arbuscular mycorrhizal fungi in soil. *Trans. Br. mycol. Soc.* 74: 407-410.

Warnock, A.J., Fitter, A.H. and Usher, M.B. (1982) The influence of a springtail *Folsomia candida* (Isecta, Collembola) on the mycorrhizal association of leek *Allium porrum* and the vesicular-arbuscular mycorrhizal endophyte *Glomus fasciculatus*. *New Phytol.* 90: 285-292.

Wilson, G.W.T., Hetrick, B.A.D. and Kitt, D.G. (1989) Suppression of vesicular-arbuscular mycorrhizal fungus spore germination by nonsterile soil. *Can. J. Bot.* 67: 18-23.

Impact of Arbuscular Mycorrhizas on
Sustainable Agriculture and Natural Ecosystems
S. Gianinazzi and H. Schüepp (eds.)
© 1994 Birkhäuser Verlag Basel/Switzerland

Approaches to the study of the extraradical mycelium of arbuscular mycorrhizal fungi

J.C. Dodd

International Institute of Biotechnology, P.O. Box 228, Canterbury, Kent CT2 7YW, UK

Introduction

One of the most neglected areas of research on arbuscular mycorrhizas is the role of the extraradical mycelial network that links colonisation of the root by arbuscular mycorrhizal fungi (AMF) with the soil matrix. This hyphal network is a key component in nutrient cycling in natural ecosystems and constitutes a major sink for carbon and other elements. Several reviews have attempted to highlight this important area (Sylvia, 1990, 1992) and this has led to an increased interest in microcosm experiments that allow separation of roots from the developing mycelium of different AMF. Various rhizobox or cuvette systems have been developed and employed which allow growth of a plant root system in one compartment (Fig. 1) whilst allowing spread of the extraradical mycelium (ERM) of AMF through a nylon mesh screen (30-40m) into a root-free compartment (Jakobsen et al., 1992a; Li et al.,1991; Schüepp et al., 1987). In this review, therefore, we will attempt to highlight recent approaches taken to improve our knowledge of the ERM in terms of its development, functioning and activity for individual AMF.

Formation of the extraradical mycelium

Few studies have attempted to follow the development of the ERM following colonisation of a seedling by an AMF, focusing almost exclusively on the spread of the fungus within the root cortex using trypan blue staining. The growth dynamics of the ERM for individual AMF needs to be studied in order to understand its function and ecological role as the interface between plant and soil. Work by Gueye et al. (1987) showed that proliferation of ERM occurred 20-25 days after planting of cowpea corresponding with increased of Phosphorus (P) inflow. More recent studies have provided a clearer picture of the architecture of the ERM that develops following colonisation of the seedling root system (Abbott et al., 1992; Friese and Allen, 1991a). Friese and Allen (1991a) used a mixed soil inoculum (dominated by spores of *Glomus spp.*) to inoculate *Artemisia tridentata* and *Oryzopsis hymenoides* from a semi-arid ecosystem in root observation chambers for 25 days. They described three major forms of AMF hyphae that enter plant roots:

(a) Germ tubes from spores - single hyphae that develop into networks of hyphae close to the root before producing multiple entry points on the root surface (Giovannetti et al., 1993a, b). These hyphae were always thin-walled and 2-3µm in diameter.

(b) Runner hyphae from adjacent colonised roots - These were observed to form single entry points on the root surface but occasionally produced 2-4. These hyphae had uneven wall thickening (1-3µm) and hyphal diameters between 10-15µm.

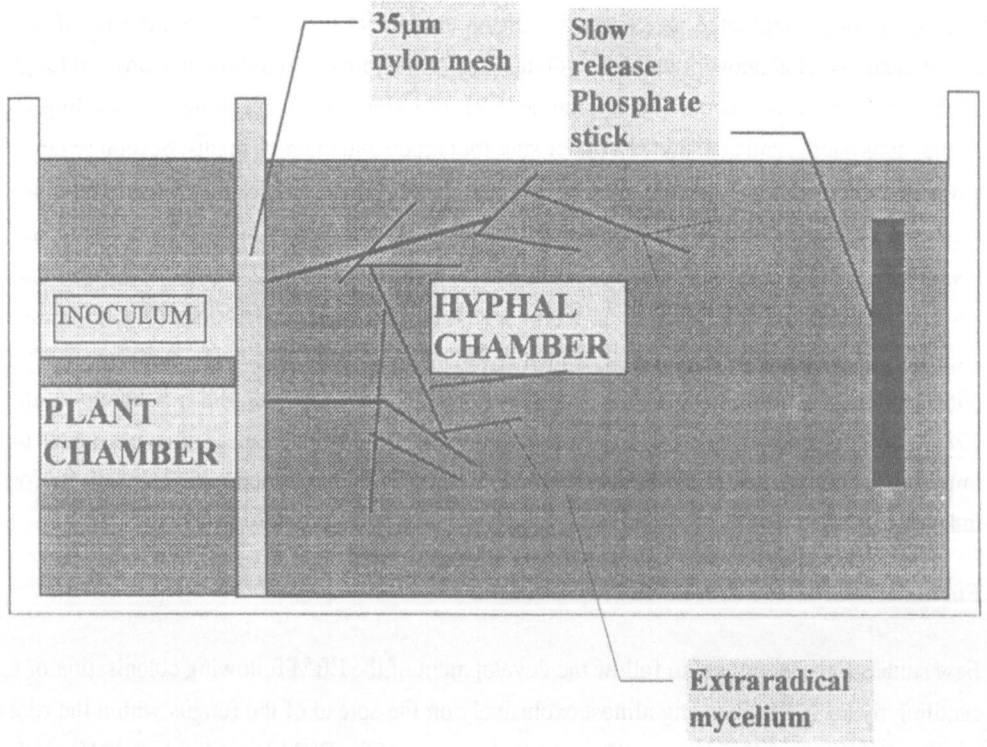

Figure 1. Rhizobox system used in experiments by Green et al., (1994)

(c) Hyphae emerging from old colonised root fragments - These were seen to be thicker-walled (3μm) and up to 20μm diameter and occasionally intertwined as they approached a root before also producing multiple entry points from branching of the hyphae close to the roots.

They also described two forms of hyphae that subsequently grew out into the soil from the root surface. The first were runner hyphae of long single strands with angular projections similar in morphology to those described above. These either ran alongside the root surface producing secondary points of entry or grew out into the soil for several cm until they encountered other roots and thereby established hyphal bridges. The second hyphal growth form constituted the absorptive hyphal network that consisted of a series of dichotomous branching orders that developed into a fan-shaped network. These were ephemeral networks which died-back 5-7 days after initiation. The branching led to proportional decreases in the hyphal diameters measured thereby leading to increases in the surface area to volume ratio of the hyphae. These hyphal networks develop septa as the cytoplasm is withdrawn allowing the ERM to tap other areas of the soil for nutrients. The authors state that this second hyphal form 'exites' the roots before spreading and branching. There is no conclusive evidence that hyphae can emerge from inside the root by direct penetration through the root epidermis. It is more likely that the initial entry points or appressoria are used to initiate the establishment of the ERM. This study, nevertheless, provides a clearer picture of the dynamics of the establishment of an arbuscular mycorrhiza in the early stages of seedling growth.

This work by Friese and Allen (1991a) probably reflects the early stages of colonisation of roots by species of *Glomus* (Dodd pers. obs.) but we have few data on other genera of AMF. Abbott et al. (1992) compared the formation of hyphae in soil by *Scutellospora calospora* (WUM 12) with a *Glomus sp.* (WUM 10) and found that the spread of the ERM of *Sc. calospora* was independent of the extent of internal colonisation of roots while that of the *Glomus sp.* was dependent on early, rapid and extensive colonisation within roots. Until recently there appears to have been a general acceptance in the literature that all AMF produce similar ERM and that the hyphae which constitute this network have similar architectures and morphologies. Figures for hyphal diameters between 5-20μm are frequently quoted with 1-5 μm given for the secondary or ephemeral hyphae (Friese and Allen, 1991a; Sieverding, 1991). These measurements of the hyphae are, however, from mixed populations of AMF or from the most commonly used species of *Glomus* used in most experimentation. A recent study (Dodd, unpublished data) of the diameters of hyphae of *Glomus manihotis* (INDO-1), *Acaulospora morrowiae* (PHIL-11A) and *Scutellospora heterogama* (ROTH) showed wide variations in the max/min and mean diameters of 50 hyphae chosen at random from within the ERM. These hyphae had been taken from the rhizosphere of the same host in the same substrate (Table I).

Table I. The mean, maximum and minimum diameters (50 measurements) of extraradical hyphae of three species of AMF grown under the same conditions on *Pueraria phaseoloides* and taken directly from the roots after 3 months' growth.

AMF	Mean diameter of hyphae (μm)	Maximum diameter of hyphae (μm)	Minimum diameter of hyphae (μm)
Acaulospora morrowiae (PHIL-11A)	5.6	9.1	3.7
Glomus manihotis (INDO-1)	7.9	18.0	3.2
Scutellospora heterogama (ROTH)			
Pigmented hyphae	7.0	9.9	4.2
Hyaline hyphae	4.5	7.2	1.6
Total	5.2	9.9	1.6

What should also be noted is the occurrence of both pigmented primary hyphae and thinner ephemeral hyaline hyphae (Fig. 2b) which branch from them in species of *Scutellospora* with pigmented spores (e.g. *Sc. heterogama*, *Sc.nigra*). To the unwary, these may easily be eliminated as saprophytic fungi when hyphal length estimations are being carried out (see later). It is also evident from this small data set that the thick-walled runner hyphae with larger than 10μm diameters, also noted by Friese and Allen (1991a), only occurred in the species of *Glomus* (Fig. 2a).

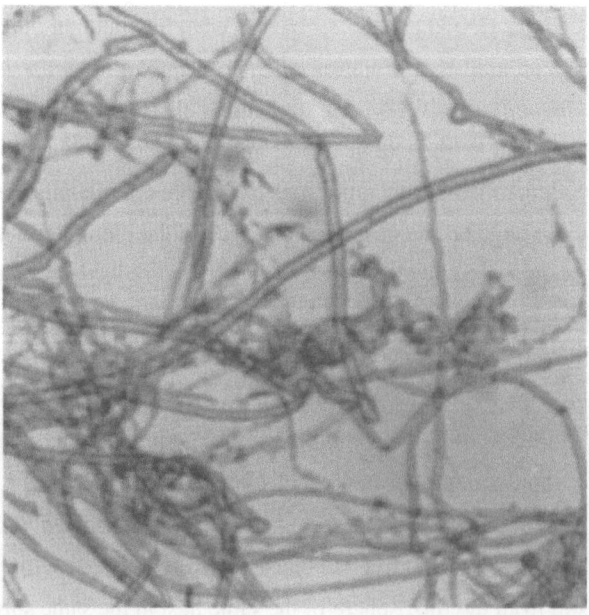

Figure 2a. Photo of extraradical mycelium of *Glomus manihotis* (INDO-1) showing thick-walled hyphae dominating (x400)

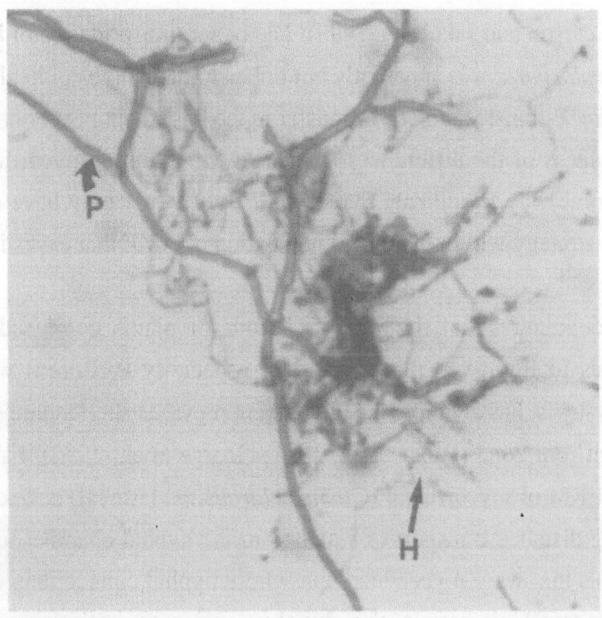

Figure 2b. Photo of extraradical mycelium of *Scutellospora heterogama* (ROTH) showing larger diameter pigmented hyphae (P) and clusters of thinner hyaline hyphae (H) arising from them (x400)

Function of the extraradical mycelium in nutrient and water uptake

The primary beneficial role to the plant of the ERM of AMF under P-deficient conditions is enhanced P-uptake and this aspect will be covered in greater detail elsewhere in this volume (see Jakobsen in chapter 11). It should be noted, however, that a comparison of the spread of ERM and the consequent P-inflow into subterranean clover by three AMF revealed both different distributions of ERM of each AMF, in the hyphal compartment of a rhizobox system, and different P-inflow rates into mycorrhizal roots and hyphae. The P-inflow with *Acaulospora laevis* (WUM 11) was approximately three times higher that of either *Glomus sp* (WUM 10) or *Sc. calospora* (WUM 12) over the 47 days of the experiment (Jakobsen et al., 1992a). In a subsequent 37 day study it was shown that the *Glomus* sp transported most ^{32}P to shoots of subterranean clover over soil-root distances below 1 cm, while *A.laevis* transported most ^{32}P to shoots over soil-root distances greater than 1 cm (Jakobsen et al., 1992b). They concluded that hyphae of *A.l aevis* had the capacity to spread faster and further in those soil conditions than hyphae of the *Glomus sp*. Further work that compared the efficiency of exchange of C and P between cucumber seedlings and three AMF, *Glomus sp.* (WUM 10), *Sc. calospora* (WUM 12) and *Glomus caledonium* (RIS 42), showed that the latter fungus had the capacity to transport as much as 11% of the applied ^{32}P compared with 1.5% for *Glomus sp.* and 0.2%

for *Sc. calospora* (Pearson and Jakobsen, 1993). The better transportation of P by *Glomus sp.*, compared with *Sc. calospora*, was apparently not linked to the lengths of hyphae produced but to improved P-uptake, P-translocation or P-transfer to the host. These experiments have greatly increased our awareness of the differences between AMF forming mycorrhizas with young seedlings and go some way to indicate that *Scutellospora* spp. may have a fundamentally different ecological strategy within the mixed population of AMF that exists in soils in natural and agro-ecosystems.

Nelsen (1987) concluded that the water relations of plants colonised by AMF were secondary responses to improved plant nutrition, especially P. Recent work, employing adapted rhizobox systems, has attempted to circumvent previous shortcomings by isolating the ERM from the plant root zone (Faber et al., 1991; George et al., 1992). Faber et al. (1991) concluded that the ERM of mycorrhizal (*Glomus claroideum* [Ames]) cowpea and sunflower was capable of long distance transport of water from the hyphal chamber to the plant. They incorporated controls that were mycorrhizal but whose hyphal connections were severed, by incorporation of a small air-gap in the rhizobox, prior to initiating water stress and measurement of water transport. A similar experimental design was employed by George et al. (1992) whereby roots and hyphae were separated into compartments. Their results indicated that the ERM of *Glomus mosseae* (Göttingen) was capable of transporting considerable amounts of P and N as NH_4^+ and NO_3^- to the plant from zones several centimetres from the root. There was, however, no evidence for significant direct water transport by hyphae to either clover or couchgrass. Discrepancies between these two similar studies may have been due to the different AMF used but plant host, soil and environmental conditions may also have been confounding factors. Work by Davies et al. (1992) also showed that the ERM of mycorrhizal pepper plants increased during drought acclimation, as measured by a soil aggregation assay (Graham et al., 1982), and that this improved drought resistance by facilitating soil water uptake. Further work still needs to be done to substantiate claims for direct water transport by ERM (Read and Boyd, 1986).

Recent work on the uptake of other nutrients, such as NH_4^+ and NO_3^-, by the ERM of individual AMF or natural populations of AMF has employed rhizobox systems and labelled isotopes. Johansen et al. (1992 a,b) showed that the ERM of cucumber seedlings colonised by *Glomus intraradices* (from Michigan State University) was capable of depleting the soil of inorganic N in the hyphal chamber. The recovery of ^{15}N in mycorrhizal plants was 38% when $^{15}NH_4^+$ was applied and 40% when $^{15}NO_3^-$ was applied. The corresponding values for non-mycorrhizal plants were 7% and 16% respectively. They concluded that hyphal transport of N had occurred over a distance of 5 cm from the applied ^{15}N sources towards the host plant. They could not conclude that NH_4^+ could be assimilated by the ERM but that it was nitrified to

NO_3^- prior to uptake. Interestingly they found that little ^{15}N was detected in the ERM of a hyphal chamber established at the other end of the rhizobox where no N was added indicating that at this stage of the symbiosis there was insignificant translocation of N from the cucumber to the ERM. This latter point is particularly relevant given the considerable interest in assessing whether the ERM can facilitate the direct transfer of nutrients between plants in natural ecosystems (Newman, 1988). Evidence for direct transfer of symbiotically-fixed N to another non-legume via the ERM is still equivocal although further work has indicated that N transfer does occur and is mycorrhiza-aided (Frey and Schüepp, 1992).

Further studies on the carbon flow to the developing ERM of 22-day-old $^{14}CO_2$-labelled cucumber plants has shown that hyphae contained 26% of the extraradical organic ^{14}C (Jakobsen and Rosendahl, 1990). In the same study the authors estimated that the fungal biomass and its respiration consumed 20% of the photoassimilated ^{14}C, 4%, of which, was allocated to the ERM in 22-day-old cucumber seedlings. The role of the ERM in carbon transfer between plants has also received attention and employed a rhizobox system incorporating a donor and receiver plant separated by 39 μm nylon mesh (Martins, 1993). Twelve-week-old *Lolium perenne* plants inoculated with a mixed dune sand inoculum had their hyphal links with the receiver plant severed or left intact before donor plants were exposed to ^{14}C for 48hr. Martins concluded that the main route for transfer between plants was via the ERM (46% of total) although what effect the total amount transferred has on the growth of the receiver plant is not clear.

Extraradical mycelium and soil aggregation

The recent upsurge in interest in sustainable agriculture and ecosystem restoration the role of AMF in soil aggregation has received further attention. A thought-provoking and fuller account of their potential role can be found elsewhere (Bethlenfalvay and Lindermann, 1992, see also Bethlenfalvay and Schüepp in chapter 10) and only recent work, specifically related to the ERM, will be covered here. Since Tisdall and Oates (1979) found a strong correlation between the amount of hyphae produced in soil and the production of water-stable aggregates (>2 mm) it has been suggested that the ERM may supply the initial framework for aggregation while bacterial polysaccharides cement the soil particles together (Sylvia, 1990). Miller and Jastrow (1990, 1992) have undertaken an intense study of a prairie ecosystem in the USA and showed that root lengths and lengths of ERM were associated with increases in the proportion of soil held as water-stable macroaggregates and the geometric mean diameter of the aggregates. They subsequently developed a model which was evaluated using data from prairie reconstructions and showed that the ERM had the strongest direct effect on the geometric mean diameter of aggregates. Further data indicated that in the reconstruction process of the prairie *Gigaspora gigantea* replaced two *Glomus* spp. as the predominant sporulator in the soil and that the former was positively associated with the length of ERM and macroaggregation (Miller and Jastrow, 1992). Thomas et al. (1993) showed a significant effect of soil hyphae on soil aggregation equivalent to that of non-mycorrhizal roots and, as stated by the authors, the energy expended in establishing an ERM is far less than that used in producing more root growth. Further studies which investigate the production and distribution of ERM of individual AMF will provide much needed information on the potential for screening inoculant AMF specifically for soil restoration (Dodd and Thomson, 1994; Sylvia and Burks, 1988).

Extraradical mycelium as inoculum

The ERM of AMF is important to the plant not only as a nutrient absorption system but also as a source of inoculum. There is often a poor correlation between spore numbers in soils and subsequent levels of colonisation and in many natural ecosystems (*e.g.* rainforests, Dodd pers. obs.), where spore numbers are low, a pre-existing network of soil hyphae is probably the chief source of inoculum (Birch, 1986; Brundrett, 1991). The reductions in levels of mycorrhizal colonisation resulting from soil disturbance indicates the importance of the pre-existing ERM as inoculum (Evans and Miller, 1990; Jasper et al., 1989a,b). Addy et al. (1993) showed that mesh (43 μm) pouches, which contained ERM produced from a maize mycorrhizal association of six weeks, could be buried in a field soil which froze in mid-winter (minimum temperature at 5 cm was -3.3°C) and still colonise bioassay plants in the following spring. Results of activity stains with fluorescein diacetate showed that some of the ERM remained viable over winter. It is probable that the existing network acts as a base from which further extension of the ERM is possible following colonisation of new root growth. Some of the early growth depressions observed in inoculation experiments with certain plant/AMF associations may be the result of the initial drain on the plant by the fungus in having to establish a new ERM. The energy needed to construct an ERM may differ between AMF and also depend on the ability of the plant itself to supply sufficient carbon. The role of tillage in agro-ecosystems needs further study if sustainable production is our future goal since disturbance of soils may be selecting for AMF which rely on spores or colonised root fragments as their important propagules in colonising new host roots and which are not the most effective at nutrient uptake.

Extraradical mycelium as food?

The ERM of AMF will come into contact with a wide range of other soil organisms in the soil which may lead to a reduction in the effectiveness of the fungal symbiont. Mycophagous Collembola are clearly important in natural ecosystems (McGonigle and Fitter, 1988) but nematodes and bacteria also come into intimate contact with the ERM. Rhizobox experiments could become more adventurous in examining the role of the ERM by including more components of the natural soil biota to examine their effects on the functioning of mycorrhizas.

Quantification of the extraradical mycelium

Methodology used in the generation of data relating to the ERM of AMF was recently covered by Sylvia (1992) where methods for the determination of hyphal lengths were discussed in detail. Since then, however, equipment such as computerised image analysis has become more widely available and can facilitate the data gathering and handling. Recent advances in our knowledge of new techniques and results gained will only be covered here.

Extraction procedures

The methods that are being used to extract hyphae from soils or other growth media are normally based on membrane filtration after blending of the sample (Hanssen et al., 1974; Jakobsen et al. 1992a). Several other techniques have been used to extract the ERM of AMF in different growth media but there has been no comparative assessment of their effectiveness (Vilariño et al. 1993). Green et al. (1994) undertook such a comparison to identify the most appropriate protocol for the presentation of the hyphae for measurement by image analysis. Four extraction techniques were tested. Cores of 25 mm diameter were taken from the hyphal chamber (Fig 1) of the rhizoboxes from regions of 0-3, 3-6 and 6-9 cm away from the plant chamber after 63 days (final harvest). Each core was divided into 4 x 5g sub-samples with each being used in one of the sample preparation procedures. The first two of these (MF1 and MF2) are based on the same membrane filter method (Hanssen et al., 1974) while the other two (EME and SFC) are extraction techniques:

1. A modification of the membrane filter method (MF1), developed by Hanssen et al. (1974), in which sub-samples of 5g were stained in 20 ml of 0.02% trypan blue in lactoglycerol at 100°C for 10 min while stirring. The volume was increased to 200 ml with dH_2O. A 40 ml aliquot was taken using a 50 ml syringe with a 4 mm calibre tip

and poured into a beaker containing 16 ml of dH_2O. This solution was then filtered through 0.45 µm nitro-cellulose filters.

2. An aqueous extraction and membrane filtration technique (MF2), used by Jakobsen et al. (1992a), in which 5g sub-samples were blended at high speed in 150 ml of dH_2O for 30s. The suspension was transferred to an Erlenmeyer flask and agitated for 60s at full speed on a magnetic stirrer. Aliquots of 30 ml were removed and pipetted into 25 µm Millipore filter holders (1.2 µm pore size). The water was removed and 2 ml of 0.05% trypan blue in lactoglycerol was added. After 5 min stain was rinsed off with dH_2O and the filters were transferred to microscope slides for examination.

3. An extraradical mycelium extraction (EME) technique, developed by Vilariño et al. (1993) in which a 5g sample of substrate was deposited in a 500 ml beaker containing 500-600 ml of dH_2O. A framework of six wires of 200 mm in diameter forming a sphere was immersed in the beaker containing the sample and spun at 900-1000 rpm for 10 min in a MSE Centaur bench centrifuge. The frame was gently washed with 120 ml dH_2O into a 200 ml beaker containing 20 ml 0.02% trypan blue and the mixture was then boiled for 10 min and filtered as above (2).

4. A sucrose flotation centrifugation (SFC) technique, developed by Schubert et al. (1987) in which 5g samples were stirred in water for 5 min The mixture was then collected on a 40 µm sieve and washed with tap water. The recovered material was then resuspended in 25 ml of 50% (w/v) sucrose solution in water in a 75 ml centrifuge tube for 2 min. The supernatant was then collected with the suspended mycelium and 20 ml fresh sucrose solution added. This step was repeated 5 times to recover the remaining mycelium. The collected material was then washed in water and resuspended in 25 ml of the sucrose solution and centrifuged in bench centrifuge at 1500 rpm for 3 min. The supernatant was then washed, stained and filtered as above.

Measurement of hyphal lengths

The resulting membrane filters and hyphae from each technique were placed on a microscope slide, covered with a few drops of polyvinyllactoglycerol (PVLG) and observed under a compound microscope at x100 for the grid-line intersect method and at x25 for image analysis. Hyphal length measurements were obtained in two ways:

1. A grid formed by a 1100 x 950 mm rectangle with 7 intercrossed lines incorporated into the eyepiece lens (x10) of a Zeiss Axioscope was used to measure hyphal lengths. Stained hyphae which were angular, aseptate in appearance and thicker than 5 µm were deemed to be hyphae of *Glomus* spp. as described by Vilariño et al. (1993) and based

on earlier research (Bethlenfalvay and Ames 1987). A total of 50 microscope fields was observed at random, counting the intersections between hyphae and the gridline and hyphal lengths were calculated from the formula of Tennant (1975). This is called the grid-line intersect method.

2. All hyphae on filters were measured and values obtained from non-mycorrhizal treatments subtracted from the corresponding mycorrhizal treatments. The image of the hyphae on the slide under the x 2.5 objective was captured with a JVC black and white video camera (UVP CCD 4722-2200/0000) attached to a Zeiss compound microscope and processed as a field of 512 x 512 pixels. The captured image was then processed through a series of steps including binary editing, line thinning and data analysis as described by Morgan et al. (1991). The software package (pc image, Foster & Findlay Associates Ltd., Newcastle-upon-Tyne, UK) and a Synapse framestore (Synoptics Ltd., Cambridge, UK) were used as a basis for the image analysis system.

Table II. Mean hyphal densities (cm·g-1 durite) of *G.geosporum* colonising *Allium porrum* extracted from different zones within the hyphal chamber using four sample preparation procedures, and quantified by microscopical assessment or image analysis after 63 days. The different letters indicate significant differences (LSD, p=0.05) between sample preparation, independent of the method of assessment and only in the 0-3 cm zone of the hyphal chamber. (From Green et al., 1994)

Sample Preparation	Distance from plant chamber	Microscopical Assessment	Image Analysis
MF1 (Hanssen et al. 1974)	0-3cm	31.8b	34.3b
	3-6cm	9.2	10.1
	6-9cm	1.6	1.5
MF2 (Jakobsen et al. 1992)	0-3cm	12.5ab	15.4ab
	3-6cm	5.4	8.2
	6-9cm	2.4	5.2
EME (Vilariño et al. 1993)	0-3cm	8.6a	11.1a
	3-6cm	2.2	2.9
	6-9cm	2.2	1.8
SFC (Schubert et al. 1987)	0-3cm	3.5a	5.3a
	3-6cm	2.1	0.5
	6-9cm	0.1	1.4

Results shown in Table II clearly show that MF1 was the best procedure for the extraction of the ERM from a sand substrate. Another membrane technique, MF2, produced comparable data. The other two methods studied were less effective at extracting hyphae but produced clean mycelium and are probably better for studies of the activity of the mycelium. For clay-based terragreen and fine sand substrates a modification of the MF2 technique was used to eliminate the build-up of particulate matter on membrane filters and thus produce samples clean enough for quantification with image analysis. This modification was found to be as accurate for quantification of extracted hyphae as MF2 but had the advantage of reducing the time taken to measure a single sample slide (Green et al., 1994).

These data suggest that an image analysis system can facilitate the collection of hyphal data with less subjectivity than the manual methods, at a faster rate and is not observer-dependent. The need for an observer to become proficient at discriminating hyphal fragments of AMF from other filamentous fungi, along with the fatigue-inducing nature of the microscopical assessment, makes image analysis a more attractive option where many samples have to be processed. The approach of subtracting the length of hyphae in control treatments from that measured in the inoculated treatments, to estimate the actual length of the ERM produced by AMF in microcosm experiments, is the compromise method of allowing for the presence of non-mycorrhizal hyphae (Green et al., 1994; Jakobsen et al., 1992a).

Using rhizobox systems it has been possible to quantify the development of the ERM of several different AMF under the same growth conditions and on the same host using sequential harvests (Gazey et al., 1992; Green et al., 1994; Jakobsen et al., 1992a). Jakobsen et al. (1992a) showed that the length of hyphae of *Sc.calospora* (WUM 12) extracted declined exponentially with increasing distances from the root up to 11 cm, whereas *A. laevis* (WUM 11) maintained a plateau of constant hyphal density 47 days after transplanting inoculated clover seedlings. A*Glomus sp.* had an intermediate pattern of spread. In a similar study, which

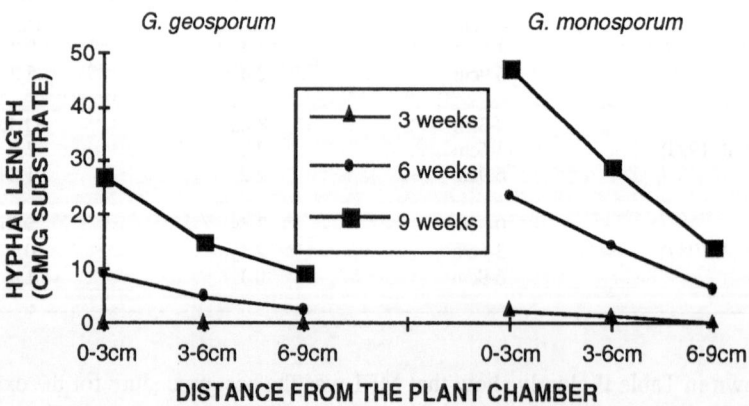

Fig. 3. The development of the extraradical mycelium of *Glomus geosporum* (Kent) and *Glomus monosporum* (Sussex) colonising roots (39% and 41% colonisation, respectively, at 9 weeks) of *Allium porrum* 3,6 and 9 weeks after sowing in rhizoboxes (Adapted from Green et al., 1994).

included a slow release phosphate source at the end of the hyphal chamber (Fig 1), Green et al. (1994) showed that *Glomus geosporum* (Kent) and *Glomus monosporum* (Sussex) produced identical patterns of spread (Fig. 3) and paralleling that observed by Jakobsen et al. (1992) for their *Glomus sp.* Green et al. (1994) also found that *G. monosporum* (Sussex) produced twice as much ERM as *G. geosporum* (Kent) per unit length of mycorrhizal root, even though similar levels of percentages of root length were colonised in the two hosts which may help explain the greater effectiveness of the former AMF under nutrient-poor conditions (Dodd and Jeffries, 1989a,b and Dodd, unpublished data).

Using a different experimental design Gazey et al. (1992) compared the production of ERM by two species of *Acaulospora* on subterranean clover at different harvests (up to 56 days after sowing), using three different levels of inoculum, and showed that *Acaulospora sp.* (WUM 18) continued to increase the length of its ERM up to the last harvest compared with *A. laevis* (WUM 11) which had declined at final harvest (Gazey et al., 1992). These results paralleled the data obtained for colonisation by both AMF in the roots at the same harvest times.

Other techniques

Frey et al. (1993) showed that they could obtain good correlations between the chitin or ergosterol content of the ERM of their *G. intraradices* isolate, colonising red clover, and the hyphal lengths extracted from a sand. The concentration of chitin and ergosterol in the ERM averaged 0.29 ng·m^{-1} and 0.24 ng·m^{-1}, respectively. Considerably higher values of both substances were obtained from colonised roots which indicated that most of the mycorrhizal biomass, for this AMF on this host, was in the root. The potential for using these biochemical parameters to characterise the biomass allocation for individual AMF needs further investigation using different hosts in microcosm experiments.

Assessing the activity of the extraradical mycelium

The previous section highlighted the importance of quantifying the spread of the ERM of individual AMF but more information on the functioning of the ERM, at a specific time, can be gained from either radiolabelling studies (see Jakobsen in chapter 11) or the use of metabolic stains. Several metabolic stains have been used in recent years to assess the activity of the ERM including the reduction of indonitrotetrazolium (INT) to show NADH diaphorase activity (Sylvia, 1988), fluorescein diacetate to reveal esterase activity (Schubert et al., 1987) and the reduction of nitro blue tetrazolium (NBT) to reveal succinate dehydrogenase activity (Hamel et al., 1990). Hamel et al. (1990) showed that FDA gave the most precise estimates of enzyme activity followed by the INT method using an isolate of *G. intraradices*. The percentage of hyphae with activity of both of these enzymes declined over 12 weeks whilst the NBT method showed an increase. The percentage of live intraradical colonisation in the roots of alfalfa or bromegrass declined more rapidly than in the ERM (Hamel et al., 1990). More recently Tisserant et al. (1993) published a technique for staining for fungal alkaline phosphatase (AP) activity in intraradical mycelium and suggested that it could be used to indicate symbiotic

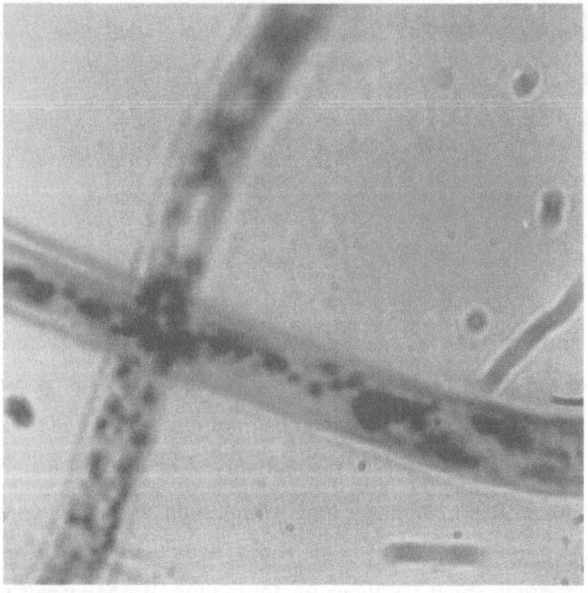

Figure 4. Extraradical hyphae of *Glomus manihotis* (INDO-1) stained for Alkaline Phosphatase (AP) activity. The black precipitate against a clear background of the slide makes quantification by computerised image analysis easy. Data can be presented as the percentage area of the internal cytoplasm occupied by black precipitate (AP activity).

efficiency. Dodd et al. (unpublished data) stained the ERM of three species of *Glomus* colonising leek and found that the method of staining for AP activity was easier and more reproducible than that using INT and was easier to quantify using image analysis given the formation of the black precipitate against a clear background (Fig. 4). Using a combination of these techniques in a rhizobox system the AP activity of *G.geosporum* (Kent) and *G. monosporum* (Sussex) colonising onion was assessed after 60 days. These treatments also received spray applications of three fungicides, at recommended field rates, 42 and 56 days after sowing. Activity was estimated as the percentage area of black precipitate occurring in the cytoplasm of the hyphae measured. The most central hyphae in ten fields of view (x400) per replicate slide were measured for AP activity after extraction of hyphae (Green et al., 1994) from the 0-3 cm zone of the hyphal chamber (Fig. 1). Table III shows that the AP activity detected in the ERM of *G.geosporum* (Kent) was approximately double that of *G.monosporum* (Sussex). Benomyl reduced the AP activity of the ERM of *G.geosporum* (Kent) but not that of *G.monosporum* (Sussex). The effect of benomyl on colonisation of roots was more dramatic (Table III). This combination of image analysis with metabolic staining allows the production of less subjective data and will allow a more thorough evaluation of the effect of different treatments, such as agrochemicals, on the complete fungal biomass in roots and soil.

Table III. The effect of spray application of benomyl on two AMF associated with *Allium cepa* grown in a rhizobox system (see Fig. 1) in a sandy loam soil. Pesticide applications were at recommended field rates at 6 and 8 weeks and data obtained at 9 weeks.

AMF Treatment	Enzyme Activity (S.E.) (% area of AP activity within cytoplasm of hyphae)	% Root Colonisation (S.E.)
Glomus geosporum, Kent		
Control	31 6.6	78 3.3
Benomyl	12 4.9	40 12.9
Glomus monosporum, Sussex		
Control	14 0.3	65 1.5
Benomyl	15 8.0	1 0.3

Identification of extraradical mycelium of AMF

Serological techniques have frequently been proposed as having the potential for detecting the hyphae of AMF in soils (Kough and Linderman, 1986). Immunofluorescence assays have been employed to detect hyphae of AMF using polyclonal antibodies (PAB's) raised against the soluble and particulate fractions of spores. Fungal hyphae were placed on a membrane filter and washed with buffered saline. The PAB was then incubated with the hyphae at 40°C for 1 hour before rinsing in buffered saline. A fluorescein-labelled goat anti-rabbit serum was incubated with the hyphae for 1 hour at 40°C and again washed in buffered saline before observation using ultraviolet illumination at a wavelength of 450-490 nm. Horn et al. (1992) produced a five-step protocol for the isolation of hyphae or spores from pot-culture substrates of two species of *Glomus* which they suggested were sufficiently pure for use as antigens for monoclonal antibody (MAB) (Hahn et al., 1994) or PAB production. Rozycka, Dodd and

Jeffries (unpublished data) recently used similar approaches to raise PAB's against particulate and soluble fractions of spores of *G. geosporum* (Kent) and *Sc. heterogama* (Roth) and the ERM of the former species and *G. manihotis* (INDO-1) (see also Hahn et al. in chapter 3). Antisera raised against spore extracts of *G. geosporum* were able to distinguish mycelium of this fungus from external mycelium of *Gi. margarita* (UPLB, Philippines) but not from mycelium of *G. monosporum* (Sussex). Preliminary work (Table IV) with the antibody (GG8) raised against mycelium of *G. geosporum* has shown that it can be used to discriminate *G.geosporum* mycelium and the ERM of *A. delicata* (EJO1) using a visual estimate of fluorescence on a scale of 0-4 (Table IV). However, mycelium from *G. geosporum* could not

Table IV. Visual estimates of the immunofluorescence (Scale 0-4) recorded for a polyclonal antibody (PAB) raised against spores (GG6) or extraradical mycelium (GG8) of *Glomus geosporum* (Kent) when tested against extraradical hyphae of different AMF (Rozycka, Jeffries and Dodd, unpublished data).

Antiserum	AMF Sample	Fluorescence reading for two dilution's of PAB's	
		1:8	1:32
GG6	*Glomus geosporum* (Kent)	2/3	1/2
	Acaulospora delicata (EJO1)	3	1/2
GG8	*Glomus geosporum* (Kent)	2/3	2
	Acaulospora delicata (EJO1)	1	0/1
	Glomus caledonium (Roth)	3	1/2
	Glomus clarum (Fin7)	2/3	1
Pre-immune		1/2	1

be distinguished from that of *G. clarum* (Fin7) or *G. caledonium* (Roth) using indirect immunofluorescence.

Further work is continuing to investigate the potential for using this PAB in competition studies in rhizobox systems to discriminate the ERM of *G. geosporum* and *A. delicata*. Variations in autofluorescence and in staining reactions have caused problems in accurately assessing the fluorescence which has made reproducible data difficult to obtain. We have concentrated on the use of fluorescein isothiocyanate conjugate in the indirect immuno-fluorescence protocols used but other fluorochromes may prove more useful e.g. B-phyco-erythrin conjugated with streptavidin (Hahn et al., 1993).

Care should be taken in the screening of antisera particularly when the immunofluorescence technique is used. Friese and Allen (1991b) reported the raising of an isolate-specific polyclonal antiserum to spores of *Gi. margarita*. It has since been shown, in a direct immunofluorescence assay, that the spores used from one of the isolates of this fungus were not viable. Thus the antiserum was only genus-specific (Allen pers. comm.). Immunofluorescence as a staining method for discriminating AMF can only be of real value if the antisera show high specificity and can be linked to a system such as a computerised image analysis to allow quantification of fluorescence rather than the observer-dependent grading techniques.

Conclusions

The recent work highlighted here shows how simple rhizobox systems along with **pure** pot-cultures of different species of AMF can be employed to provide more fundamental data on the function of the ERM of AMF at different stages of the development of the mycorrhizal symbiosis. New approaches used in some of the studies e.g. computerised image analysis, metabolic stains, serological identification of AMF as well as the developing molecular techniques will provide us with new data on the ecological function of individual AMF and populations in different ecosystems. Data obtained in the last few years clearly indicate that different AMF may have unique strategies for scavenging the soil matrix for nutrients. The study of mycelial development in rhizobox systems may provide an ideal screening procedure if AMF, with the capacity to produce extensive ERM quickly, are needed for ecosystem restoration projects or for reducing further soil erosion (Sylvia and Burks, 1988). The approach of dissecting the population of AMF from a particular soil by using pure cultures in microcosm experiments should be pursued because there is a generally held belief that all AMF can be grouped together in their functionality. This opinion has arisen chiefly because most studies employ a species of *Glomus*, and no other species from the other genera, in their studies. This is understandable given the relative ease in initiating and maintaining in pure-culture many species of *Glomus* compared with species of *Acaulospora* or *Scutellospora*. It is clear from the above work that we need further information on the activity of the ERM of different species of AMF across the range of morphospecies currently defined. The use of rhizobox systems supplies an experimental unit where both the extra- and intraradical hyphae of AMF can be studied simultaneously. Future work should, therefore, maximise the data to be obtained from such experiments rather than concentrating on one factor alone. The development of the newer technologies, outlined above, will make data acquisition easier and less subjective. The comparison of data between different groups working in this area will then become more reliable and meaningful.

References

Abbott, L.K., Robson, A.D., Jasper, D.A. and Gazey, C. (1992) What is the role of VA mycorrhizal hyphae in soil? In: *Mycorrhizas in Ecosystems* (eds) D.J. Read, D.H. Lewis, A.H. Fitter and I.J. Alexander. CAB International, Wallingford, Oxon, UK. pp. 37-41.

Addy, H.D., Schaffer, G., Miller, M.H. and Peterson, R.L. (1993) Persistence of the extra-radical mycelium of VAM fungi in frozen soil over winter. In: *Abstracts of the International Symposium on Management of Mycorrhizas in Agriculture, Horticulture and Forestry*. University of W. Australia, Perth, W. Australia, 1992.

Bethlenfalvay, G.J. and Ames, R.N. (1987) Comparison of two methods for quantifying extraradical mycelium of vesicular-arbuscular mycorrhizal fungi. *Soil Sci. Soc. Am. J.* 51: 834-837.

Bethlenfalvay, G.J. and Linderman, R.G. (1992) *Mycorrhizae in Sustainable Agriculture*. ASA Special Publication 54, Madison, Wisconsin, USA. 124p.

Birch, C.P.D. (1986) Development of VA mycorrhizal infection in seedlings in semi-natural grassland turf. In: *Physiology and Genetical Aspects of Mycorrhizae* (eds) V. Gianinazzi-Pearson and S. Gianinazzi. INRA, Paris. pp. 233-237.

Brundrett, M. (1991) Mycorrhizas in natural ecosysterms. *Adv. Ecol. Res.* 21: 171-313.

Davies, F.T., Potter, J.R. and Linderman, R.G. (1992) Mycorrhiza and repeated drought exposure affect drought resistance and extraradical hyphae development of pepper plants independent of plant size and nutrient content. *J. Plant Physiol.* 139: 289-294.

Dodd, J.C. and Jeffries, P. (1989a) Effect of herbicides on three vesicular-arbuscular mycorrhizal fungi associated with winter wheat (*Triticum aestivum* L.). *Biol. Fertil. Soils* 7: 113-119.

Dodd, J.C. and Jeffries, P. (1989b) Effect of fungicides on three vesicular-arbuscular mycorrhizal fungi associated with winter wheat (*Triticum aestivum* L.). *Biol. Fertil. Soils* 7:120-128.

Dodd, J.C. and Thomson, B.D. (1994) The screening and selection of inoculant arbuscular mycorrhizal and ectomycorrhizal fungi. *Plant Soil.* 159: 149-158.

Evans, D.G. and Miller, M.H. (1990) The role of the external mycelial network in the effect of soil disturbance upon vesicular-arbuscular mycorrhizal colonization of maize. *New Phytol.* 114: 65-71.

Faber, B.A., Zasoski, R.J. and Munns, D.N. (1991) A method for measuring hyphal nutrient and water uptake in mycorrhizal plants. *Can. J. Bot.* 69: 87-94.

Frey, B. and Schuepp, H. (1992) Transfer of symbiotically fixed nitrogen from berseem (*Trifolium alexandrium* L.) to maize via vesicular-arbuscular mycorrhizal hyphae. *New Phytol.* 122: 447-454.

Frey, B., Vilariño, A., Schüepp, H. and Arines, J. (1993) Chitin and ergosterol content of extraradical and intraradical mycelium of the AM fungus *Glomus intraradices*. Abstract presented at 9th North American Conference on Mycorrhizae. Guelph, Ontario, Canada (1993).

Friese, C.F. and Allen, M.F. (1991a) The spread of VA mycorrhizal fungal hyphae in the soil: Inoculum types and external hyphal architecture. *Mycologia* 83: 409-418.

Friese, C.F. and Allen, M.F. (1991b) Tracking the fates of exotic and local VA mycorrhizal fungi: methods and patterns. *Agric. Eco. Environ.* 34: 87-96.

Gazey, C., Abbott, L.K. and Robson, A.D, (1992) The rate of development of mycorrhizas affects the onset of sporulation and production of external hyphae by two species of *Acaulospora*. *Mycol.Res.* 96: 643-650.

George, E., Haussler, K-U, Vetterlein, D., Gorgus, E. and Marschner, H. (1992) Water and nutrient translocation by hyphae of *Glomus mosseae*. *Can. J. Bot.* 70: 2130-2137.

Giovannetti, M., Avio, L., Sbrana, C. and Citernesi, A.S. (1993a) Factors affecting appressorium development in the vesicular-arbuscular mycorrhizal fungus *Glomus mosseae* (Nicol. & Gerd.) Gerd. & Trappe. *New Phytol.* 123:114-122.

Giovannetti, M., Sbrana, C., Avio, L., Citernesi, A.S. and Logi, C. (1993b) Differential morphogenesis in arbuscular mycorrhizal fungi during pre-infection stages. *New Phytol.* 125: 587-593.

Graham, J.H., Linderman, R.G. and Menge, J.A. (1982) Development of external hyphae by different isolates of mycorrhizal *Glomus* spp. in relation to root colonisation and growth of Troyer citrange. *New Phytol.* 91: 183-189.

Green, D.G., Vilarino, A., Alty, P., Jeffries, P. and Dodd, J.C. (1994) Quantification of mycelial development of arbuscular mycorrhizal fungi using image analysis. *Mycorhiza* (In press).

Gueye, M., Diem, H.G. and Dommergues, Y.R. (1987) Variation in N2 fixation, N and P contents of mycorrhizal *Vigna unguiculata* in relation to the progressive development of extra-radical hyphae of *Glomus mosseae*. *MIRCEN J.* 3: 75-86.

Hahn, A., Bonfante, P., Horn, K., Pausch, F. and Hock, B. (1994) Production of monoclonal antibodies from arbuscular mycorrhizal fungi by an improved immunization and screening procedure. *Mycorrhiza* (In press).

Hamel, C., Fyles, H. and Smith, D.L. (1990) Measurement of development of endomycorrhizal mycelium using three different vital stains. *New Phytol.* 115: 297-302.

Hanssen, J.F., Thingstad, T.F. and Gohsøyr, J. (1974) Evaluation of hyphal lengths and fungal biomass in soil by a membrane filter method. *Oikos* 25:102-107.

Horn, K., Hahn, A., Pausch, F. and Hock, B. (1992) Isolation of pure spore and hyphal fractions from vesicular-arbuscular mycorhizal fungi. *J. Plant Physiol.* 141: 28-32.

Jakobsen, I. and Rosendahl, L. (1990) Carbon flow into soil and external hyphae from roots of mycorrhizal cucumber plants. *New Phytol.* 115: 77-83.

Jakobsen, I., Abbott, L.K. and Robson, A.D. (1992a) External hyphae of vesicular-arbuscular mycorrhizal fungi associated with *Trifolium subterraneum* L. 1. Spread of hyphae and phosphorus inflow into roots. *New Phytol.* 120: 371-380.

Jakobsen, I., Abbott, L.K. and Robson, A.D. (1992b) External hyphae of vesicular-arbuscular mycorrhizal fungi associated with *Trifolium subterraneum* L. 2. Hyphal transport of ^{32}P over defined distances. *New Phytol.* 120: 509-516.

Jasper, D.A., Abbott, L.K. and Robson, A.D. (1989a) Soil disturbance reduces the infectivity of external hyphae of VA mycorrhizal fungi. *New Phytol.* 112: 93-99.

Jasper, D.A., Abbott, L.K. and Robson, A.D. (1989b) Hyphae of a VA mycorrhizal fungus maintain infectivity in dry soil, except when the soil is disturbed. *New Phytol.* 112: 101-107.

Johansen, A., Jakobsen, I. and Jensen, E.S. (1992) Hyphal transport of ^{15}N-labelled nitrogen by a vesicular-arbuscular mycorrhizal fungus and its effect on depletion of inorganic soil N. *New Phytol.* 122: 281-288.

Johansen, A., Jakobsen, I. and Jensen, E.S. (1993a) External hyphae of vesicular-arbuscular mycorrhizal fungi associated with *Trifolium subterraneum* L. 3. Hyphal transport of ^{32}P and ^{15}N. *New Phytol.* 124: 61-68.

Johansen, A., Jakobsen, I. and Jensen, E.S. (1993b) Hyphal transport by a vesicular-arbuscular mycorrhizal fungus of N applied to the soil as ammonium or nitrate. *Biol. Fertil. Soils* 16: 66-70.

Kough, J.L. and Linderman, R.G. (1986) Monitoring extra-matrical hyphae of a vesicular-arbuscular mycorrhizal fungus with an immunofluorescent assay and the soil aggregation technique. *Soil Biol. Biochem.* 18: 307-313.

Li X.-L, George E, Marschner H (1991) Phosphorus depletion and pH decrease at the root-soil and hyphae-soil interfaces of VA mycorrhizal white clover fertilised with ammonium. *Plant Soil* 119: 397-404.

Martins, M.A. (1993) The role of the external mycelium of arbuscular mycorrhizal fungi in the carbon transfer process between plants. *Mycol. Res.* 97: 807-810.

McGonigle, T.P. and Fitter, A.H. (1988) Ecological consequences of arthropod grazing on VA mycorrhizal fungi. *Proc. Royal Soc. Edin.* 94B: 25-32.

Miller, R.M. and Jastrow, J.D. (1990) Hierarchy of root and mycorrhizal fungal interactions with soil aggregation. *Soil Biol. Biochem.* 22: 579-584.

Miller, R.M. and Jastrow, J.D. (1992) The role of mycorrhizal fungi in soil conservation. In: *Mycorrhizae in Sustainable Agriculture.* (eds) G.J. Bethlenfalvay and R.G. Linderman ASA Special Publication 54, Madison, Wisconsin, USA. pp. 29-44.

Morgan, P., Cooper , C.J., Battersby, N.S., Lee, S.A., Lewis, S.T., Machin, T.M., Graham, S.C. andWatkinson, R.J. (1991) Automated image analysis method to determine fungal biomass in soils and on solid matrices. *Soil Biol. Biochem.* 23: 609-616.

Nelsen, C.E. (1987) Water relations of vesicular-arbuscular mycorrhizal systems. In: *Ecophysiology of VA mycorrhizal plants.* (ed) G.R. Safir. CRC Press, Boca Raton, Fla. pp. 71-91.

Newman, E.I. (1988) Mycorrhizal links between plants: their functioning and ecological significance. *Adv. Ecol. Res.* 18: 243-270.

Pearson, J.N and Jakobsen, I. (1993) Symbiotic exchange of carbon and phosphorus between cucumber and three arbuscular mycorrhizal fungi. *New Phytol.* 124: 481-488.

Read, D.J. and Boyd, R. (1986) Water relations of mycorrhizal fungi and their host plants. In: *Water, fungi and plants* (eds) P.G. Ayres and L. Boddy. Cambridge University Press, Cambridge, UK. pp. 287-303.

Schubert, A., Marzachí, C., Mazzitelli, M., Cravero, M.C. and Bonfante-Fasolo, P. (1987) Development of total and viable extraradical mycelium in the vesicular-arbuscular mycorrhizal fungus *Glomus clarum* Nicol. & Schenck. *New Phytol.* 107: 183-190

Schüepp, H., Miller, D.D. and Bodmer, M. (1987) A new technique for monitoring hyphal growth of vesicular-arbuscular mycorrhizal fungi through soil. *Trans. Brit. Mycol. Soc.* 89: 429-435.

Sieverding, E. (1991) *Vesicular-Arbuscular Management in Tropical Agrosystems.* Technical Cooperation (GTZ)-Federal Republic of Germany, Eschborn. 371p.

Sylvia, D.M. (1988) Activity of external mycelium of vesicular-arbuscular mycorrhizal fungi. *Soil Biol. Biochem.* 20: 39-43.

Sylvia, D.M. (1990) Distribution, Structure and Function of external hyphae of vesicular-arbuscular mycorrhizal fungi. In: *Rhizosphere Dynamics* (eds) J.E. Box and L.H. Hammond. Westview Press, Boulder, CO, pp.144-167.

Sylvia, D.M. (1992) Quantification of external hyphae of vesicular-arbuscular mycorrhizal fungi. In: *Methods in microbiology (Techniques for the study of mycorrhiza, vol 24)* (eds) J.R. Norris, D.J. Read, and A.K. Varma Academic press, London, pp. 53-65.

Sylvia, D.M. and Burks, J.N. (1988) Selection of a vesicular-arbuscular mycorrhizal fungus for practical inoculation of *Uniola paniculata*. *Mycologia* 80: 565 - 568.

Tennant, D. (1975) A test of a modified line intersect method of estimating root length. *J.Ecol.* 63: 995-1001.

Thomas, R.S., Franson, R.L. and Bethlenfalvay G.J. (1993) Separation of vesicular-arbuscular mycorrhizal fungus and root effects on soil aggregation. *Soil Sci. Soc. Am. J.* 57: 77-81.

Tisdall, J.M. and Oades, J.M. (1979) Stabliization of soil aggregates by the root segments of rryegrass. *Aust. J. Soil Sci.* 17: 429-441.

Tisserant, B., Gianinazzi-Pearson, V., Gianinazzi, S. and Gollotte, A. (1993) *In planta* histochemical staining of fungal alkaline phosphatase activity for analysis of efficient arbuscular mycorrhizal infections. *Mycol. Res.* 97: 245-250.

Vilariño, A., Arines, J. and Schüepp, H. (1993) Extraction of vesicular-arbuscular mycorrhizal mycelium from sand samples. *Soil Biol. Biochem.* 25: 99-100.

Water relations and alleviation of drought stress in mycorrhizal plants

M. Sánchez-Díaz and M. Honrubia[1]

Departamento de Fisiología Vegetal, Universidad de Navarra, 31080 Pamplona, Spain
[1]Departamento de Biología Vegetal, Laboratorio de Micología, Facultad de Biología, Campus de Espinardo, Universidad de Murcia. 30100 Murcia, Spain

Introduction

Changes in water relations and increased mineral uptake have been the two major reported effects of arbuscular mycorrhizal infection on host plants (Cooper, 1984; Safir, 1987). Some authors have suggested that mycorrhizas may be even more important to plant growth under dry conditions than when soil moisture is plentiful (Allen and Allen, 1986; Nelsen, 1987; Sánchez-Díaz et al., 1990). However, the mechanisms whereby mycorrhizas may increase host drought resistance have not been elucidated. One problem is to distinguish between nutritional advantages and those conferred by improved water uptake. Another important cosideration is that mechanisms which contribute to improved plant water and nutritional status, do not necessarily increase drought tolerance (Nelsen, 1987).

Most experimental work on arbuscular mycorrhizas is carried out under "optimum" conditions but, quite often, field studies fail to validate the dramatic results observed under such conditions (usually glasshouse studies in pot cultures) (Fitter, 1985). This has led many ecologists to suggest that mycorrhizas are not important in the field. Alternatively, as suggested by Allen and Allen (1986), mycorrhizas may have advantageous effects only during stress periods. In their study, effects of mycorrhizas on stomatal resistance were observed only during drought. Allen and Allen (1986) and Allen (1991) have proposed that often the critical mycorrhizal response is during these "ecological crunches" (Weins, 1977) and that these events represent the agents of selection during which mycorrhizas are particularly important to the plant. Therefore, under many natural or man caused stress situations (excess or deficiency of nutrients or water, irradiation, competition, infection by pathogens, toxicity by heavy metal contaminated soils, etc.) (see other reviews in this book and Barea, 1991; Bethlenfalvay and Linderman, 1992; Barea et al., 1993), mycorrhizas can play an important role in the maintenance of plant populations and the yield of crops. In this contribution we will be concentrating on the effects of arbuscular mycorrhizas infection on the plant water relations and the ecology of arbuscular mycorrhizal fungi (AMF) in semi-arid lands.

The water relations of the arbuscular mycorrhizal symbiosis

The effects of mycorrhizal infection on the drought resistance of plants are a separate subject from changes in plant water relations due to mycorrhizas in well-watered plants.

1. Changes in well-watered plants due to mycorrhizal infection Most studies show that mycorrhizal infection results in increased stomatal conductance and transpiration rate in well-watered plants (Fitter, 1988). There are several explanations for this (Cooper, 1984; Nelsen, 1987; Ayres and West, 1993):

a) **Direct flow via hyphae**. Extraradical hyphae may be responsible for the increased water uptake. Hyphae with a diameter of 2-5 μ can penetrate soil pores inaccessible to root hairs (10-20 μ diameter). Total hyphae length can reach 50 m cm^{-3} soil (Allen, 1991). In *Bouteloua gracilis* infected by *Glomus fasciculatum*, leaf resistance was reduced by 50% and transpiration increased by 100% without there being any change in leaf or root water potentials (Allen, 1982). Mycorrhizal and non-mycorrhizal plants had similar leaf areas and root lengths, although roots of infected hosts had fewer and shorter root hairs. Allen estimated that the rate of transport from extraradical hyphae to the root was 2.8 x 10^{-5} mg s^{-1} per entry point, this being sufficient to maintain normal water relations. It has been suggested by others, however, that direct water uptake to such a significant level is not hydraulically possible (Fitter, 1985).

b) **Modified phosphorus content** Altered water relations resulting from AMF infection may be a function of enhanced plant phosphorus status and, thus, a secondary consequence of colonization (Safir et al., 1971, 1972; Nelsen and Safir, 1982). Fitter (1988) showed that uninfected non-fertilized *Trifolium pratense* had low stomatal conductances and leaf phosphorus concentrations, whereas plants subjected to high phosphorus additions, and hence those with greatest leaf phosphorus concentrations, exhibited high stomatal conductances. In contrast, Augé et al. (1986a) observed higher conductance in mycorrhizal rose plants fertilized with low levels of phosphorus than in plants subjected to high phosphorus treatment. They concluded that leaf phosphorus concentrations did not affect conductance but, rather, infection was the important factor. However, as Fitter (1988) observed, the leaf phosphorus concentrations were generally low, despite the level of phosphorus nutrition, and treatments were not sufficiently different from each other to effect a nutritional response. Endophyte infection levels were lower in the high-phosphorus than in the low-phosphorus treatment group justifying the conclusions by Augé et al. (1986a) that increased conductance was correlated with infection levels.

c) **Altered hormonal balance** As water relations can partially be modified by the plant hormonal status, several workers have investigated the role of the mycorrhizas in altering their balance. Allen et al. (1980) examined the effect of arbuscular mycorrhizal infection on *Bouteloua gracilis*, a range grass, grown in axenic culture. They determined that in 50-day-old plants the cytokinin activity was 57 and 111% higher in leaves and roots, respectively, in arbuscular mycorrhizal than in non-mycorrhizal plants. In a further study, Allen et al. (1982) found significantly higher gibberellin activity in leaves and somewhat lower in roots, due to infection. In addition, there were concomitant reductions in abscisic acid levels in the leaves, with only slight changes in the roots. Levy and Krikun (1980) suggested that their observations made on AMF mediated water relations of citrus seedlings could be due to altered hormonal balance. Using the same host, Dixon et al. (1988) demonstrated that AMF infected seedlings had higher cytokinin concentrations in xylem exudate than non infected ones. Edriss et al. (1984) showed that three AMF could greatly increase leaves cytokinin levels in *Citrus aurantium* provided with low and intermediate levels of P, and that one of them also did so at a very high P level. In a similar study, Dixon et al. (1985) examined the concentrations of cytokinins in leaves of *Citrus limon* inoculated with five endophytes. All arbuscular mycorrhizal plants were larger than the controls. In the plants in which two of the endophytes induced the greatest growth, significantly higher zeatin levels were found. Recently, Drüge and Schönbeck (1992) studying the effect of AMF infection on transpiration, photosynthesis and growth of flax (*Linum usitatissimum* L.) in relation to cytokinin levels found that during the beginning of the mycorrhizal infection zeatin riboside levels were temporarily decreased in roots and increased in shoots when compared to non-mycorrhizal plants. However, when the symbiosis had established, colonized roots revealed significantly higher zeatin riboside levels than those of non-mycorrhizal plants. Significant growth responses of shoots and roots due to mycorrhizal infection were preceded by higher zeatin riboside levels in the respective organs. The authors concluded that the enhanced internal cytokinin levels are involved in the improved photosynthesis and growth of mycorrhizal flax.

Although abscisic acid (ABA) is a typical stress hormone, Danneberg et al. (1992) have found that ABA-levels were higher in AMF colonized roots of maize than in non-infected controls. The role of hormones in water relations and in physiology of arbuscular mycorrhizal symbiosis is interesting, although it is to be expected that AMF can increase levels of growth-regulating compounds in plants as many bacteria and even plant pathogenic fungi can do (Ayres and West, 1993). Many pathogens produce auxins, cytokinins, gibberellins and ethylene, so it is not necessarily true that an increase in the concentration of any one compound is beneficial, unless that compound is limiting physiological activity in the plant (Hayman, 1983). Furthermore the ratio of the different compounds may be the important factor controlling growth (Horsfall and Cowling, 1978).

d) Other mechanisms Each of the above mechanisms is probably important at some time and under some circumstances, the balance between them reflecting the dynamic nature of the mycorrhizal association. Indeed, further mechanisms may be operative. Thus, Kothari et al. (1990) suggested that higher water uptake rates in mycorrhizal plants could be a result of anatomical alterations in infected roots; for example, more differentiated metaxylem vessels and changes in root exodermis formation. Interestingly, it has been suggested that increases in the hydraulic conductance of roots from rusted barley and *V. faba* are associated with a reduction in root cortical thickness (Tissera and Ayres, 1988).

2. Drought resistance of mycorrhizal plants AMF may increase drought resistance of plants by means of several mechanisms, including increased water uptake due to hyphal extraction of soil water (Hardie, 1985), increased stomatal sensitivity to leaf-air vapor pressure deficit (Huang et al., 1985), regulated stomatal conductance in response to hormonal signals (Allen et al., 1982), or by lowered leaf osmotic potential for greater turgor maintenance (Augé et al., 1986b).

AMF effect on plant water status during drought has also been associated with improved host nutrition, particularly P (Nelsen and Safir, 1982; Fitter, 1988). However, several authors have reported that drought resistance of mycorrhizal plants is independent of plant P concentration (Sweatt and Davies, 1984; Augé et al., 1986b; Bethlenfalvay et al., 1988; Peña et al., 1988; Sánchez-Díaz et al., 1990). Figure 1 (Peña et al., 1988) shows that during drought, in the *Medicago-Rhizobium-Glomus* symbiosis, nodule activity was significantly higher in infected than in non infected plants; the higher activity could not be explained by improved P uptake caused by the fungus since P-concentrations in mycorrhizal plants were always lower than in phosphorus fertilized plants.

Other factors associated with AMF colonization such as changes in leaf elasticity and increased rooting length and depth (Ellis et al., 1985; Kothari et al., 1990) may also influence drought resistance. Complicating any explanation of increased water fluxes in AMF infected plants under water stress, is the possible interaction with signals that communicate changes in soil water status

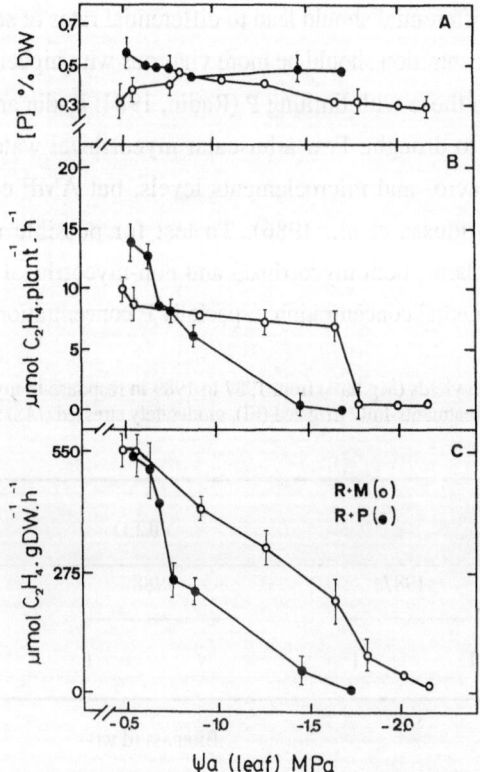

Figure 1. Phosphorus content in percentage of dry weight (A); acetylene reduction activity per plant (B) and per nodule dry weight (C), as a function of leaf water potential. R+M= plants inoculated with *R. meliloti* and *G. mosseae* and irrigated with N-and P-free nutrient solution. R+P= plants inoculated with *R. meliloti* and irrigated with N-free nutrient solution containing P. (From Peña et al., 1988.)

from root to shoot (Zhang and Davies, 1989). Thus, it has been suggested that AMF may be associated with altered levels of abscisic acid (Allen et al., 1982), although firm evidence is lacking. Augé and Duan (1991) have found that roses growing in a split-root system where only one side of the root system was colonized by *Glomus intraradices* and water stressed, while the corresponding side remained uninfected and watered, the stomatal conductance of these plants was reduced to 80% of the control plants (both root halves watered) within 11 days after the beginning of the stress. In contrast, if the whole split-root system was water stressed, only after 17 days the stomatal conductance was reduced below that of the controls in plants. Most significantly, no treatment-related differences in leaf water potential or relative water content were observed during the period of water stress. Although arbuscular mycorrhizal infection may have led to localized differences in soil drying and root water relations, it is possible that infection may have interfered directly with ABA concentrations in root or shoot.

A major problem with regard to many water relations studies has been that AMF infected plants were of different size and tissue P content than non-infected ones. If plants are grown in containers of equal size, a plant size differential should lead to differential rates of soil water depletion. Also, plants with optimum P concentration should be more vigorous with higher photosynthetic rates and stomatal conductances than those with limiting P (Radin, 1984; Radin and Eidentock, 1986), and might respond differently to drought. Few arbuscular mycorrhizal water relations studies have documented leaf tissue macro- and microelements levels, but AMF could influence levels of elements other than P (Bildusas et al., 1986). To test for possible mechanisms of drought resistance in mycorrhizal plants, both mycorrhizal and non-mycorrhizal control plants should be equal in size and tissue elemental concentration, especially P concentration (Davies et al., 1992).

Table I. Maize biomass and grain yields (Mg ha^{-1}) from 1987 to 1989 in response to mycorrhizal inoculation (I) or not (N) and water-management treatments-fully irrigated (FI), moderately stressed (MS) and severely stressed (SS). (From Sylvia et al., 1993.)

Water trt.	YIELD					
	1987		1988		1989	
	N	I	N	I	N	I
	Biomass (d wt)					
SS	8.02	9.84	11.79	12.11	11.31	13.38
MS	14.08	15.68	13.96	14.69	14.44	15.19
FI	19.69	19.91	17.32	17.03	15.77	19.26
Water (W)	**		**		**	
Inoc. (I)	**		NS		**	
W x I	**		NS		**	
	Grain					
SS	3.92	5.22	5.06	5.43	6.78	7.56
MS	7.71	8.95	7.29	8.38	7.90	8.22
FI	10.73	11.11	10.23	10.50	9.11	10.74
Water (W)	**		**		**	
Inoc. (I)	**		NS		**	
W x I	NS		NS		NS	

Elucidating the role of arbuscular mycorrhizas in drought resistance is further complicated by the interaction of mycorrhizal fungi with prolonged or temporary drought (Augé et al., 1986b). Furthermore, most experiments with AMF have been conducted in controlled greenhouse or growth chamber environments. There is relatively sparse information on the function of AMF in field environments (Sylvia and Williams, 1992). Fitter's (1986) field studies suggested that AMF improve the drought resistance of plants. White et al. (1992) showed, in a study in the Red Desert of Wyoming, that irrigation schedule was more important than irrigation rate for enhancing establishment of functional AMF biomass.

Sylvia et al. (1993) conducted field trials, over three seasons, to directly test the effect of AMF on water-stressed corn (*Zea mays* L.). Grain and total above ground biomass yields increased with irrigation and a positive response to mycorrhizal inoculation was constant across irrigation levels for both grain yield and total biomass (Table I). Also, due to the smaller size of water-stressed plants, but a consistent growth response to inoculation across water treatment, the response of corn to AMF inoculation increased with increasing water stress.

Ecological aspects of arbuscular mycorrhizas in semi-arid lands

The occurrence, distribution and ecological importance of AMF have been studied in a variety of plant communities (Daniels Hetrick, 1984; Read, 1991,1992) and particularly in arid and semi-arid lands (Díaz and Honrubia, 1994; Dodd and Krikun, 1984; Ho, 1987; López-Sánchez and Honrubia, 1992; Allen and Allen, 1992; Honrubia et al., unpublished data).

Several *Glomales* species have been reported from semi-arid western Mediterranean area (Roldán-Fajardo, 1985; Salamanca, 1991; Díaz and Honrubia, 1993a; Honrubia et al., unpublished data), which shows a floristic richness of these fungi in xeric conditions. The genus *Glomus* is the best represented in terms of species variability, but *Entrophospora infrequens* (Hall) Ames and Schneider seems to be one of the most widely distributed species in this area (unpublished observations), together with *Glomus geosporum* (Nicolson and Gerdemann) Walker and *G. mosseae* (Nicolson and Gerdemann) Gerdemann and Trappe.

According to Allen and Allen (1992), the spore density and biodiversity of AMF are related to the size of the individual plants, but also, as we have previously demonstrated, a seasonal distribution of the amount of AMF propagules (spores, external mycelium, infected roots) can be observed throughout the year and related to the plant species and communities. In semi-arid Mediterranean areas (Fig. 2) we observed (López-Sánchez and Honrubia, 1992; Díaz and Honrubia, 1994) that although in general, the maximum spore density in the field is reached during the fruit-bearing period of the host plant; it remains high in autumn and falls to a minimum in winter, tending to increase in spring. These observations agree with those of Miller (1978), Reeves et al. (1979) and Allen et al. (1984), who noted similar tendencies in the behaviour patterns of AMF.

On the other hand, root infection reaches its maximum when the plant flowers, and then decreases to a minimum in summer, which agrees with the observations of Rabatin (1979) and Giovannetti (1985). In semi-arid lands, some plant species such as *Anthyllis cytisoides* L. and

174

Salsola genistoides Juss. can sprout and flower in autumn, during which period an increase in root infection is observed (López-Sánchez and Honrubia, 1992).

Figure 2. Landscape of western semi-arid Mediterranean areas

When plants are developed in xeric conditions, the species of the *Fabaceae, Asteraceae* and *Poaceae* have the highest levels of mycorrhizal infection, whereas *Chenopodiaceae* have the lowest. *Anthyllis cytisoides* L. (*Fabaceae*), *Brachypodium retusum* (Pers.) Beauv. and *Lygeum spartum* L. (*Poaceae*), which are widely distributed and well adapted to semi-arid habitats in western Mediterranean area, seem to be very dependent on AMF infection in natural soils (López-Sánchez et al., 1992; Díaz and Honrubia, 1993b). Species of the *Chenopodiaceae* family such as *Salsola vermiculata* L. and *S. genistoides* Juss. behave as mycorrhizal independent or mycorrhizal facultative respectively. The latter species can reach high mycorrhizal infection levels when growing together with another mycorrhizal dependent plant, such as *Artemisia herba-alba* Asso (*Asteraceae*) and in poor soil deficient in P; but while growing alone and in soils with a high P level, *S. genistoides* Juss. only presents traces of mycorrhizal infection (López-Sánchez and Honrubia, 1992). This behaviour seems to be usual in this family, as the observations of Miller et al. (1983) suggest. *Cupressaceae* species which naturally grow in semiarid mediterranean areas, for example, *Tetraclinis articulata,* have also been observed to form arbuscular mycorrhizas (Díaz and Honrubia, 1993c).

Depth distribution of propagules of the AMF in eroded semi-arid lands should be a very common strategy for their survival, and results that we have obtained recently confirm this (Honrubia et al., unpublished data; Molina-Niñirola et al., unpublished data). The study was carried out in four sites in the Thermomediterranean area with a *Chamaeropo humili - Rhamnetum lycioides* O. Bolòs climatic vegetation. Several natural conditions, refering to the orientation, slope and vegetation, in marle soils, were considered. Site 1 had a N-NW orientation, *Pinus halepensis* Mill. plantation of 14 years-old, with a herbaceous community of *Bradrypodium retusum* (Pers.) Beauv. Site 2 had a similar orientation but without tree vegetation and with a very dense cover, over 90%, formed by *Lygeum spartum* L., *Stipa tenacissima* L. and *Thymus hyemalis* Lge. Site 3 had a S orientation, very steep slope and a high erosion degree; the vegetation was formed by *Salsola genistoides* Juss. Finally, site 4 had a N-NE orientation, *Pinus halepensis* Miller plantation of about 30 years-old and with an abundant vegetation cover formed by *B. retusum* (Pers.) Beauv., *Rosmarinus officinalis* L.

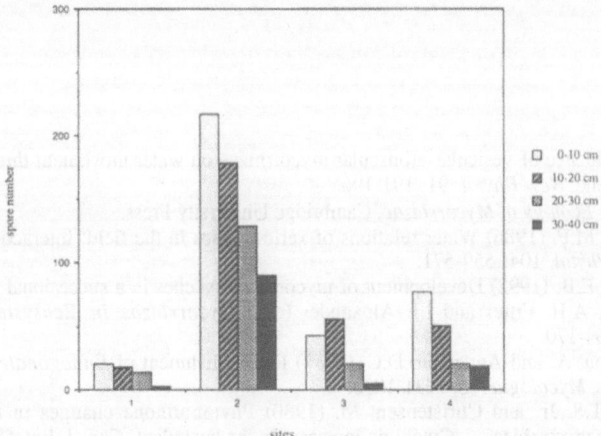

Figure 3. Number of spores of AMF detected at four different depths in four sites in the Thermomediterranean area

Four depths were considered up to 40 cm. In all the sites, spores of AMF were found even at 40 cm depth (Fig. 3). The highest numbers of spores were observed at site 2, which has a very well established dense plant community of camephytes and hemicryptophytes. The number of spores at 40 cm in this site were higher than those obtained in the surface layer of all the other three sites. Biodiversity of AMF was also highest at this site. It is interesting to observe that in site 3 the number of spores found in the surface layer were lower than those between 10 and 20 cm depth. Site 3 had high erosion problems because of the steep slope and a low vegetation cover. These characteristics could explain why the maximum number of spores were found at the second level of depth and not in the surface layer. Its also interesting to note that in sites 1 and 4 with an

176

ectomycorrhizal component, lower number of spores of AMF were found. It is particularly appropriate to compare sites 1 and 2 because of their proximity, orientation and slope similarities. Density of spores in depth and biodiversity of AMF seem to be related to the vegetation cover and the stability of the plant community in semi-arid habitats.

The external mycelium of AMF should play an important role in water uptake of plants, particularly in semi-arid habitats; probably not for assuring high transpiration rates, but for maintaining minimal needs to preserve physiological activities and permit survival (Read, 1992). As George et al. (1992) suggested, the diameter of the AMF hyphae needed to maintain a water transport rate significant to the total plant water uptake should be much greater than that of *Glomus mosseae* (Nicolson and Gerdemann) Gerdemann and Trappe. But water uptake contribution through the external mycelium become progressively more beneficial for the host-plant in water stress conditions (Read and Boyd, 1986; Allen and Allen, 1986).

In conclusion more studies on ecophysiological aspects are needed to understand the role played by the mycelium of AMF in the water uptake process of the plants growing in drought stress conditions.

References

Allen, M.F. (1982) Influence of vesicular-arbuscular mycorrhizae on water movement through *Bouteloua gracilis* (H. B.K.) lag ex Steud. *New Phytol* 91: 191-196.

Allen, M.F. (1991) *The Ecology of Mycorrhizae.* Cambridge University Press.

Allen, E.B. and Allen, M.F. (1986) Water relations of xeric grasses in the field: interactions of mycorrhizas and competition. *New Phytol.* 104: 559-571.

Allen, M.F. and Allen, E.B. (1992) Development of mycorrhizal patches in a succesional arid ecosystem. In: D.J. Read, D.H. Lewis, A.H. Fitter and I.J. Alexander (eds) *Mycorrhizas in Ecosystems.* Cambridge: CAB International, pp. 164-170.

Allen, M.F., MacMahon, A. and Anderson, D.C. (1984) Reestablishment of *Endogonales* on Mount St. Helens: survival of residuals. *Mycologia* 76: 1031-1038.

Allen, M.F., Moore, T.S. Jr. and Christensen, M. (1980) Phytohormone changes in *Bouteloua gracilis* by vesicular-arbuscular mycorrhizae. I. Cytokinin increases in the host plant. *Can .J .Bot.* 58: 371-374.

Allen, M.F., Moore, T.S. Jr. and Christensen, M. (1982) Phytohormone changes in *Bouteloua gracilis* infected by vesicular-arbuscular mycorrhizae. II. Altered levels of gibberellin-like substances and abscisic acid in the host plant. *Can J. Bot.* 60: 468-471.

Augé, R.M. and Duan, X. (1991) Mycorrhizal fungi and non hydraulic root signals of soil drying. *Plant Physiol.* 97: 821-824.

Augé, R.M., Schekel, K.A. and Wample, R.L. (1986a) Greater leaf conductance of well-watered VA mycorrhizal rose plants is not related to phosphorus nutrition. *New Phytol.* 103: 107-116.

Augé, R.M., Schekel, K.A. and Wample, R.L. (1986b) Osmotic adjustment in leaves of VA mycorrhizal and non mycorrhizal rose plants in response to drought stress. *Plant Physiol* 82: 765-770.

Ayres, P.G. and West, H.M. (1993) Stress responses in plants infected by pathogenic and mutualistic fungi. In: L. Fowden, T. Mansfield and J. Stoddart (eds) *Plant Adaptation to Environmental Stress.* London: Chapman and Hall, pp. 295-311.

Barea, J.M. (1991) Vesicular-arbuscular mycorrhizae as modifiers of soil fertility. In: *Advances in Soil Science.* New York: Springer-Verlag, pp. 1-40.

Barea, J.M., Azcón, R. and Azcón-Aguilar, C. (1993) Mycorrhiza and crops. In: *Advances in Plant Pathology.* Academic Press Limited pp. 167-189.

Bethlenfalvay, G.J., Brown, M.S., Ames, R.N. and Thomas, R.S. (1988) Effects of drought on host and endophyte development in mycorrhizal soybeans in relation to water use and phosphate uptake. *Physiol .Plant* 72: 565-571.

Bethlenfalvay, G.J. and Linderman, R.G. (eds.) (1992): *Mycorrhizae in Sustainable Agriculture*. Proceedings of a symposium sponsored by the Soil Science Society of America, American Society of Agronomy and the Crop Science Society of America. Madison, Wisconsin.

Bildusas, I.J., Dixon, R.K., Plfeger, F.L. and Stewart, E.L. (1986) Growth, nutrition and gas exchange of *Bromus inermis* inoculated with *Glomus fasciculatum* . *New Phytol.* 102: 303-311.

Cooper, K.M. (1984) Physiology of VA mycorrhizal associations. In: C. Powell and J. Bagyaraj (eds) *VA Mycorrhiza* . Boca Raton: CRC Press, pp. 155-186.

Daniels Hetrick, B.A. (1984) Ecology of VA mycorrhizal fungi. In: C. Powell and J. Bagyaraj (eds) *VA Mycorrhiza* . Boca Raton: CRC Press, pp. 35-55.

Danneberg, G., Latus, C., Zimmer, W., Hundeshagen, B., Schneider-Poetsch, H.J. and Bothe, H. (1992) Influence of vesicular-arbuscular mycorrhiza on phytohormone balances in maize (*Zea mays* L.). *J. Plant Physiol.* 141: 33-39.

Davies, F.T. Jr., Potter, J.R. and Linderman, R.G. (1992) Mycorrhiza and repeated drought exposure affect drought resistance and extraradical hyphae development of pepper plants independent of plant size and nutrient content. *J Plant Physiol* . 139: 289-294.

Díaz, G. and Honrubia, M. (1994) A mycorrhizal survey of plants growing on mine wastes in Southeast Spain. Arid Soil Research and Rehabilitation 8: 59-68.

Díaz, G. and Honrubia, M. (1993a) Notes on Glomales from Spanish semiarid lands. Nova Hedwigia 57: 159-168.

Díaz, G. and Honrubia, M. (1993b) Respuestas de crecimiento del albardín (*Lygeum spartum* L.) a la inoculación con hongos micorrícicos y a la fertilización fosforada. Cryptogamie, Mycologie 14(2): 117-125.

Díaz, G. and Honrubia, M. (1993c) Arbuscular mycorrhizae on *Tetraclinis articulata* (Cupressaceae): development of mycorrhizal colonization and effect of fertilization and inoculation. *Agronomie* 13: 267-274.

Dixon, R.K., Garret, H.E. and Cox, G.S. (1985) Cytokinins in leaves of mycorrhizal citrus. In: Molina, R. (editor). Proceedings 6th North American Conference on Mycorrhizae. Forest Research Laboratory, Oregon State University Corvallis.

Dixon, R.K., Garret, H.E. and Cox, G.S. (1988) Cytokinins in the root pressure exudate of *Citrus jambhiri* Lush. colonized by vesicular-arbuscular mycorrhiza. *Tree Physiol.* 4: 9-18.

Dodd, J.C. and Krikun, J. (1984) Observations on endogonaceous spores in Negev desert. *Trans. Br. Mycol. Soc.* .82: 536-540.

Drüge, U. and Schönbeck, F. (1992) Effect of vesicular-arbuscular mycorrhizal infection on transpiration, photosynthesis and growth of flax (*Linum usitatissimum* L.) in relation to cytokinin levels. *J. Plant. Physiol.* 141: 40-48.

Edriss, M.H., Davis, R.M. and Burger, D.W. (1984) Influence of mycorrhizal fungi on cytokinin production in sour orange. *J. Am. Soc. Hortic. Sci.* 109: 587-590.

Ellis, J.R., Larsen, H.J. and Boosalis, M.G. (1985) Drought resistance of wheat plants inoculated with vesicular-arbuscular mycorrhizae. *Plant and Soil* 86: 369-378.

Fitter, A.H. (1986) Effect of benomyl on leaf phosphorus concentration in alpine grasslands: a test for mycorrhizal benefit. *New Phytol,* 103: 767-776.

Fitter, A.H. (1985) Functioning of vesicular-arbuscular mycorrhizas under field conditions. *New Phytol.* 99: 257-265.

Fitter, A.H. (1988) Water relations of red clover *Trifolium pratense* L. as affected by VA mycorrhizal infection and phosphorus supply before and during drought. *J. Exp. Bot.* 39: 595-603.

George, E., Häussler, K., Kothari, S.K., Li, X.L. and Marschner, M. (1992) Contribution of mycorrhizal hyphae to nutrient and water uptake of plants. In: D.J. Read, D.H. Lewis, A.H. Fitter and I.J. Alexander (eds) *Mycorrhizas in Ecosystems*. Cambridge: CAB International, pp. 42-47.

Giovannetti, M. (1985) Seasonal variations of vesicular-arbuscular and endogonaceous spores in maritime sand dunes. *Trans. Br. Mycol. Soc.* 84: 678-684.

Hardie, K. (1985) The effect of removal of extraradical hyphae on water uptake by vesicular-arbuscular mycorrhizal plants. *New Phytol.* 101: 677-684.

Hayman, D.S. (1983) The physiology of vesicular-arbuscular endomycorrhizal symbiosis. *Can J Bot* 61: 944-963.

Ho, I. (1987) Vesicular-arbuscular mycorrhizae of halophytic grasses in the Alvord desert of Oregon. *Northwest Science* 61: 148-151.

Horsfall, J.G. and Cowling, E.B. (1978) Plant D isease. III. How Plants Suffer from Disease. New York: Academic Press.

Huang, R.S., Smith, W.K. and Yost, R.S. (1985) Influence of vesicular-arbuscular mycorrhiza on growth, water relations and leaf orientation in *Leucaena leucocephala* (Lam) de Wit. *New Phytol.* 99: 229-243.

Kothari, S.K., Marschner, H. and George, E. (1990) Effect of VA mycorrhizal fungi and rhizosphere microorganisms on root and shoot morphology, growth and water relations in maize. *New Phytol.* 116: 303-311.

Levy, Y. and Krikun, J. (1980) Effect of vesicular-arbuscular mycorrhiza on *Citrus jambhiri* water relations. *New Phytol.* 85: 25-32.

López-Sánchez, E., Díaz, G. and Honrubia, M. (1992) Influence of vesicular-arbuscular mycorrhizal infection and P addition on growth and P nutrition of *Anthyllis cytisoides* L. and *Brachypodium retusum* (Pers.) Beauv *Mycorrhiza* 2: 41-45.

López-Sánchez, E. and Honrubia, M. (1992) Seasonal variation of vesicular-arbuscular mycorrhizal in eroded soils from Southern Spain. *Mycorrhiza* 2: 33-39.

Miller, R.M. (1978) Some occurrences of vesicular-arbuscular mycorrhizae in natural and disturbed ecosystems of the Red Desert. *Can. J. Bot.* 57: 619-623.

Miller, R.M., Moorman, T.B. and Schmidt, S.K. (1993) Interspecific plant association effects on vesicular-arbuscular mycorrhiza occurence in *Atriplex confertifolia. New Phytol.* 95: 241-246.

Molina-Niñirola, C., Cano, A., Honrubia, M., Díaz, G. and Torres, P. Estudios sobre la ecología de micorrizas en el semiárido del Sudeste español. Unpublished data.

Nelsen, C.R. and Safir, G.R. (1982) Increased drought tolerance of mycorrhizal onion plants caused by improved phosphorus nutrition. *Planta* 154: 407-413.

Nelsen, C.h.E. (1987) The water relations of vesicular-arbuscular mycorrhizal systems. In: Safir, G.R. (editor): *Ecophysiology of VA Mycorrhizal Plants.* Boca Raton: CRC Press, pp. 71-91.

Peña, J.I., Sánchez-Díaz, M., Aguirreolea, J. and Becana, M. (1988) Increased stress tolerance of nodule activity in the *Medicago-Rhizobium-Glomus* symbiosis under drought. *J. Plant Physiol.* 133: 79-83.

Rabatin, S.C. (1979) Seasonal and edaphic variation in vesicular-arbuscular mycorrhizal infection of grasses by *Glomus tenuis. New Phytol.* 83: 95-102.

Radin, J.W. (1984) Stomatal responses to water stress and to abscisic acid in phosphorus-deficient cotton plants. *Plant Physiol.* 76: 392-394.

Radin, J.W. and Eidenbock, M.P. (1986) Carbon accumulation during photosynthesis in leaves of nitrogen- and phosphorus-stressed cotton. *Plant Physiol.* 82: 869-871.

Read, D.J. (1991) Mycorrhizas in ecosystems. *Experientia* 47: 376-391.

Read, D.J. (1992) The mycorrhizal mycelium. In: Allen, M.F., (editor): *Mycorrhizal Functioning. An Integrative Plant-fungal Process .* New York: Chapman and Hall, pp. 102-133.

Read, D.J. and Boyd, R. (1986) Water relations of mycorrhizal fungi and their host plants. In: Ayres, P., Boddy, L., (eds) *Water, Fungi and Plants.* Cambridge: Cambridge University Press, pp. 287-303.

Reeves, F.B., Wagner, D., Moorman, T. and Kiel, J. (1979) The role of endomycorrhizae in revegetation practices in the semi-arid west. I. A comparison of incidence of mycorrhizae in severely disturbed natural environments. *Am. J. Bot.* 66: 6-13.

Roldán-Fajardo, B.E. (1985) Micorrizas VA en cultivos arbóreos: almendro, naranjo y olivo. (Ph D. Thesis). 258 pp. Granada: Univ. of Granada.

Safir, G.R. (1987) *Ecophysiology of VA Mycorrhizal Plants.* Boca Raton: CRC Press.

Safir, G.R., Boyer, J.S. and Gerdemann, J.W. (1971) Mycorrhizal enhancement of water transport in soybean. *Science* 172: 581-583.

Safir, G.R., Boyer, J.S. and Gerdemann, J.W. (1972) Nutrient status and mycorrhizal enhancement of water transport in soybean. *Plant Physiol.* 49: 700-703.

Salamanca, P. (1991) Estudio sobre la simbiosis microbio-planta (micorrizas y *Rhizobium*-leguminosas) en la revegetación de suelos en zonas áridas. (Ph D Thesis) 170 pp. Granada: Univ. of Granada.

Sánchez-Díaz, M., Pardo, M., Antolín, M., Peña, J. and Aguirreolea, J. (1990) Effect of water stress on photosynthetic activity in the *Medicago-Rhizobium-Glomus* symbiosis. *Plant Science* 71: 215-221.

Sweatt, M.R. and Davies, F.T. Jr. (1984) Mycorrhizae, water relations, growth and nutrient uptake of geraniums grown under moderately high phosphorus regimes. *J. Amer. Soc. Hort. Sci.* 109: 210-213.

Sylvia, D.M., Hammond, L.C., Bennett, J.M., Haas, J.A. and Linda, S.B. (1993) Field response of maize to a VAM fungus and water management. *Agron. J.* 85: 193-198.

Sylvia, D.M. and Williams, S.E. (1992) Vesicular-arbuscular mycorrhizal and environmental stress. In: G.J. Bethlenfalvay, R.G. Linderman (eds) *Mycorrhizae in Sustainable Agriculture,* pp. 101-124.

Tissera, P. and Ayres, P.G. (1988) Hydraulic conductance and anatomy of roots of *Vicia faba* L plants infected by *Uromyces viciae-fabae* (Pers.) Shroet. *Physiol. Mol. Plant Pathol.* 32: 192-207.

Weins, J.A. (1977) On competition and variable environments. *American Scientist* 65: 590-597.

White, J.A., De Puit, E.J., Smith, J.L. and Williams, S.E. (1992) Vesicular-arbuscular mycorrhizal fungi and irrigated mined land reclamation in Southwestern Wyoming. *Soil Sci. Soc. Am. J.* 56: 1466-1471.

Zhang, J. and Davies, W.J. (1989) Abscisic acid produced in dehydrating roots may enable the plant to measure the water status of the soil. *Plant Cell Environ.* 12: 73-81.

Impact of Arbuscular Mycorrhizas on
Sustainable Agriculture and Natural Ecosystems
S. Gianinazzi and H. Schüepp (eds.)
© 1994 Birkhäuser Verlag Basel/Switzerland

Impact of arbuscular mycorrhizal fungi on plant uptake of heavy metals and radionuclides from soil

K. Haselwandter, C. Leyval[1] and F.E. Sanders[2]

Institut für Mikrobiologie, Universität Innsbruck, Technikerstr. 25, A-6020 Innsbruck, Austria
[1]*Centre de Pedologie Biologique, C.N.R.S., 17, rue N.D. des Pauvres, B.P.5, F-54501 Vandoeuvre-Les-Nancy Cedex, France*
[2]*Department of Pure and Applied Biology, University of Leeds, Leeds LS2 9JT, UK*

Introduction

As a result of environmental pollution, a number of cations are deposited into soil ecosystems where they may remain for a long period of time. Among these elements are heavy metals and radionuclides.

Soil microorganisms are known to play a key role in the mobilization and immobilization of cations, of heavy metals (Birch and Bachofen, 1990a; Wnorowski, 1991) and radionuclides (Bunzl and Schimmack, 1988; Birch and Bachofen, 1990b). Mycorrhizal fungi are ubiquitous soil microorganisms providing a direct physical link between bulk soil and plant root surfaces (Miller and Jastrow, 1992). It is therefore important to assess the influence of these fungi on the mobility of such elements in soil, and on their transfer from the soil into the plant. The arbuscular mycorrhizal effect on heavy metal and radionuclide uptake by plants is illustrated by reference to selected studies.

Heavy metals

Metal compounds transported in the atmosphere originate from natural (terrestrial, marine, volcanic and biogenic) as well as from anthropogenic (combustion, industrial) sources. In soils heavy metals are present (10%) as native content of some soil parent material, but most (90%) arrive in soils by dry and wet atmospheric deposition (e.g. zinc smelters) and as a result of agronomic practices, including fertilizer and sewage sludge application (Adriano, 1986). Some of these metals, such as Cu, Co and Mo, are essential trace elements for animals, microorganisms and plants, while others are not (Cd, Pb). However, at high concentrations all of them become toxic. Their bioavailability and toxicity to microorganims including mycorrhizal fungi, plants and animals is influenced by many factors, particularly pH, temperature, redox potential, cation exchange capacity of the solid phase and competition between ions (Leyval et al., 1994; Schmitt and Sticher, 1991). Metal toxicity levels are also metal specific. For example, Cd and Zn have common concentration ranges in soils of 0.1 - 1 mg kg^{-1} and 3 - 50 mg kg^{-1} dry weight, and maximum tolerable concentrations of 3 mg kg^{-1} and 300 mg kg^{-1}, respectively (Stoeppler, 1991).

Arbuscular mycorrhizal fungi (AMF) commonly occurring in natural as well as agricultural soils are known to enhance plant uptake of nutrients, especially phosphorus (Kothari et al., 1990; Pearson and Jakobsen, 1993). They have also been shown to increase plant uptake of metal ions present at trace concentrations, such as Zn (Faber et al., 1990; Pacovsky, 1986), Cu (Gildon and Tinker, 1983b; Manjunath and Habte, 1988) and Co (Killham, 1985; Rogers and Williams, 1986).

Killham and Firestone (1983) have determined the influence of AMF infection by *Glomus fasciculatum* on heavy metal uptake and growth of the perennial grass *Ehrharta calycina* on a sandy loam treated to simulate acidic and heavy metal depositions. Cu, Ni, Pb, Zn, Fe and Co were applied in simulated acid rain (pH 3.0, 4.0, and 5.6) at deposition rates approximating to those from smelter effluents. Metal concentrations in shoots of mycorrhizal plants were greater than those of non-mycorrhizal controls exposed to simulated acid rain, with differences increasing with increasing acidity and heavy metal content of the treatment solutions.

When Gildon and Tinker (1983a) analyzed the effect of heavy metals on the extent and development of AMF infection of onions by *Glomus mosseae*, they found that the infection intensity was reduced progressively by increasing additions of Cu or Zn. The two metals had surprisingly similar effects. However, clover plants growing on areas which had been heavily contaminated with metal showed levels of infection of around 30%, whereas much lower levels of heavy metal contamination had completely prevented infection in onions in their pot experiments. According to the infection intensity achieved (= percentage of root length colonized), *G. mosseae* isolated from the contaminated soil was much more tolerant of Zn and Cd than the isolate from uncontaminated soil. In addition, the shoot dry weight and P concentration were significantly enhanced by the metal tolerant fungus in comparison to the intolerant strain. This study was one of the first to show that different strains of AMF may have sharply different qualities.

Recently, Leyval et al. (1991) and Weissenhorn et al. (1993) have confirmed the results of Gildon and Tinker (1983a). Soils polluted through atmospheric deposition from a smelter, through application of sewage sludge or soluble salt amendments were found to contain AMF ecotypes more tolerant of Cd than a reference strain of *G. mosseae*, isolated from a non-contaminated soil. This tolerance to Cd was observed at the level of spore germination (Weissenhorn et al., 1993) and at the level of mycorrhizal colonization (Fig. 1).

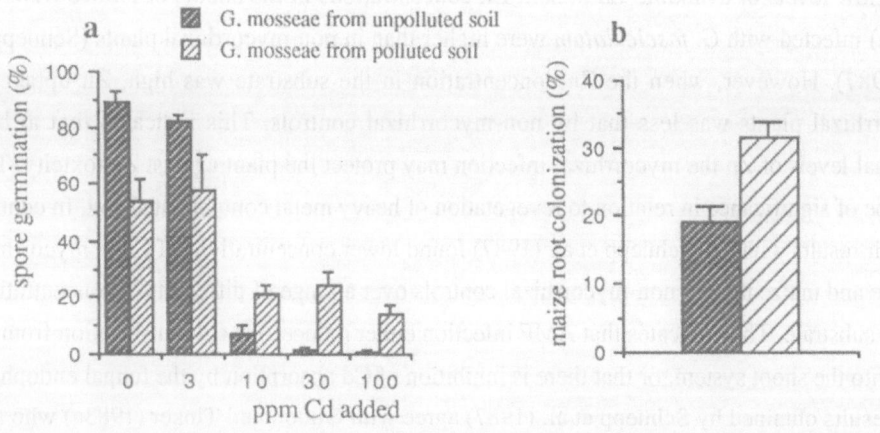

Figure 1. Effect of Cd on mycorrhizal spore germination in sand (a) and on mycorrhizal colonization of maize roots in a Cd polluted soil (b): Comparison of tow *Glomus mosseae* isolates from unpolluted and heavy metal-polluted soil. Maize was inoculated with the same number of spores for both fungi and cultivated for 6 weeks. The polluted soil contained 18 ppm of total and 0.4 ppm of $Ca(NO_3)_2$-extractable Cd. Mycorrhizal colonization of roots was estimated by the method of Trouvelot et al. (1986).

Griffioen et al. (1994) reported high levels of mycorrhizal colonization of heavy metal-tolerant *Agrostis capillaris* growing on soils contaminated by Zn and Cd in the vicinity of a Zn refinery. By contrast, in the area surrounding an old copper mine they found a negative correlation between AMF infection of *A. capillaris* and the total copper content of the soils. Unfortunately, they did not attempt to relate their results to the bioavailability of the metals and the physico-chemical properties of the different soils.

Dueck et al. (1986) examined the effect of *G. fasciculatum* and Zn on two grasses (*Festuca rubra* and *Calamagrostis epigejos*) occurring in coastal dunes downwind of a blast furnace complex and which are becoming increasingly polluted. They found that AMF infection can alleviate the negative effect of Zn on plant growth. As expected, both Zn and AMF influenced plant growth. While Zn inhibited biomass production, AMF infection stimulated the growth of roots and shoots, especially in the case of *F. rubra*. With respect to the growth of roots, both species showed a significant interaction between the effects of Zn and AMF. Apparently, the negative effect of Zn on the growth of roots was alleviated or even absent in the AMF + Zn treatment. The negative effect of Zn on the shoot, however, was not decreased by AMF infection. Zn concentrations in the roots and the shoots of both species were significantly increased by Zn application, but were not reduced by AMF infection. Only in the roots of *F. rubra* was there a significant interaction between the effects of applied Zn and AMF infection. AMF led to an increase in the Zn concentration in the roots when external concentration was low, but had no effect at higher external concentrations.

At low levels of available Zn in soil, Zn concentrations in the shoots of lettuce (*Lactuca sativa*) infected with *G. fasciculatum* were higher than in non-mycorrhizal plants (Schüepp et al., 1987). However, when the Zn concentration in the substrate was high, Zn uptake by mycorrhizal plants was less that by non-mycorrhizal controls. This indicated that at high external levels of Zn the mycorrhizal infection may protect the plant against Zn toxicity. This may be of significance in relation to revegetation of heavy metal contaminated soil. In contrast to their results with Zn, Schüepp et al. (1987) found lower concentrations of Cd in mycorrhizal lettuce and maize than in non-mycorrhizal controls over a range of different Cd concentrations in the substrate. This indicates that AMF infection either reduces the Cd translocation from the root into the shoot system, or that there is inhibition of Cd absorption by the fungal endophyte. The results obtained by Schüepp et al. (1987) agree with Gildon and Tinker (1983a) who also found lower Cd concentration in the shoot and higher Cd tolerance of the host when the clover plants were AMF infected. In soybean, too, AMF could reduce Zn, Cd and Mn concentrations in the leaves when the plants grew in soil with high metal concentrations (Heggo and Angle, 1990).

In a pot experiment using a clay loam soil amended with different amounts of sewage sludge, inoculation with a reference strain of *G. mosseae* did not change maize shoot dry weights. In the pots with high level of heavy metals, Cu, Zn and Pb uptake was lower by mycorrhizal than by non-mycorrhizal plants. Mycorrhizal infection seemed to increase Zn and Cu uptake at low metal concentrations (Fig. 2, from Leyval et al., 1991) as observed previously by Schüepp et al. (1987).

Somewhat different results were obtained in pot experiments using a silty loam soil polluted through atmospheric deposition from a smelter and also limed (Weissenhorn et al., 1993). In such soil indigenous mycorrhizal fungi increased maize growth and significantly decreased in particular Cd, but also Cu, Zn and Mn concentrations in maize roots and shoots (Weissenhorn, 1994). Inoculation of the same soil after gamma-irradiation with a *G. mosseae* reference strain or a Cd tolerant *G. mosseae* isolate had no effect on maize growth and metal uptake by roots and shoots (Weissenhorn, 1994). Difference in plant biomass, which was four times greater in the second experiment due to different light intensities in the growth chamber, might have contributed to the different results. These results show that the influence of AMF on plant

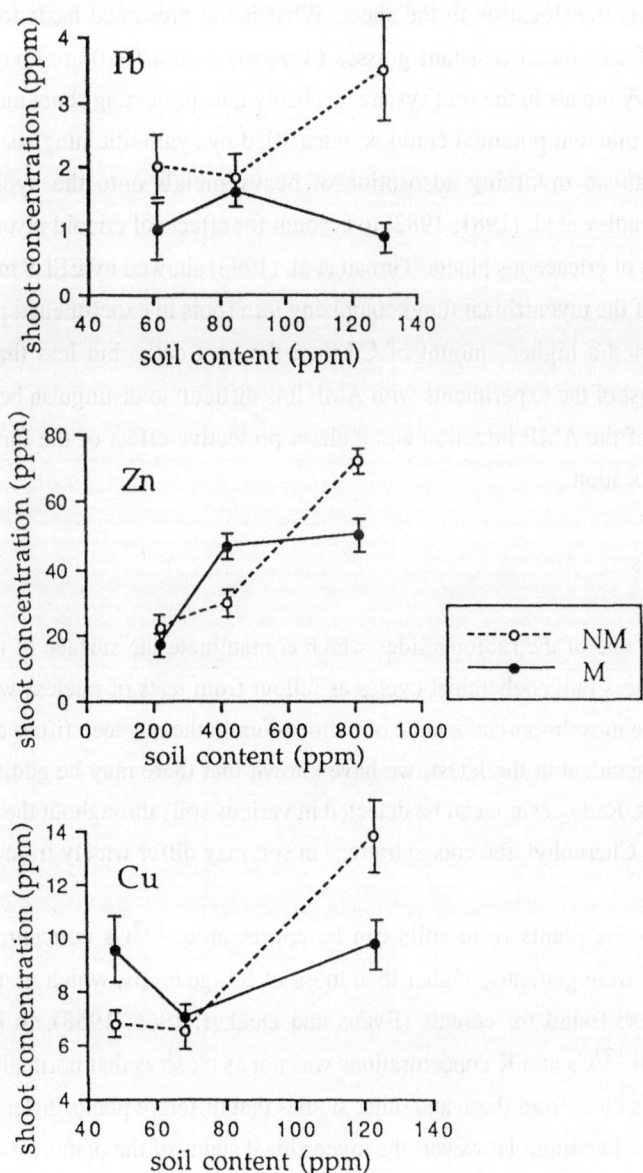

Figure 2. Heavy metal (Cu, Zn and Pb) concentrations of maize shoots (4th, 5th and 6th leaves) grown on a soil amended with three levels of sewage sludge. NM non-mycorrhizal plants, M mycorrhizal plants inoculated with *G. mosseae*. (From Leyval et al., 1991)

uptake depends on plant growth conditions, on the fungal partner, on the metal and its availability in the soils as well as on other edaphic properties, and is therefore difficult to predict.

It is interesting to note that in soils with low metal concentration the AMF infection increased the uptake of an essential micronutrient such as Zn, but decreased the uptake of the highly toxic Cd, or at least its translocation to the shoot. What is the presumed basis for this protective mechanism ? Heavy metal resistant grasses (*Agrostis stolonifera*) are known to be able to accumulate heavy metals in the root system, probably thus protecting shoot metabolism (Wu et al., 1975). This inherent potential could be intensified by symbiotic fungi associated with the roots. A mechanism involving adsorption of heavy metals onto the hyphal surface was suggested by Bradley et al. (1981; 1982) to account for effects of ericoid mycorrhiza on heavy metal resistance of ericaceous plants. Turnau et al. (1993) showed by EELS microanalysis that the cytoplasm of the mycorrhizal fungi colonizing fern roots in experimental plots treated with Cd dust contained a higher amount of Cd than the host cells, but less than the substrate. However, in most of the experiments with AMF it is difficult to distinguish between a biomass dilution effect of the AMF infection and a direct protective effect of the fungi against metal uptake or translocation.

Radionuclides

Radiocesium is one of the radionuclides which contaminate the surface of the earth and the biosphere. It enters biogeochemical cycles as fallout from tests of nuclear weapons, and this was probably the most important source of radiocesium in the nineteen fifties and sixties. Since the Chernobyl accident at the latest, we have known that there may be additional sources of ^{137}Cs and ^{134}Cs. Radiocesium can be detected in various soils throughout the world, although, especially since Chernobyl, the concentrations in soil may differ widely from area to area, and from site to site.

^{137}Cs uptake by plants from soils can be appreciable. ^{137}Cs concentrations found in vegetable crops were generally higher than those of forage crops, which in turn were usually higher than those found for cereals (Evans and Dekker, 1966; 1968). In these plants, the relation between ^{137}Cs and K concentrations was not as close as that normally found between ^{90}Sr and Ca. It is clear from these and other studies that different plants differ in their potential to accumulate radiocesium. However, the mycorrhizal status of the plants investigated was not determined, and a possible influence of mycorrhizal infection on radiocesium uptake was neglected.

A possible interaction between AMF infection and the uptake of radiocesium (^{134}Cs) by grasses was first reported by McGraw et al. (1979). AMF infection of *Paspalum notatum* roots by two of ten species from the genera *Glomus, Gigaspora, Acaulospora* and *Sclerocystis* led to a twofold increase of the radioactivity in leaf tissue 48 h after injection of ^{134}Cs into soil; this indicates that some AMF may enhance the uptake of radiocesium.

Rogers and Williams (1986) studied in more detail the influence of AMF on the uptake of ^{137}Cs and ^{60}Co by *Melilotus officinalis* and *Sorghum sudanense*. While the Co concentration in clover was not significantly different, that of Cs was 2.0 and 1.7 times that in the control at the first (65 days) and second (93 days) harvest. In the case of *Sorghum*, the Cs concentration was greater in the AMF than in the non-mycorrhizal grass, but not significantly; the Co concentration was 2.5 times higher than that in the control at the first harvest (85 days), but did not differ significantly at the second harvest (119 days). In these experiments three different AMF inocula were used, which most likely have influenced the extent to which the radionuclide uptake was affected. The predominance of *Glomus microcarpum* in one inoculum and its complete absence from another indicate that this species may be stimulatory to Cs accumulation by clover. On the other hand, some experiments have shown that AMF infection may have reduced significantly the radiocesium content of *Festuca ovina* (cf. Haselwandter and Berreck, 1994). Three and five weeks after application of ^{137}Cs the shoot tissue radioactivity of mycorrhizal plants was about half that of the non-mycorrhizal controls.

Uptake of Sr also appears to be affected by mycorrhizal infection. When soybean was infected by *G. mosseae* root absorption of ^{90}Sr was enhanced (Jackson et al., 1973).

Concluding remarks and outlook

Fungi can interact with toxic metals in numerous ways (Gadd, 1993; Winkelmann and Winge, 1994). Undoubtedly, heavy metal and radionuclide transfer from soil into plants is mediated by microorganisms including AMF. However, there are still many questions relating to this topic which the studies carried out so far have not been able to answer. These questions should be addressed in future investigations, in the hope of obtaining conclusive answers, and of explaining the many controversial results described in the literature, of which a few have been mentioned above.

A first question which could be raised is: Do species and/or strains of AMF differ in their effects on plant uptake of heavy metals and radionuclides? It is well established that ectomycorrhizal fungi accumulate ^{137}Cs in a species-specific manner, albeit dependent upon the extent to which the biosphere is contaminated (see e.g. Haselwandter et al., 1988; or review by Haselwandter and Berreck, 1994). Heavy metals are also known to be accumulated by fungi, including ectomycorrhizal fungi (see e.g. Gast et al., 1988; Wilkins, 1991) and AMF (Turnau et al., 1993), in a species and heavy metal specific manner. In general, it is not the same species which sequester high levels of radiocesium and heavy metals. For the reasons outlined above it has to be assumed that different AMF species - or even strains of the same species, as pointed out by Gildon and Tinker (1983a) - differ in their efficacy in mediating and regulating heavy metal and/or radionuclide uptake by plants.

A second question is: Does the percentage of root length colonized affect the uptake of heavy metals and/or radionuclides? The infection intensity under which the uptake studies have been carried out was not quantified in all the studies referred to above. However, not only the fungal species forming AMF infection might be crucial in determining its effect upon heavy metal or radionuclide uptake, but also the infection intensity, or the proportion of the root system which is mycorrhizal. Furthermore, vital staining (Gianinazzi and Gianinazzi-Pearson, 1992) which could provide valuable further information on the activity of the AMF within the roots, was not attempted in the studies mentioned above.

To what extent AMF infection may contribute to increase or decrease heavy metal uptake by plants *in situ* under normal agricultural practices is another question. In different field situations with polluted and unpolluted soils, no correlation was observed between levels of mycorrhizal root colonization and soil and plant metal concentrations (Weissenhorn et al., 1994). By contrast, levels of mycorrhizal colonization were well correlated with plant P content and soil pH.

Another point to be raised concerns the host species under investigation. As clearly pointed out by Schüepp et al. (1987), different plant species differ in their potential to accumulate heavy metals. Hence when the effect of arbuscular mycorrhiza on the uptake of metals is being evaluated, the test plant(s) to be used should be carefully chosen.

It is always necessary to consider the level of availability of a given metal when the AMF effect upon its uptake is to be analyzed. In general, the availability of a metal in soil is determined by a number of physico-chemical in addition to biological soil characteristics. Ions of trace elements can be present in soil at concentrations so low that plant growth may be limited. On the other hand, the same elements may be present at concentrations so high as to limit plant growth through toxicity. Different mechanisms may regulate the uptake or exclusion by roots or AMF of a given metal, depending on its external concentration.

This can be illustrated using the example of ericoid mycorrhizal fungi with their potential to produce hydroxamate siderophores. *Hymenoscyphus ericae*, a typical ericoid mycorrhizal fungus, produces ferricrocin as principal siderophore when Fe-availability is low (Haselwandter et al., 1992). When Fe is present at high concentrations, synthesis of the compound ceases (Dobernigg and Haselwandter, 1992). The plant, however, seems still to be protected by mycorrhizal infection against Fe-toxicity (Shaw et al., 1990). Specific metal binding macromolecules (metallothioneins) are produced by plants and certain fungi in the presence of high metal concentrations (Rauser, 1990). This indicates that, at different levels of metal availability, different mechanisms could be governing the interaction between mycorrhizal infection and metal uptake. Such mechanisms may reside in the AMF, the plant or concern the effectiveness of the association between host and fungal symbiont. A significant role for other rhizosphere microorganisms should not be overlooked.

Finally, it must be emphasized that, under natural circumstances, a given metal will be subject to interactions with other cations, an important consideration when analyzing the control exerted by AMF on metal uptake by plants. Such ionic interactions may greatly influence plant uptake of heavy metals and/or radionuclides.

There are clearly enormous difficulties in interpreting the results of experiments on the effects of AMF on the uptake of metals by plants from soil, since these are usually complicated by growth effects due to greater uptake of P by mycorrhizal compared to non-mycorrhizal plants. Growth responses to AMF, including effects on dry matter partitioning within the plant, may lead to lower internal concentrations of a metal in various plant parts due to internal dilution, and therefore reduced toxicity, even if arbuscular mycorrhizas do not directly affect the uptake of that metal. Effects of AMF on P uptake may also mask any beneficial effects of AMF in enhancing the uptake of trace elements when their availability is low.

An experimental methodology which avoids some of these difficulties involves measurement in mycorrhizal and non-mycorrhizal plants of rates of metal uptake expressed per unit of root biomass or length. Such measurements allow direct assessment of mycorrhizal effects on acquisition. Determination of the rates at which the metals are accumulated in the root or translocated to the shoot can reveal any preferential retention in the fungal tissue. To calculate rates involves measurements at intervals over a period of time, leading to larger and more complicated experiments, but the present authors would argue these are essential if, at least, some of the current confusion is to be resolved.

In conclusion, all the mechanisms discussed above may have practical implications, for example with regard to limitation of plant growth, when heavy metals are present either in trace concentrations or at high and hence toxic concentrations. In both cases, AMF infection may also play a key role with regard to the potential of a given plant species to colonize a particular habitat. Furthermore, knowledge of the regulation of cation uptake by roots and AMF, in particular of radionuclides, may and will help to control the transfer of radionuclides, such as ^{137}Cs, from the soil into the plant, and hence through the food chain. At present, it appears impossible to draw firm general conclusions or to predict the precise role of a given AMF in a specific soil with regard to the mobilization or immobilization of a toxic metal. This may be because the information on a possible impact of AMF on ion uptake by plants is based on experiments with a very limited number of taxa of both AMF and host plants. This is a feature consistent with other areas of mycorrhiza research (Klironomos and Kendrick, 1993).

References

Adriano, D.C. (1986) *Trace Elements in the Terrestrial Environment* Springer, New York, USA, pp 533.

Birch, L.D. and Bachofen, R. (1990a) Complexing agents from microorganisms. *Experientia* 46: 827-834.

Birch, L.D. and Bachofen, R. (1990b) Effects of microorganisms on the environmental mobility of radionuclides. In: J.M. Bollag and G. Stotzky (eds):*Soil Biochemistry,* Vol. 6. Marcel Dekker, New York, pp. 483-527.

Bradley, R., Burt, A.J. and Read, D.J. (1981) Mycorrhizal infection and resistance to heavy metal toxicity in *Calluna vulgaris. Nature* 292: 335-337.

Bradley, R., Burt, A.J. and Read, D.J. (1982) The biology of mycorrhiza in the Ericaceae. VIII. The role of mycorrhizal infection in heavy metal resistance. *New Phytol.* 91: 197-209.

Bunzl, K. and Schimmack, W. (1988) Effect of microbial biomass reduction by gamma-irradiation on the sorption of ^{137}Cs, ^{85}Sr, ^{139}Ce, ^{57}Co, ^{109}Cd, ^{65}Zn, ^{103}Ru, ^{95m}Tc and ^{131}I by soils. *Radiat. Environ. Biophys.* 27: 165-176.

Dobernigg, B. and Haselwandter, K. (1992) Effect of ferric iron on the release of siderophores by ericoid mycorrhizal fungi. In: D.J. Read, D.H. Lewis, A.H. Fitter and I.J. Alexander (eds) *Mycorrhizas in Ecosystems.* CAB International, Wallingford, Oxon, pp. 252-257.

Dueck, Th.A., Visser, P., Ernst, W.H.O. and Schat, H. (1986) Vesicular-arbuscular mycorrhizae decrease zinc-toxicity to grasses growing in zinc-polluted soil. *Soil Biol. Biochem.* 18: 331- 333.

Evans, E.J. and Dekker, A.J. (1966) Plant uptake of Cs-137 from nine Canadian soils. *Can. J. Soil Sci.* 46: 167-176.

Evans, E.J. and Dekker, A.J. (1968) Comparative Cs-137 content of agricultural crops grown in a contaminated soil. *Can. J. Plant Sci.* 48: 183-188

Faber, B.A., Zazoski, RJ., Burau, R.G. and Uriu, K. (1990) Zinc uptake by corn as affected by vesicular-arbuscular mycorrhizae. *Plant and Soil* 129: 121-130.

Gadd, G.M. (1993) Interactions of fungi with toxic metals. *New Phytol.* 124: 25-60.

Gast, G.H., Jansen, E., Bierling, J. and Haanstra, L. (1988) Heavy metals and their relationship with soil characteristics. *Chemosphere* 17: 798-799.

Gianinazzi, S. and Gianinazzi-Pearson, V. (1992) Cytology, histochemistry and immunocytochemistry as tools for studying structure and function in endomycorrhiza. In: J.R. Norris, D.J. Read and A.K. Varma (eds) *Methods in Microbiology,* Vol. 24, Academic Press, London, pp. 109-139.

Gildon, A. and Tinker, P.B. (1983a) Interactions of vesicular- arbuscular mycorrhizal infection and heavy metals in plants. I. The effects of heavy metals on the development of vesicular- arbuscular mycorrhizas. *New Phytol.* 95: 247-261.

Gildon, A. and Tinker, P.B. (1983b) Interactions of vesicular- arbuscular mycorrhizal infection and heavy metals in plants. II. The effect of infection on uptake of copper. *New Phytol.* 95: 263- 268.

Griffioen, W.A.J., Ietswaart, J.H. and Ernst, W.H.O. (1994) Mycorrhizal infection of an Agrostis capillaris population on a copper contaminated soil. *Plant and Soil* 158: 83-89.

Haselwandter, K. and Berreck, M. (1994) Accumulation of radionuclides in fungi. In: G. Winkelmann and D. Winge (eds) *Metal Ions in Fungi,* Marcel Dekker, New York, pp. 259-278.

Haselwandter, K., Berreck, M. and Brunner, P. (1988) Fungi as bioindicators of radiocaesium contamination: Pre- and post- Chernobyl activities. *Trans. Br. mycol. Soc.* 90: 171-174.

Haselwandter, K., Dobernigg, B., Beck, W., Jung, G., Cansier, A. and Winkelmann, G. (1992) Isolation and identification of hydroxamate siderophores of ericoid mycorrhizal fungi. *BioMetals* 5: 51- 56.

Heggo, A. and Angle, J.S. (1990) Effects of vesicular-arbuscular mycorrhizal fungi on heavy metal uptake by soybeans. *Soil Biol. Biochem.* 22: 865-869.

Jackson, N.E., Miller, R.H. and Franklin, R.E. (1973) The influence of vesicular-arbuscular mycorrhizae on uptake of 90Sr from soil by soybeans. *Soil Biol. Biochem.* 5: 205-212

Killham, K. (1985) Vesicular-arbuscular mycorrhizal mediation of trace and minor element uptake in perennial grasses: relation to livestock herbage. In: A.H. Fitter, D. Atkinson, D.J. Read and M.B. Usher (eds) *Ecological Interaction in soil: Plants, Microbes and Animals.* Blackwell, Oxford, pp 225-232.

Killham, K. and Firestone, M.K. (1983) Vesicular-arbuscular mycorrhizal mediation of grass response to acidic and heavy metal depositions. *Plant and Soil* 72: 39-48.

Klironomos, J.N. and Kendrick, W.B. (1993) Research on mycorrhizas: Trends in the past 40 years as expressed in the 'MYCOLIT'database. *New Phytol.* 125: 595-600.

Kothari, S.K., Marschner, H. and Romheld, V. (1990) Direct and indirect effects of VA mycorrhizal fungi and rhizosphere microorganisms on acquisition of mineral nutrients by maize (Zea mays L.) in a calcareous soil. *New Phytol.* 116: 637-646.

Leyval, C., Berthelin, J., Schontz, D., Weissenhorn, I. and Morel, J.L. (1991) Influence of endomycorrhizas on maize uptake of Pb, Cu, and Cd applied as mineral salts and sewage sludge. In: J.G. Farmer (ed) *Heavy Metals in the Environment*, CEP Consultants Ltd., pp. 204-207.

Leyval, C., Weissenhorn, I., Glashoff, A. and Berthelin, J. (1994) Influence des metaux lourds sur la germination des spores de champignos endomycorrhizien a arbuscules dans les sols. *Acta Botanica Gallica* (in press).

Manjunath, A. and Habte, M. (1988) Development of vesicular- arbuscular mycorrhizal infection and the uptake of immobile nutrients in Leucaena leucocephala. *Plant and Soil* 106: 97-103.

McGraw, A.-C., Gamble, J.F. and Schenck, N.C. (1979) Vesicular- arbuscular mycorrhizal uptake of cesium-134 in two tropical pasture grass species. *Phytopathology* 69: 1038.

Miller, R.M. and Jastrow, J.D. (1992) The application of VA mycorrhizae to ecosystem restoration and reclamation. In: M.F. Allen (ed) *Mycorrhizal Functioning*, Chapman and Hall, New York, pp. 438-467.

Pacovsky, R.S. (1986) Micronutrient uptake and distribution in mycorrhizal of phosphorus fertilized soybean. *Plant and Soil* 95: 379- 388.

Pearson, J.N. and Jakobsen, I. (1993) The relative contribution of hyphae and roots to phosphorus uptake by arbuscular mycorrhizal plants, measured by dual labelling with ^{32}P and ^{33}P. *New Phytol.* 124: 489-494.

Rauser, W.E. (1990) Phytochelatins. *Ann. Rev. Biochem.* 59: 61-86.

Rogers, R.D. and Williams, S.E. (1986) Vesicular-arbuscular mycorrhiza: Influence on plant uptake of cesium and cobalt. *Soil Biol. Biochem.* 18: 371-376.

Schmitt, H.W. and Sticher, H. (1991) Heavy metal compounds in the soil. In: E. Merian (ed) *Metals and their Compounds in the Environment*, VCH Verlagsgesellschaft Weinheim, Germany, pp. 312- 331.

Schüepp, H., Dehn, B. and Sticher, H. (1987) Interaktionen zwischen VA-Mykorrhizen und Schwermetall-belastungen. *Angew. Botanik* 61: 85-96.

Shaw, G., Leake, J.R., Baker, A.J.M. and Read, D.J. (1990) The biology of mycorrhiza in the *Ericaceae*. XVII. The role of mycorrhizal infection in the regulation of iron uptake by ericaceous plants. *New Phytol.* 115: 251-258.

Stoeppler, M. (1991) Cadmium. In: E. Merian (ed) *Metals and their Compounds in the Environment*, VCH Verlagsgesellschaft Weinheim, Germany, pp. 803-851.

Trouvelot, A., Kough, J.L. and Gianinazzi-Pearson, V. (1986) Mesure du taux de mycorhization VA d'un système radiculaire. Recherche de méthodes d'estimation ayant une signification fonctionnelle. In: V. Gianinazzi-Pearson and S. Gianinazzi (eds) *Physiological and Genetical Aspects of Mycorrhizae*, INRA, Paris, pp. 217-221

Turnau, K., Kottke, I. and Oberwinkler, F. (1993) Element localization in mycorrhizal roots of *Pteridium aquilinum* (L.) Kuhn collected from experimental plots treated with cadmium dust. *New Phytol.* 123: 313-324.

Weissenhorn, I. (1994) Arbuscular mycorrhizae in heavy metal polluted soils: metal tolerance and role in metal transfer to plants. PhD thesis, Universite de Nancy I, 166 pp.

Weissenhorn, I., Leyval, C. and Berthelin, J. (1993) Cd-tolerant arbuscular mycorrhizal (AM) fungi from heavy-metal polluted soil. *Plant and Soil* 157: 247-256.

Weissenhorn, I. Leyval, C. and Berthelin, J. (1994) Bioavailability of heavy metals and arbuscular mycorrhiza (AM) in a soil polluted by atmospheric deposition from a smelter. *Biol. Fertil. Soils* (submitted).

Wilkins, D.A. (1991) The influence of sheathing (ecto-)mycorrhizas of trees on the uptake and toxicity of metals. *Agric. Ecosyst. Environ.* 35: 245-260.

Winkelmann, G. and Winge, D. (1994) *Metal Ions in Fungi.* Marcel Dekker, New York.

Wnorowski, A.U. (1991) Selection of bacterial and fungal strains for bioaccumulation of heavy metals from aqueous solutions. *Wat. Sci. Tech.* 23: 309-318.

Wu, L., Thurman, D.A. and Bradshaw, A.D. (1975) The uptake of copper and its effect upon respiratory processes of roots of copper tolerant and non-tolerant classes of Agrostis stolonifera. *New Phytol.* 75: 225-229.

Biocontrol of plant pathogens using arbuscular mycorrhizal fungi

J.E. Hooker, M. Jaizme-Vega[1] and D. Atkinson[2]

Soil Biology Unit, Department of Land Resources, SAC, Mill of Craibstone, Aberdeen, AB2 9TQ, UK
[1]CITA, Departmento Protección Vegetal, Apartado 60, La Laguna, E-38080, Tenerife, Islas Canarias
[2]SAC, West Mains Road, Edinburgh, EH9 3JG, UK

Introduction

Mycorrhizal fungi colonise the roots of over 90% of plant species, to the mutual benefit of both the plant host and fungus. These symbiotic fungi are present in most terrestrial ecosystems and play a major role in both the growth of plants and important ecosystem processes. Although there are several different types by far the most common are the arbuscular mycorrhizas which are formed by most plant species, including the majority of commercially important crop and horticultural plants. The hyphae of the symbiotic fungi penetrate roots of susceptible plants forming specialised structures known as arbuscles, and also sometimes vesicles; mycorrhizal hyphae form an extensive mycelium within and outside the root which extends into the soil, often several centimetres. These hyphae often bear spores but in some species of arbuscular mycorrhical fungi (AMF) these often also occur intra-radically. Each structure has a different function. Arbuscles are the structures where exchanges of carbon to the fungus and nutrients and water to the host plant take place and vesicles are considered to be primarily for storage. The external mycelium serves as an acquisition and transfer system and the spores a means of survival. The association between the AMF and plants thus brings major advantages to both partners. The fungus obtains a source of carbon derived from plant photosynthesis and so does not have to compete for scarce rhizophere carbon in order to survive and proliferate and the host plant receives several benefits.

Without doubt, the most well known beneficial consequence of the extensive interactions which occur between AMF and their host plants is improved plant phosphate nutrition. The responses of plants, deficient in phosphorous, to inoculation and subsequent colonisation are well documented with very significant effects on plant growth and phosphate accumulation reported (see Harley and Smith, 1983 and Jakobsen in this volume). Effects on phosphate uptake are particularly pronounced because it is rapidly fixed in soils and thus immobilised and unavailable to plant roots.

The benefits of colonisation by AMF are though not limited to phosphorous nutrition alone. Other benefits reported are improved nitrogen (Barea et al., 1991) and trace element (Cooper and Tinker, 1978) content, increased tolerance to drought (Allen and Allen, 1986), reduced heavy metal uptake (Haselwandter et al., in this volume), effects on soil structure (Sutton and Sheppard, 1976). In addition it has been suggested that colonisation may increase resistance to plant pathogens. The extent to which this occurs is the subject of this paper.

Research on the potential for adding organisms to or manipulating organisms already present in agricultural systems in order to enhance a plant's protection against pathogens ie biological control, has received increased attention in recent years. The main driving force for this attention is the increasing public, and in some countries governmental, concern that current chemical practices used in intensive agricultural systems are harmful to the environment and unsustainable. The aim of this paper is to review the evidence from existing research for the role of AMF in the protection of plants against pathogens and suggest practical methods for managing systems in order to optimise their benefits.

Biological control by AMF

The role of AMF in the control of plant pathogens has been the subject of several reviews (e.g. Schenck, 1987; Jalali and Jalali, 1991; Caron, 1989; Sharmaet al., 1992). What is apparent from

Table I: Influence of colonisation by AMF on disease caused by fungal pathogens in plants

Pathogen	Host plant	Effect	Reference
Aphanomyces eutieches	Pea	-	Rosendahl (1985)
Fusarium oxysporum	Onion	-	Dehne (1982)
Phytopthora	Citrus	+	Davis and Menge (1980)
Fusarium solani (1983)	Soybean	-	Zambolim and Schenck
Phytopthera cinnamomic	Avocado	-	Davis et al (1978)
Sclerotium rolfsii	Peanut	-	Krishna and Bagyaraj (1983)
Thielaviopsis basicola (1975)	Tobacco	-	Baltruschat and Schönbeck

- indicates a reduction in disease	+ indicates an increase in disease

Table II: Influence of colonisation by AMF or nematode infection of plants

Nematode	Host plant	Effect	Reference
Paratylenchus penetrans	Pigeon pea	+	Elliott et al (1984)
Rotylenchus reniformis (1982)	Tomato	-	Sitaramaiah and Sikora
Paratylenchus brachyurus (1982)	Cotton	-	Hussey and Roncardori
Meloidogyne hapla	Onion	-	MacGuidwin et al (1985)
Meloidogyne incognita (1987)	Tamarillo		Cooper and Grandisum

\-   indicates a reduction in disease   +   indicates an increase in disease

these and our own review of the literature is that there is no single response and the degree of protection offered by the symbiosis, if at all, varies upon plant species, AMF species and cultural and environmental conditions (see for example references cited in Tables I and II). However, in the great majority of cases a reduction in disease is reported.

It is possible to summarise the evidence obtained from research to date as follows:

1   Colonisation by AMF may offer protection against soil-borne fungal pathogens.
2   Infection of roots by nematodes may be reduced when they are colonised by AMF.
3   Increased nutrient concentrations in the leaves of AMF colonised plants may lead to enhanced development but not increased incidence of foliar pathogens.

The available evidence is therefore that AMF can provide protection against pathogens.

Mechanisms of action

The mechanisms by which AMF control plant pathogens are not well understood. One of the problems is identifying from among the many changes which occur to plants as a result of parasitic colonisation those which influence resistance to pathogens. These changes include nutrient status, partitioning of photosynthate to roots, distribution of growth, the formation of new chemical compounds, altered root anatomy, modified root exudation and root system morphology.

Single mechanisms are often sought in order to explain observable events but in reality in biological systems such as these we are considering it is more likely that a range of different factors are responsible, each providing a contribution to the observed control. Arbuscular mycorrhizal effects will also be influenced by plant growing conditions which will themselves influence infection by the pathogen. Unlike chemical control the effects of biological control will be determined by interactions between a wide range of factors. In addition, the relative importance of these factors will change depending upon the plant pathogen, the AMF species and the environmental conditions prevailing at a specific point in time. The major factors which seem likely to be involved are:

Nutrient status: This is usually enhanced in arbuscular mycorrhizal plants and studies by both Davis (1980) and Graham and Menge (1982) suggested that the control of soil-borne pathogens by AMF could be replaced by adding phosphate. However, the study of Caron et al (1986), comparing the effects of both colonisation and added phosphate on the development of a soil-borne pathogen, suggests a mechanism independent of phosphorous status.

Biochemical changes in plant tissues: AMF have been reported to increase concentrations of fungal chitinases within roots (Dehne et al., 1978). High levels of amino acids, especially arginine, which occur in arbuscular mycorrhizal plants have been found to suppress sporulation in *Thielaviopsis* (Baltruschat and Schönbeck, 1975). Increases in peroxidases and phytoalexins have also been reported for arbuscular mycorrhizal plants and it has been suggested that this confers resistance (Morandi et al, 1984). In addition is is known that there are proteins which are synthesised only in roots colonised by AMF (Schellenbaum et al., 1992; Berta et al.,1994).

Anatomical changes: Increased lignification of root endodermal cells as a direct result of colonisation has been demonstrated (Dehne and Schönbeck, 1979) and it is possible that his makes penetration by the pathogen more difficult.

Stress alleviation: Arbuscular mycorrhizal plants are better able to tolerate abiotic stress conditions such as nutrient shortage, drought and heavy metals. These stress conditions often predispose plants to disease. By preventing such stress they may also increase the resistance of the plant to pathogens.

Microbial changes in the rhizosphere: There is considerable evidence that microbial changes occur in the rhizosphere of plants as a result of colonisation. Changes occur in both the composition and size of rhizosphere microbial populations (eg Ames et al, 1984; Meyer and Linderman, 1986) with effects dependent on AMF species (Secilia and Bagyaraj, 1987). Direct effects of soil leachates derived from the rhizosphere soil of arbuscular mycorrhizal plants on the morphogenesis of fungal pathogens have also been reported. Meyer and

Linderman (1986) identified significant reductions in sporangium and zoospore numbers in the presence of leachates derived from the rhizosphere soil of plants colonised by *Glomus fasciculatum* compared to those present when in contact with leachates from non-colonised plants. It is not clear from this study, however, whether the active agent(s) were derived from the plant roots, AMF or microorganisms in the rhizosphere or mycorrhizosphere. Further studies by Caron et al. (1986) have shown reductions in pathogen populations in the mycorrhizosphere of tomatoes, and an associated reduction in host plant disease due to *Fusarium*.

Induced changes to root system morphology: The very extensive changes to root system morphology usually induced by colonisation have not usually been considered in the context of plant protection. Although variable effects have been reported the most frequent consequence of colonisation is a large increase in branching, resulting in a root system containing a relatively larger proportion of higher order roots (eg Berta et al., 1990; Hooker et al., 1992; Tisserant et al., 1992; Berta et al., 1994) and a reduction in root system

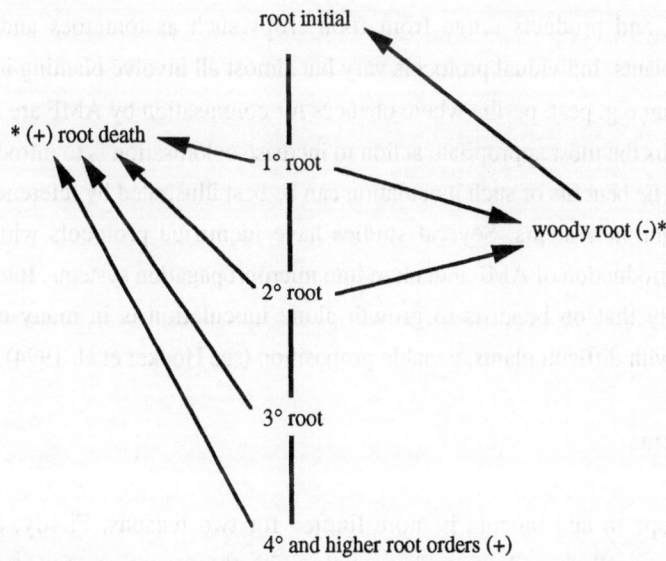

Figure 1. Effect of colonisation on root morphology and likely influence on susceptibility to root pathogens. Root length (+) increased or (-) decreased in mycorrhizal plants. * likely to influence susceptibility to pathogens.

longevity (Hooker and Atkinson, unpublished data). Given the magnitude of the changes induced there is likely to be a significant impact on infection by and disease progression of soil-borne pathogens (Fig. 1).

Appropriate management

The existing and growing evidence for the involvement of AMF in providing protection against plant pathogens is important in the current political and ethical environment and the associated demands for a reduction in the application of chemicals to agricultural systems. Although further research is vital to improve our understanding of the role of the symbiosis in plant protection it is possible to make some management recommendations using existing knowledge.

It is possible to place agricultural plant production systems into two distinct categories, either closed (such as in horticultural systems) and open within which the majority of arable crops are produced. Both provide different opportunities for optimising the arbuscular mycorrhizal symbiosis.

Closed systems

These involve the propagation of either seed, cutting or micropropagated plantlets in glasshouses, and products range from food crops such as tomatoes and cucumber to ornamental plants. Individual protocols vary but almost all involve planting into a relatively sterile medium e.g. peat, perlite where chances for colonisation by AMF are small. In these closed systems the most appropriate action to increase colonisation is to introduce AMF into the system. The benefits of such inoculation can be best illustrated by reference to studies of micropropagation systems. Several studies have identified protocols which permit the successful introduction of AMF inoculum into micropropagation systems. It has been shown unequivocally that on benefits to growth alone inoculation is in many circumstances, particularly with difficult plants, a viable proposition (see Hooker et al. 1994).

Open systems

Here the scope to add inocula is more limited for two reasons. Firstly, the land areas involved are usually very large and secondly, with the exception of systems where soil sterilization takes place, introduced inocula would have to compete with indigenous fungi.

However, if we accept that AMF do contribute to plant protection then it follows that a reduction in their presence could increase disease. So in these 'open systems' the strategy most likely to lead to success is to alter management practices so as to favour AMF.

Tillage: Tillage has been shown to have a significant influence on arbuscular mycorrhizal colonisation, with low tillage favouring higher colonisation (Yocum et al., 1985; Mulligan et al., 1985; Vivekanander and Fixen, 1991), and it has been demonstrated that it is the disruption of the hyphal network that is responsible (Jasper et al., 1989; Evans and Miller, 1990). The use of a no-till or low-till management would thus be likely to increase colonisation.

Crop rotations: Because AMF are obligate symbionts, fallow (Thompson, 1987) or a rotation with a non-mycorrhizal plant species (Harinkumar and Bagyaraj, 1989) such as members of the *Brassicaceae* or *Chenopodiaceae* deplete the inoculum potential of the soil. However, if the potential for colonisation was included as a criteria for selection of cover crops then the inoculum potential of the soil could be increased during the winter months.

Fertilisers: In general colonisation by AMF reduces as the supply of soluble phosphate increases. Thus, current practices which often involve the addition of large quantities of soluble phosphorous to crops are likely to result in a reduction in colonisation to a level where they may not be effective (Gianinazzi-Pearson et al., 1985). From a nutritional point of view this may not be a concern if fertiliser is plentiful but it may also reduce protection from disease. The use of slow release, organic and insoluble forms of phosphorous fertiliser offer alternatives which are unlikely to reduce colonisation to the same extent.

Fungicides: These compounds can, not surprisingly, have an adverse influence on AMF. In general, reductions in colonisation and or activity are a consequence of their application (e.g. Nesheim and Linn, 1969). However, a number of fungicides have been reported to have no or very little effect on colonisation or function and some such as the systemic fungicides fosetyl-alluminium and metalaxyl are specific to Oomycetous fungi only and can even enhance the symbiosis and stimulate growth (S. Gianinazzi, personal communication).

Plant breeding: It is now known that different cultivars can vary in their level of dependency on the arbuscular mycorrhizal symbiosis (O'Bannon et al., 1980). In wheat it has been demonstrated that modern varieties show less of a response to inoculation than their ancestors (Manske, 1990). This suggests that breeding, which has in the past selected varieties which produce high yields with high fertiliser inputs, may have selected out varieties with an increased potential for colonisation. The response to AMF should therefore be a criteria in selection for plants in low-input systems.

198

The future

It is important that AMF should not be seen as the single and final solution for the control of plant pathogens, but as part of an integrated approach. In 'open systems' management is likely to be the way forward, with practices shaped by research aimed at a greater understanding of the role of AMF in plant-soil systems. However, in 'closed systems' inoculation is likely to be the way forward in many cases. For many years difficulties associated with inoculum production have prevented large scale use. However, inoculum is now available commercially throughout Western Europe, Japan, USA and South America.

It is certain that only by adopting these approaches is it likely that the benefits of AMF to agricultural production will be optimised, leading to a reduction in chemical inputs and increased sustainability.

References

Allen, E.B. and Allen, M.F. (1986) Water relations of xeric grasses in the field: interactions of mycorrhizas and competition. *New Phytol.* 104: 559-571.

Ames, R.N., Reid, C.P.P. and Ingham, E.R. (1984) Rhizosphere bacterial population responses to root colonisation by a vesicular-arbuscular mycorrhizal fungus. *New Phytol.* 96: 555-563.

Baltruschat, H. and Schönbeck, F. (1975). Studies on the influence of endotrophic mycorrhizae on the injection of tobacco by *Thielaviopsis basicola. Phytopatholo.Z.* 84: 172-188.

Barea, J.M., Azcón-Aquilar, C. and Azcón, R. (1991) The role of VA mycorrhizas in improving plant N acquisition from soil as assessed with ^{15}N. In: C. Flitton (ed) *The Use of Stable Isotopes in Plant Nutrition, Soil Fertility and Environmental Studies.* Joint IAEA, FAO Division, Vienna, 209-216.

Berta, G., Fusconi, A., Trotta, A. and Scannerini, S. (1990) Morphogeneric modifications induced by the mycorrhizal fungus *Glomus* strain E3 in the root system of *Allium porrum* L. *New Phytol.* 114: 207-215.

Berta, G., Trotta, A., Fusconi, A., Hooker, J.E., Monro, M., Atkinson, D., Giovannetti, M., Marini, S., Loreti, F., Tisserant, B., Gianinazzi-Pearson, V. and Gianinazzi, S. (1994) The effects of arbuscular mycorrhizal infection on plant growth. Root system morphology and soluble protein content in *Prunus cerasifera L.Tree Physiol.* (in press)

Caron, M. (1989) Potential use of mycorrhizae in control of soil-borne diseases. *Can. J. Plant. Pathol.* 11: 177-179.

Caron, M., Fortin, J.A. and Richard, C. (1986) Effect of phosphorous concentration and *Glomus intraradices* on *Fusarium* crown and root rot of tomatoes. *Phytopathology* 76: 942-946.

Cooper, K.M. and Tinker, P.B. (1978) Translocation and transfer of nutrients in vesicular-arbuscular mycorrhizas. II. Uptake and translocation of phosphorous, zinc and sulphur. *New Phytol.* 81: 43-52.

Cooper, K.M. and Grandison, G.S. (1987) Effects of vesicular-arbuscular mycorrhizal fungi on infection of tamarillo (*Cyphomandra betacea*) by *Meloidogyne incognita* in fumigated soil. *Plant Disease,* 71: 1101-1106.

Davis, R.M. (1980) Influence of *Glomus fasciculatus* on *Thielaviopsis basicola* root rot of citrus. *Plant Disease,* 64: 839-840.

Davis, R.M. and Menge, J.A. (1980) Influence of *Glomus fasciculatus* and soil phosphorous on phytopthora root rot of citrus. *Phytopathology,* 70: 447-452.

Davis, R.M., Menge, J.A. and Zentmeyer, G.A. (1978) Influence of vesicular-arbuscular mycorrhizae on Phytopthora root rot of three crop plants. *Phytopathology* 68: 1614-1617.

Dehne, H.W., Schönbeck, F. and Baltruschat, H. (1978) Untersuchungen zum Einfluss der endotrophen Mycorrhiza auf Pflanzenkrankheiten 3. Chitinase - aktivitat und Ornithinzyklus (The influence of endotrophic mycorrhiza on plant diseases III. Chitinase activity and the ornithine cycle). *Z .Pflkrank.* 85: 66-678.

Dehne, H.W. and Schönbeck, F. (1979) Untersuchungen zum Einfluss der endotrophen Mykorrhizza Pflanzenkrankheiten II. Phenolstoffwechsel und Lignifizierung (Studies on the influence of endotrophic mycorrhizae on plant disease II. Phenol metabolism and lignification). *Phytopathol .Z*. 95: 210-216.

Dehne, H.W. (1982) Interation between vesicular-arbuscular mycorrhizal fungi and plant pathogens. *Phytopathology*, 72: 1114-1119.

Elliott, A.P., Bird, G.W. and Safir, G.R. (1984) Joint influence of *Pratylenchus penetrans* (Nematoda) and *Glomus fasciculatum* (Phycomyceta) on the ontogeny of *Phaseolus vulgaris.. Nematropica* 14: 111-119.

Evans, D.G. and Miller, M.H. (1990) The role of the external mycelial network in the effect of soil disturbance upon vesicular-arbuscular mycorrhizal colonisation of maize. *New Phytol*. 114: 65-71.

Gianinazzi-Pearson, V., Trouvelot, A. and Gianinazzi, S. (1985) Evaluation of the infectivity and effectiveness of indigenous vesicular-arbuscular fungal populations in some agricultural soils in Burgundy. *Can. J. Bot.*. 63: 1521-1524.

Graham, J.H. and Menge, J.A. (1982) Influence of vesicular-arbuscular mycorrhizal and soil phosphorous on tale-all disease of wheat. *Phytopathology* 72: 95-98.

Harinikumar, K.M. and Bagyaraj, D.J. (1989) Effect of cropping sequence, fertilisers and farmyard manure on vesicular-arbuscular mycorrhizal fungi in different crops over three consecutive seasons. *Biol. Fert. Soils* 7: 173-175.

Harley, J.L. and Smith, S.E. (1983) *Mycorrhizal symbiosis*. Academic Press, London and New York, 483.

Hooker, J.E., Munro, M. and Atkinson, D. (1992) Vesicular-arbuscular mycorrhizal fungi induced alteration in poplar root system morphology. *Plant and Soil*, 145: 207-214.

Hooker, J.E, Gianinazzi, S., Vestberg, M., Barea, J.M. and Atkinson, D. (1994) The application of arbuscular mycorrhizal fungi to micropropagation systems: an opportunity to reduce inputs. *Ag Sci Fin* . (in press).

Hussey, R.S. and Roncadori, R.W. (1982) Vesicular-arbuscular mycorrhizae may limit nematode activity and improve plant growth. *Plant Disease*, 66: 9-14.

Jalali, B.L. and Jalali, I. (1991) Mycorrhiza in plant disease control. In: *Handbook of Applied Mycology, Soil and Plants*, Vol 1, K. Arora, B. Rai, K.G. Mokerji and G.R. Knudsen (eds), Marcel Dekker, New York, pp 131-154.

Jasper, D.A, Abbott, L.K and Robson, A.D (1989) Hyphae of a vesicular-arbuscular mycorrhizal fungus maintain infectivity in dry soil, except when the soil is disturbed. *New Phytol*. 112: 101-107.

Krishna, K.R. and Bagyaraj, D.J. (1983) Interaction between *Glomus fasciculatum* and *Sclerutium rolfsii* in peanut. *Can. J. Bot*. 61: 2349-2351.

MacGuidwin, A.E., Bird, G.W. and Safir, G.R. (1985) Influence of *Glomus fasciculatum* on *Meloidogyne hapla* infecting *Allium cepa*. *J Nematol*. 17: 389-395.

Manske, G.G.B. (1990). Genetical analysis of the efficiency of VA mycorrhiza with spring wheat. *Agric Ecosystems and Environment* 29: 273-280.

Meyer, J.R. and Linderman, R.G. (1986) Selective influence on populations of rhizosphere or rhizoplane bacteria and actinomycetes by mycorrhizas formed by *Glomus fasciculatum*. *Soil Biol Biochem*. 18: 191-196.

Morandi, D., Bailey, J.A. and Gianinazzi-Pearson, V. (1984) Isoflavanoid accumulation in soybean roots infected with vesicular-arbuscular mycorrhizal fungi. *Physiol. Pl. Path*. 24: 356-364.

Mulligan, M.F., Smucker, A.J.M. and Safir, G.F. (1985) Tillage modifications of dry edible bean root colonisation by VAM fungi. *Agron J*. 77: 140-144.

Nesheim, O.N. and Linn, M.B. (1969). Deleterious effect of certain fungitoxicants on the formation of mycorrhizae on corn by *Endogone fasciculat* and on corn root development. *Phytopathology*, 59: 297-300 pp.

O'Bannon, J.H., Evans, D.W. and Peaden, R.N. (1980) Alfalfa varietal response to seven isolates of vesicular-arbuscular mycorrhizal fungi. *Can. J. Pl .Sci.*. 60: 859-863.

Rosendahl, S. (1985) Interactions between the vesicular-arbuscular mycorrhizal fungus *Glomus fasciculatum* and *Aphanomyces eutieches* root rot of peas. *Phytopathol. Zeitschrift*, 114, 31-41.

Schellenbaum, L., Gianinazzi, S., and Gianinazzi-Pearson V. (1992). Comparison of acid soluble protein synthesis in roots of endomycorrhizal wild type *Pisum sativum* and corresponding isogenic mutants. *J. Plant Physiol*. 141: 2-6.

Schenck, N.C. (1987) Vesicular-arbuscular mycorrhizal fungi and the control of fungal root diseases. In: *Innovative Approaches to Plant Disease Control*. I. Chet (ed), John Wiley & Sons, New York.

Secilia, J. and Bagyaraj, D.J. (1987) Bacteria and actinomycetes associated with pot cultures of vesicular-arbuscular mycorrhizas. *Can. J. Microbiol.*. 33: 1069-1073.

Sharma, A.K., Johri, B.N. and Gianinazzi, S. (1992) Vesicular-arbuscular mycorrhizae in relation to plant disease. *World Journal of Microbiology and Biotechnology*, 8: 550-563.

Sitaramaiah, K. and Sikora, R.A. (1982) Effect of the mycorrhizal fungus *Glomus fasciculatus* on the host-parasite relationship of *Rotylenchus reniformis* in tomato. *Nematol.* 28: 412-419.

Sutton, J.C. and Sheppard, B.R. (1976) Aggregation of sand-dune soil by endomycorrhizal fungi. *Can. J. Bot.*54: 326-333.

Thompson, J.P. (1987). Decline of vesicular-arbuscular mycorrhizae in long fallow disorder of field crops and its expression in phosphorous deficiency of sunflower. *Aust. J. Agric. Res.* 38: 847-867.

Tisserant, B., Schellenbaum, L., Gianinazzi-Pearson, V., Gianinazzi, S. and Berta, G. (1992) Influence of infection by an endomycorrhizal fungus on root development and architecture in *Platanus acerifolia. Allionia* 30: 171-184.

Vivekanandan, M and Fixen, P E (1991) Cropping systems on mycorrhizal colonisation, early growth and phosphorous uptake of corn. *Soil Sci. Soc. Am..* 55: 136-140 .

Yocum, D.H., Larsen, H.J. and Boosalis, M.G. (1985) The effects of tillage treatments and a fallow season on VA mycorrhizae of winter wheat. In: Molina, R (ed) *Proc 6th North American Conference on Mycorrhizal.* Forest Research Laboratory, Corvallis, Oregon, 297 pp.

Zambolim, L. and Schenck, N.C. (1983) Reduction of the effects of pathogenic root rot infecting fungi on soybean by the mycorrhizal fungus *Glomus mosseae. Phytopathology* 73: 1402-1405.

Management of positive interactions of arbuscular mycorrhizal fungi with essential groups of soil microorganisms

G. Puppi, R. Azcón[1,] and G. Höflich[2]

Dipartimento di Biologia Vegetale Università "La Sapienza", Roma, Italy
[1]*Estación Esperimental del Zaidin, CSIC, Prof. Abareda 1, Granada, Spain*
[2]*Zentrum für Agrarlandschafts- und Landnutzungsforschung, Eberswalder Str. 84, Müncheberg, Germany*

Introduction

The optimal development of crops often demands a high input of mineral fertilizers (especially nitrogen) and pesticides. The use of these substances is not only costly and energy demanding, but also pollutes soils and ground water. Plant development may be improved by the combination of mycorrhizal fungi and rhizosphere micro-organisms acting in coordination at the root-soil interface (Linderman, 1992).

Mycorrhizal fungi are key components of soil microbiota and the regulation of mycorrhizal formation and function is influenced by soil microorganisms (Azcon-Aguilar and Barea, 1992). Interactions between mycorrhizal fungi and soil microorganisms involve nutrient cycling with impact on plant growth and nutrition. However manipulation of these beneficial combinations of microorganisms depends on the understanding of the ecosystem in order to apply a suitable selection of these microorganisms (Azcon, 1989; Azcon et al., 1991). Consequence of microbial activity in the rhizosphere can be the alteration in plant nutrient availability as a result of biochemical reactions and other effects derived from the production of plant growth regulating substances, vitamins and enzymes (Ames et al. 1987, Barea 1986). AMF are subject both to beneficial and to detrimental reactions due to rhizosphere microorganisms (Barea and Azcon-Aguilar, 1982).

Environmental factors may affect the rate of spore germination and subsequent development of the fungal mycelium around the roots (Azcon, 1987; Azcon-Aguilar and Barea, 1985; Mayo et al., 1986). The first entry point is a critical stage in mycorrhizal development (Mosse and Hepper, 1975). Microbial compounds produced by soil microorganisms, that increase root cell permeability, such as plant hormones, are involved in the formation of the symbiosis (Azcon et al. 1978; Azcon-Aguilar et al. 1986).

As a consequence of mycorrhiza formation, changes in the nutritional and physiological plant status occur (Smith and Gianinazzi-Pearson, 1988). The root exudates, modified by the mycorrhizal condition, in turn affect soil microbial populations (Ames et al., 1984; Graham et al., 1981). The term mycorrhizosphere describes the influence of mycorrhizas on the rhizosphere zone (Rambelli, 1973; Linderman, 1988).

Arbuscular mycorrhiza - *Rhizobium* interactions

Rhizobium bacteria in symbiosis can, by means of their enzyme nitrogenase and energy supplied by photosynthates from plants, satisfy the nitrogen demand of their hosts up to 90% from the atmosphere, depending on plant species and environmental factors (Merbach 1982). The nitrogen thus fixed is partly released into the soil through dead roots and compounds exuded from the nodules and can be used by non-leguminous crops grown in mixture with leguminous plants (Höflich et al., 1990). After the decomposition of the roots, the nitrogen is available to the following crop.

The occurrence of arbuscular mycorrhizas in a nodulated root system implies the existence of tripartite symbiosis. In this integrated system the activity of each member influence the subsistence and development of the other associated part. In fact, N supply by N_2 fixation could be critical to maintain a balanced nutritional and physiological status in the host plant. This, in turn, is important for mycorrhiza formation and function (Hayman, 1983). On the other hand, the high P demand of the N_2 fixation process is supplied by the arbuscular mycorrhizal fungus (AMF) (Barea et al., 1987). Nodulation and arbuscular mycorrhiza formation appear as interactive processes. Physiological and biochemical bases are involved in plant-AMF-*Rhizobium* interactions (Kucey and Paul, 1982). Nevertheless more localized effects at the root or nodule level also occur since both microorganisms interact at the precolonization stages. As rhizosphere inhabitants, direct interactions between them can take place. For example, *Rhizobium spp.* were able to improve the mycelial growth from *Glomus mosseae* spores under axenic conditions (Gonzalez, 1988) and extracellular polysaccharides from *Rhizobium meliloti* enhanced the mycorrhiza formation by *Medicago sativa* (Azcon-Aguilar et al., 1980). Cell free supernatants of *Rhizobium* cultures, containing plant hormones, also increased the mycorrhizal colonization of *M. sativa* by *G. mosseae* to an extent similar to that of solutions containing auxins, gibberellins and cytokinins (Azcon et al., 1978; Lynch, 1976).

It was assumed that *Rhizobium* and AMF do not compete for infection sites, but recently Ruiz-Lozano and Azcon (1993) reported that in *Cicer* plants the most nodulating strain of *Bradyrhizobium* tested reduced the mycorrhizal development of *G. fasciculatum*. Other authors attributed this competitive behaviour to the limited photosynthetic rate in the host (Bethlenfalvay et al., 1983). When this occurs, AMF usually show a competitive advantage for the carbohydrates over *Rhizobium* (Bayne et al., 1984). Accordingly the reduction of light intensity and thus photosynthetic carbon supply affected more strongly root colonization by *Bradyrhizobium* than that by AMF (Ruiz-Lozano and Azcon, unpublished results).

Besides the possibility of an indirect interaction operating through plant nutrition, some results suggest that arbuscular mycorrhizal effects on *Rhizobium* activity preceded any effect on plant growth (Smith and Daft, 1977; Smith et al., 1979; Asimi et al., 1980). Time course experiments with matched P-fertilized and mycorrhizal plants showed higher nitrogenase activity in roots from mycorrhizal plants (Waidyanatha et al., 1979). These results provide evidence for a differential P demand for N_2-fixing processes and plant growth; the functioning nodules have higher P requirements. In a recent study Azcon and Barea (1992) found that mycorrhizal *Medicago sativa* did not reach the same nodulation level and N_2 fixation rate compared with plants supplied with high levels of phosphate. In spite of this limitation the results do suggest the capability of AMF to use soil N that is less available to non-mycorrhizal plants. This new aspect of the involvement of AMF on N-assimilation and N uptake from soil must be taken into account for the right interpretation of interactive processes between AMF and N_2-fixing microorganisms.

Beneficial effects of dual symbioses may vary according to the combination host-endophytes (Azcon et al., 1991). If any detrimental environmental conditions, as water limitation, salinity or another type of stress occur, the tripartite symbiosis maintains higher photosynthetic and nodule activity (Sanchez-Diaz et al., 1990), N-fixation (^{15}N) (Azcon et al., 1988), transpiration and conductance (Bethlenfalvay et al., 1987) than P-fed non mycorrhizal plants. This tolerance of tripartite symbiosis to stress has a great practical and ecological relevance.

Arbuscular mycorrhiza - *Frankia* interactions

Atmospheric N_2 can be also fixed by actinomycete endophytes living in root nodules of certain host plants. These actinorhizal plants are mostly trees or shrubs common to early successional stages in nutrient-poor, marginal or disturbed habitats (Torrey, 1978; Rose, 1980), such as alder (*Alnus* spp.), *Casuarina*, *Elaeagnus*, *Myrica*, etc.; most of these species are actually used to revegetate marginal sites. Roots of these plants can be associated with two, or even three, symbionts: the nodule-forming, N_2-fixing actinomycete *Frankia* and either ectomycorrhizal or arbuscular mycorrhizal fungi. Since the first observations on such tripartite associations in the late 70's, very few papers have dealt with this topic, despite the obvious interest of such symbioses. Most of the available information was recently reviewed by Cervantes and Rodriguez-Barrueco (1992).

Improved growth of dual-inoculated compared with single-inoculated plants, is commonly reported (Rose and Youngberg, 1981; Gardner et al., 1984; Chatarpaul et al., 1989; Fragga-Beddiar and Le Tacon, 1990; Jha et al., 1993). The superior growth was even higher after outplanting (Visser et al., 1991; Lumini et al., 1994).

AMF infection seems to increase nodule dry weight (Rose and Youngberg, 1981; Fragga-Beddiar and Le Tacon, 1990; Jha et al., 1993); but Russo (1989) could observe such effects only at intermediate P levels. Visser et al. (1991) correlated the higher shoot and root productivity of *Elaeagnus commutata* and *Shepherdia canadensis*, when inoculated both with *Frankia* and AMF with nodule status. A similar correlation was observed also by Isopi et al. on alder (unpublished results).

Concerning the effect of AMF on nitrogen fixation in actinomycete nodules, the fungi may help to satisfy the large amount of phosphorus required by the nitrogen fixing activity. *Alnus acuminata* seedlings inoculated with *Frankia* and *Glomus intraradix* showed maximum nitrogenase activity at lower levels of phosphorus than required by uninoculated plants (Russo, 1989). Fragga-Beddiar and Le Tacon (1990) observed that mycorrhizal infection increased plant and nodule dry weight as well as nodule number, in *Alnus glutinosa* seedlings inoculated with *Glomus fasciculatum*, more than phosphorus addition. More recently Jha et al. (1993) observed in *Alnus nepalensis*, inoculated with *Frankia* and *Glomus mosseae* singularly and in combination over a range of P treatments, an increase in nodule dry weight and nitrogenase activity, independent of the amount of phosphorus available, in dual inoculated plants. *Frankia*, on the other hand, increased mycorrhizal colonization.

Increased mycorrhizal infection in *Frankia* nodulated plants was observed also by Rose and Youngberg (1981) and Chatarpaul et al. (1989), while Isopi et al. (unpublished results) found that such effect varied according to the mycorrhizal strain. Meyer and Linderman (1986b) have shown that *Glomus fasciculatum* infection influences the population of actinomycetes in the rhizosphere and it is thus possible that non-nutritional mechanisms could also be involved in the synergistic effect between *Frankia* and AMF.

Arbuscular mycorrhizas associative diazotrophic bacteria interactions

Much attention has been devoted in the last decade to the associative diazotrophic bacteria, which are specialized rhizospheric organisms (see Garbaye, 1991), strictly dependant upon compounds released by the plant. In some instances the association is so close to a symbiosis and both *Azospirillum* and *Acetobacter* can actually penetrate the root of gramineous plant species. The real contribution of such associations is however still much questioned, since the N-fixing efficiency of these organisms is not very high, unless sufficient energy substrates are available. Mycorrhizas are though now regarded as a tool to improve diazotrophs-gramineous plants association.

The first observations on *Azospirillum* and AMF inoculations were made by Barea et al. (1983), and these were followed by Subba Rao et al. (1985a, 1985b), Pacovsky et al. (1985), Pacovsky (1989), Tilak and Singh (1988), all of them reporting positive interactions on plant growth. In contrast, although changes in host physiology were detected, differences were not found in total dry weight of *Sorghum bicolor* plants inoculated either with *Glomus fasciculatum*, with a strain of *Azospirillum brasiliense* or with both endophytes (Pacovsky 1988).

More recently, positive effects of dual colonization of non-leguminous plant roots by both AMF and diazotrophic bacteria other than *Azospirillum* have been investigated. Dual inoculation could be specially advantageous in the case of *Acetobacter diazotrophicus* (Paula et al. (1992) since this bacterium seems to fix atmospheric N also in the presence of nitrate (Teixera et al., 1987). It may be interesting to note that *A. diazotrophicus* has not been isolated from soil and the bacteria seem to be chiefly transmitted from plant to plant through vegetative propagation (Cavalcante and Döbereiner, 1988).

Selection of AMF strains for the improvement of crop yields and diazotrophs efficiency should consider intersymbiont compatibility in addition to host-plant compatibility. In recent experiments, four AMF, *Glomus constrictum*, *G. fasciculatum*, *G. occultum* and *Scutellospora persica*, isolated from italian sand dunes, have been inoculated singularly or in combination with *A. diazotrophicus* and other plant growth promoting rhizobacteria (PGPR) (*Azospirillum brasiliense*, *Herbaspirillum seropedicae*, *Pseudomonas cepacica*), on sweet sorghum seedlings (Isopi et al., 1993). Growth effects varied according to the fungal strain used, and were improved by the presence of other "helper" bacteria.

The successful spreading and maintenance of infection might be ensured by the fact that newly formed AMF spores bear *Acetobacter* propagules. The presence of hyphosphere bacteria on the surface and within spores of AMF has been reported by many authors (Mosse, 1962; Varma et al., 1981; Tilak et al., 1989; Vancura et al., 1990; Klyuchnikov and Kozhevin, 1990) and the ability of *A. diazotrophicus* to successfully colonize sorghum plants, using AMF spores as vectors has been observed (Paula et al., 1991). Such inocula may be at a too low concentration to be really effective (Del Gallo, personal communication), but the topic is of obvious interest.

Arbuscular mycorrhizas - plant growth promoting rhizobacteria interactions

PGPR is the term used to describe the root colonizing bacteria (Suslow, 1982). They can be beneficial to plants and can be used as inoculants. The most appreciated effects of PGPR are those specifically based on the production of biologically active compounds (hormones, chelators, siderophores, enzymes, vitamins, etc.) being occasionally plant pathogens antagonists or having implications in the nutrient cycling, i.e. N_2 free fixers, P-solubilizers, and on the organic matter turnover. The activities of these soil microorganisms, synergistically interacting with mycorrhizal fungi may be beneficial for plant development and growth. The effect of these bacteria on the morphology, geometry and physiology of the root system can also affect mycorrhizal formation and response. Meyer and Linderman (1986a) reported not only additive plant growth but also mutual colonization enhancement in *Pseudomonas* sp. and *Glomus* dual inoculation. Soil microorganisms can improve saprophytic independent growth of mycorrhizal fungi in the preinfection stage (Azcon-Aguilar et al., 1986). Mycelial growth from axenic germinated spores of *Glomus mosseae* was highly improved by a PGPR isolate (Azcon, 1987).

Changes in the root physiology and exudation when plants become mycorrhizal altered the microbial population in the rhizosphere soil (Ames et al., 1984; Bagyaraj and Menge, 1978) with positive effects on plant growth. Root colonization by AMF may be enhanced in interactions with some microorganisms and particularly with some PGPR. The result of greater colonization is an increased fungal structure formation in the cortex that provides more area for metabolic interchanges between associated organisms. In an indirect or direct way, microbial associates to mycorrhizal roots may affect the hyphal development in the soil, increasing the possibility of nutrient or microbial metabolites' acquisition from the soil.

Other microorganisms characterized as PGPR are siderophores producers (Kloepper et al., 1980). Siderophores can affect plants by chelating Fe. These compounds play an important role in plant growth and disease biocontrol (Nielands and Leong, 1986). The siderophores sequester iron, making it unavailable to the other soil microorganisms, which are unable to produce the chelator or lack the iron assimilation system for ferric siderophores. Siderophores and arbuscular mycorrhizal interactions may be used successfully for plant growth enhancement.

Arbuscular mycorrhiza - phosphate-solubilizing microorganisms interactions

Early work by Barea et al. (1975) and Azcon et al. (1976) on phosphate-solubilizing microorganisms revealed the beneficial effect of dual inoculation associated with AMF. The effectiveness of such co-inoculations are possibly due to mycorrhizal hyphae reaching microhabitats where microbial phosphate solubilization takes place and roots cannot reach.

Several *in vitro* experiments have shown that many soil microorganisms can solubilize phosphate ions from sparingly soluble inorganic and organic P sources. However, this microbial process may be limited in soil by reduced carbohydrate availability to the bacterium, and difficulties for translocation of the phosphate ions to the root. The possible P adsorption on clay and minerals, can be avoided if these ions are taken up by mycorrhizal hyphae present in the microenvironment. When adsorption was avoided using a soil-less medium (Piccini and Azcon, 1988) the synergistic action was evident. Besides the described mechanisms, some of the observed interactions were based on the production of plant hormones which affect the cell-root permeability (Azcon et al., 1978; Toro et al., unpublished).

Possibilities of using symbiotic microorganisms with special reference to arbuscular mycorrhiza - *Rhizobium* interactions

As Linderman (1992) has stated, the challenge of agrobiological studies is to characterize rhizosphere microbial relationships and to optimize all components so that they function coordinately in the system. The results available suggest that plant growth can be optimized to obtain maximum yields using specific combinations of selected microorganisms that function in a synergistic form. This biological potential must be used in sustainable agriculture and crop management strategy.

The management strategy will involve the selection of suitable mycorrhizal isolates in combination with beneficial microbes producing biologically active compounds or favouring nutrient cycling (Ames et al., 1987; Höflich et al., 1992). The selection can be made for compatibility and combined efficiency regarding a particular soil-plant and environmental system (Azcon et al., 1991).

The diversity of AMF and *Rhizobium* bacteria is closely related to the diversity of plant communities. It is reduced when natural ecosystems are converted into agroecosystems or when the management is intensified (Sieverding, 1990; Rabatin and Stinner, 1989). *Glomus* spp. are relatively tolerant towards agricultural practices.

In the rotation of crops, both groups of organisms are promoted by leguminous plants and reduced during fallow periods (Johnson and Pfleger, 1992). In mixtures of grasses and leguminous plants, the nutrient flow between the plants can stimulate the development of rhizosphere microorganisms of both species (Brown et al., 1992). Brassicaceans and Chenopodiaceans do not act as hosts for AMF and thus reduce mycorrhizal colonization of the following crop. However, the use of mycorrhizal plants such as maize or sunflower, as cover crops or in intercropping can have positive effects (Johnson and Pfleger. 1992).

Soil management can also influence the development of AMF. Compacted soils reduce the root growth, soil aeration and the establishment of rhizosphere microorganisms and frequent tillage can inhibit the development of AMF (Johnson and Pfleger, 1992). Mineral and organic fertilizers also exert an influence on the microflora, according to soil nutrient content, composition (N, P, K, Mg ratio) and added quantity. Optimal nutrient and pH status appropriate to the plant species are important. An unbalanced P fertilization or excess P reduces AMF (Johnson and Pfleger, 1992). Soil nutrients, if necessary by additional fertilization, assist the development of young plants during the first stages of growth, and are a prerequisite for the establishment of the symbioses. In the presence of effective *Rhizobium* bacteria, as a rule no N fertilizer is necessary for leguminous crops (Höflich, 1993). The survival of *Rhizobium* bacteria in the soil is supported by organic substances rich in C (Giddens et al., 1982).

Biocides can influence the rhizosphere microflora either directly or indirectly through their influence on the host-plant and on the colonization by pests (Trappe et al., 1984); but herbicides and insecticides in general have no negative effect on *Rhizobium* or AMF symbioses, if the prescriptions for application are observed (Johnson and Pfleger, 1992; Wache, 1987). Light deficiency inhibits infection with AMF, nodulation, N_2 fixation and the efficiency of the symbioses (Reinhard et al., 1992; Höflich, 1993). Light intensity and duration influence the photosynthetic efficiency of plants and the consequent supply of assimilates to the N_2-fixing nodules and the production of root exudates necessary for the colonization of roots with rhizosphere microorganisms.

Biotic factors can also influence the survival of growth promoting microorganisms in the soil and their efficiency. Possible causes are: competition for available nutrients; bacteriophages disintegrating *Rhizobium* bacteria and reducing their infectivity (Evans et al., 1979); parasitic bacteria (Keya and Alexander, 1975) or actinomycetes and fungi producing antibiotics (Gibson and Newton, 1981) can limit the survival of the symbionts. Protozoans, *Meloidogyne spp.* and *Heterodera spp.* have also antagonistic effects towards symbionts (Pena-Cabriales and Alexander, 1979; Trabulsi, 1980; McGrinnity and Kapusta, 1980) and virus infections reducing the positive effect of soil microorganisms.

Selection of growth stimulating microorganisms

Agricultural soils often contain insufficient or ineffective microorganisms. Repeated attempts have therefore been made to inoculate the soil with microorganisms with the desired traits in order to improve the development of plants. A reproducible growth stimulation by inoculation requires :

microorganisms with the following characteristics:

- high specific efficiency (i.e. phytohormone formation, N_2 fixation, acquisition of soil nutrients, protection from pathogens and stress)
- high affinity to a spectrum of host plants as broad as possible
- competitivity towards other rhizosphere microorganisms deleterious for plant growth
- good adaptation to different ecological conditions
- reproducible stimulation of the plant development under field conditions.

and effective technological solutions for:

- production of inoculum
- inoculation.

Due to the broad genetic diversity it should be possible to select strains with the necessary traits from the natural populations. Techniques of molecular genetics should also enable the development of strains of microorganisms with the desired propertie. However, the selection of microorganisms with the desired metabolic capacities, such as N_2 fixation or protection against pathogens, is only the first step towards a successful inoculation. The results of laboratory and pot experiments must be demonstrated in the field. There is, for example, frequently no direct relation between nitrogenase activity, nodule number or the extend of colonization of the roots by AMF and plant yield (McGonigle, 1988; Höflich et al., 1993). The deciding factor of success is reproducible efficacy of the symbiosis under natural conditions. Important criteria are the promotion of shoot and root growth, especially during the development of the young plants, the protection from pests, and the survival of the inoculated microorganisms in the rhizosphere.

Inoculum production and application

Type and quality of the preparations have a decisive influence on the success of an inoculation with microorganisms. Preparations with carrier (peat, lime, compost, polyacrylamide) are more favourable for the survival of bacteria during storage, transport and in the soil than liquid or lyophilized preparations (Hamatova, 1980; Gibson and

Newton, 1981; Jung, 1982).

The efficiency of mixed preparations is internationally disputed, since different microorganisms may interact negatively (Milto, 1982). For a combined inoculation, the different preparations should preferably be mixed immediately before the application.

For the cultivation of AMF inocula, a substrate with a carrier (peat-bentonite-mixture) and maize as host plant gave good results, the milled inoculum consisting of root fragments, hyphae and spores in the carrier mixture (Höflich and Glante, 1991). An AMF inoculum with peat-bentonite as carrier is suitable for seed and seed row inoculation; in addition, a combined inoculation with *Rhizobium* bacteria is possible. For further information on AMF inocula, the reader is referred to Jeffries and Dodd (1991). A combined inoculation of AMF with growth stimulating bacteria can possibly favour the inoculum production (Singh, 1992).

An early and effective colonization of the roots by the inoculated microorganisms is favoured by application of the preparations immediately before sowing or by application of the granulated inoculum into the seed row. Granules improve the survival of the microorganisms in case of drought during sowing. A reliable titre for bacteria is 10^6 cfu per seed. It may be lower in the case of effective bacteria or favourable ecological conditions. Microorganisms with low plant affinity cannot survive in the rhizosphere even when applied in high quantities. While *Rhizobium* preparations are already used in the plant production, the application of AMF preparations is still in test stage. The amount of inoculum necessary for an agricultural application of AMF is rather high, while it is more easily applicable in tree nurseries, horticulture and in the recultivation of devastated soils.

Conclusions

Beneficial effects of mycorrhizas on plant growth are well established in controlled conditions and field trials. Nevertheless, although frequently reported as mycorrhizal effects, it should not be excluded that other microorganisms may be involved in such positive interactions. Mycorrhizas are commonly defined as plant-fungus associations; in functional terms however they may better be regarded as a valve regulating plant-soil interactions, possibly increasing carbon sink to roots and carbon efflux to soil (Garbaye, 1991). Their management in consideration of other soil microorganisms is a challenge and an opportunity for sustainable plant-soil systems (Staley et al., 1992).

Such a holistic approach implies a profound knowledge of the interrelationships among plants and associated fungi and bacteria at the physiological, genetic and molecular level; it implies also a deep knowledge of the site (soil type and processes, vegetational history) where the microbial inoculum is to be applied in order to make predictions possible and reduce management inputs. Not everything may be controlled, but something may be foreseen. The real problem is how to render more productive protective ecosystems.

Further research is necessary, both at basic and applied level, i.e. regarding the maintenance of inoculated organisms in the field. It is worth noting, however, that the number of field experiments is every year increasing. It is therefore to be hoped that, notwithstanding the difficulties, and even some failures and contradictions, this body of experience will soon allow a wider utilization of mycorrhizas and associated organisms in plant generation.

References

Ames, R.N., Mihara, K.L. and Bethlenfalvay, G.J. (1987) The establishment of microorganisms in vesicular-arbuscular mycorrhizal and control treatments. *Biology and Fertility of Soils* 3: 217-223.

Ames, R.N., Reid, C.P.P. and Ingham, E.R. (1984) Rhizosphere bacteria population responses to root colonization by a vesicular-arbuscular mycorrhizal fungus. *New Phytol.* 96: 555-563.

Asimi, S., Gianinazzi-Pearson, V. and Gianinazzi, S. (1980) Influence of increasing soil phosphorus levels on interaction between VA mycorrhzae and *Rhizobium* in soybeans. *Can. J. Bot.* 58: 2200-2205.

Azcon, R. (1987) Germination and hyphal growth of *Glomus mosseae* in vitro: effects of rhizosphere bacteria and cell-free culture media. *Soil Biol.Biochem.* 19: 417-419.

Azcon, R. (1989) Selective interaction between free-living rhizosphere bacteria and vesicular-arbuscular mycorrhizal fungi. *Soil Biol.Biochem.* 21: 639-644.

Azcon, R. and Barea, J.M. (1992) Nodulation, N_2 fixation (^{15}N) and nutrition relationships in mycorrhizal or phosphate amended alfalfa plants. *Symbiosis* 12: 33-41.

Azcon, R., Azcon-Aguilar, C. and Barea, J.M. (1978) Effects of plant hormones present in bacterial cultures on the formation and responses to VA mycorrhiza. *New Phytol.* 80: 359-369.

Azcon, R., Barea, J.M. and Hayman, D.S. (1976) Utilization of rock phosphate in alkaline soils by plants inoculated with mycorrhizal fungi and phosphate-solubilizing bacteria. *Soil Biol.Biochem.* 8: 135-138.

Azcon, R., El-Atrach, F. and Barea, J.M. (1988) Influence of mycorrhiza vs. soluble phosphate on growth, nodulation, and N_2 fixation (^{15}N) in alfalfa under different levels of water potential. *Biology and Fertility of Soils* 7: 28-31.

Azcon, R., Rubio, R. and Barea, J.M. (1991) Selective interactions between different species of mycorrhizal fungi and *Rhizobium meliloti* strains, and their effects on growth, N_2-fixation (^{15}N) and nutrition of *Medicago sativa* L. *New Phytol.* 117: 399-404.

Azcon-Aguilar, C. and Barea, J.M. (1985) Effect of soil micro-organisms on the formation of vesicular-arbuscular mycorrhizas. *Trans. Brit. mycol. Soc.* 84: 536-537.

Azcon-Aguilar, C. and Barea, J.M. (1992) Interactions between mycorrhizal fungi and other rhizosphere microorganisms. In: M.F. Allen (ed) *Mycorrhizal Functioning.* Chapman and Hall, New York, pp.163-198.

Azcon-Aguilar, C., Barea, J.M. and Olivares, J. (1980) Effects of *Rhizobium* polysaccharides on VA mycorrhiza formation. 2nd Intern.Symp. on Microbial Ecology, University of Warwick, Coventry, U.K., Abstract No. 187.

Azcon-Aguilar, C., Diaz-Rodriguez, R.M. and Barea, J.M. (1986) Effect of soil microorganisms on spore germination and growth of the vesicular-arbuscular mycorrhizal fungus *Glomus mosseae. Trans.Brit. mycol.Soc.* 86: 337-340.

Bagyaraj, D. and Menge, J.A. (1978) interaction between a VA mycorrhiza and *Azotobacter* and their effects on rhizosphere microflora and plant growth. *New Phytol.* 80: 567-573.

Barea, J.M. (1986) Importance of hormones and root exudates in mycorrhizal phenomena. In: V.Gianinazzi-Pearson and S.Gianinazzi (eds) *Physiological and Genetical aspects of Mycorrhizae.* INRA, Paris, pp. 177-187.

212

Barea, J.M. and Azcon-Aguilar, C. (1982) Production of plant growth-regulating substances by the vesicular-arbuscular mycorrhizal fungus *Glomus mosseae*. *Appl. Environ. Microbiol.* 43: 810-813.

Barea, J.M., Azcon, R. and Hayman, D.S. (1975) Possible synergustic interactions between *Endogone* and phosphate-solubilizing bacteria in low-phosphate soils. In: F.E. Sanders, B. Mosse, and P.B. Tinker (eds) *Endomycorrhizas*. Acad.Press, London, pp. 409-417.

Barea, J.M., Azcon-Aguilar, C. and Azcon, R. (1987) Vesicular-arbuscular mycorrhiza improve both symbiotic N_2-fixation and N uptake from soil as assessed with a N technique under field conditions. *New Phytol.* 106: 717-725.

Barea, J.M., Bonis, A.F. and Olivares, J. (1983) Interactions between *Azospirillum* and VA mycorrhiza and their effects on growth and nutrition of maize and ryegrass. *Soil Biol. Biochem.* 15: 706-709.

Bayne, H.B., Brown, M.S. and Bethlenfalvay, G.J. (1984) Defoliation effects on mycorrhizal colonization, nitrogen fixation and photosynthesis. *Physiol. Plant.* 62: 576-580.

Bethlenfalvay, G.J., Bayne, H.G. and Pacovsky, R.S. (1983) Parasitic and mutualistic associations between a mycorrhizal fungus and soybean: The effect of phosphorus on host plant-endophyte interactions. *Physiol. Plant.* 57: 543-548.

Bethlenfalvay, G.J., Brown, M.S. and Newton, W.E. (1987) Photosynthetic water- and nutrient-use efficiency in a mycorrhizal legume. In: *Mycorrhizae in the next decade*. Practical applications and research priorities. Proc. 7th NACOM, Gainesville, Florida, pp. 231-233.

Brown, M.S., Ferrera-Cerrato, R. and Bethlenfalvay, G.J. (1992) Mycorrhiza-mediated nutrient distribution between associated soybean and corn plants evaluated by the Diagnosis and Recommendation Integrated System (DRIS). *Symbiosis* 12: 83-94.

Cavalcante, V.A. and Doberainer, J. (1988) A new acid-tolerant nitrogen-fixing bacterium associated with sugarcane. *Plant and Soil* 108: 23-31.

Cervantes, E. and Rodriguez-Barrueco, C. (1992) Relationships between the mycorrhizal and actinorhizal symbioses in non-legumes. In: J.R. Norris, D.J. Read and A.K. Varma (eds) *Techniques for the study of mycorrhizae*. Methods in Microbiology 24: 317-432.

Chatarpaul, L., Chakravarty, P. and Subramaniam, P. (1989) Studies in tretapartite symbioses. Role of ecto-and endomycorrhizal fungi and *Frankia* on growth performance of *Alnus incana*. *Plant and Soil* 118: 145-150

Evans, J., Barnett, Y.M. and Vincent, J.M. (1979) Effect of a bacteriophage on colonization and nodulation of clover roots by paired strains of *Rhizobium trifolii*. *Can. J. Microbiol.* 25: 974-978.

Fraga-Beddiar, A. and LeTacon, F. (1990) Interaction between a VA Mycorrhizal Fungus and *Frankia* Associated with Alder (*Alnus glutinosa* (L.) Gaertn.). *Symbiosis* 9: 247-258.

Garbaye, J. (1991) Biological interaction in the mycorrhizosphere. *Experentia* 47: 370-375.

Gardner, I.C., Clelland, D.M. and Scott, A. (1984) Mycorrhizal improvment in nonleguminous nitrogen fixing associations with particular reference of *Hippophae rhamnoides*. *Plant and Soil* 78: 189-199.

Gibson, A.H. and Newton, W.E. (1981) *Current perpectives in nitrogen fixation*. Proc. 5th Intern. Symp. on Nitrogen Fixation. Canberra.

Giddens, J.E., Duningan, E.P. and Weaver, R.W. (1982) *Legume inoculation in the Southeastern USA*. Southern Cooperative Series Bull. Spec. Volume 283

Gonzalez, S.B. (1988) *Ecologia y biotecnologia de micorrizas en leguminosas (soja y alfalfa)*. Ph.D. Thesis, University of Granada.

Graham, J.H., Leonard, R.T. and Menge, J.A. (1981) Membrane mediated decrease in root exudation responsible for phosphorus inhibition of vesicular-arbuscular mycorrhiza formation. *Plant Physiol.* 68: 548-552.

Hamatova, E. (1980) Production of inoculants and Rhizobium-collection in Czechoslovakia. In: *Current perpectives in nitrogen fixation*. Proc. 5th Intern. Symp. on Nitrogen Fixation, Canberra, p.516.

Hayman, D.S. (1983) The physiology of vesicular-arbuscular endomycorrhizal symbiosis. *Can. J Bot.* 61: 944-963.

Höflich, G.(1993) Effect of variety and environmental factors on the phyto-effectivity of bacterial inoculations in peas. *Zentralbl. Mikrobiol.* 148: 315-324.

Höflich, G. and Glante, F. (1991) Inokulumanzucht und Inokulation ertragswirksamer VA-Mykorrhizapilze. *Zentralbl. Mikrobiol.* 146: 247-252.

Höflich, G., Glante, F., Liste, H.-H., Weise, I, Ruppel, S. and Scholz-Seidel, C. (1992) Phytoeffective combination effects of symbiotic and associative microorganisms on legumes. *Symbiosis* 14: 427-438.

Höflich, G., Glante, F., Kühn, G. und Hickisch, B. (1993) Phyto-effective symbiosis in pea. *Zentralbl.Mikrobiol.* 148: 48-54.

Höflich, G., Kühn, G., Meinsen, C., Schuppenies, R., Schäfer, E. und Stitz, K. (1990) Lösungen zur verstärkten Nutzung der biologischen Luftstickstoffbindung in Leguminosengrasgemischen. *Arch. Acker-*

Pflanzenbau Bodenkd. **34**: 701-707.

Isopi, R., Fabbri, P, Pennelli, B., Del Gallo, M. and Puppi, G. (1993) Effect of VA mycorrhizas and *Acetobacter diazotrophicus* on *Sorghum vulgare* L. *Giorn. Bot. It.* **127**: 529.

Jeffries, P. and Dodd, C.J. (1991) The use of mycorrhizal inoculants in forestry and agriculture. In: D.K. Arora, B. Rai, K.G. Mukerji and G.R. Knudsen (eds) *Handbook of applied mycology*. Vol.1. Soil and Plants. Marcel Dekker, New York, pp.77-129.

Jha, D.K., Sharma, G.D. and Mishra, R.R. (1993) Mineral nutrition in the tripartite interaction between *Frankia*, *Glomus* and *Alnus* at different soil phosphorus regimes. *New Phytol.* **123**: 307-311.

Johnson, N.C. and Pfleger, F.L. (1992) Vesicular-arbuscular mycorrhizae and cultural stresses. In: G.J. Bethlenfalvay and R.G. Linderman (eds) *Mycorrhizae in Sustainable Agriculture*, ASA Special Publication 54, Madison, WI, USA, pp.71-99.

Jung, G. (1982) Polymer-entrapped *Rhizobium* as an inoculant for legumes. *Plant and Soil.* **65**:219-231

Keya, S.O. and Alexander, M. (1975) Regulation of parasitism by host density: The *Bdellovibrio-Rhizobium* interrelationships. *Soil Biol. Biochem.* **7**: 231-237.

Kloepper, J.W., Leong, J. Teintze, M. and Schroth, M.N. (1980) Enhanced plant growth by siderophores produced by plant growth promoting rhizobacteria. *Nature* **286**: 885-886.

Klyuchnikov, A.A. and Kozhevin, P.A. (1990) Dynamics of *Pseudomonas fluorescens* and *Azospirillum brasiliense* populations during the formation of the vesicular-arbuscular mycorrhiza (VAM) fungi. *Plant Physiol.* **79**: 562-563.

Kucey, R.M.N. and Paul, E.A. (1982) Carbon flow photosynthesis, and N_2 fixation in mycorrhizal and nodulated faba beans (*Vicia faba* L.) *Soil Biol. Biochem.* **14**: 407-412.

Linderman, R.G. (1988) Mycorrhizal interaction with the rhizosphere microflora: The mycorrhizosphere effect. *Phytopatology* **78**: 488-505.

Linderman, R.G. (1992) VA mycorrhizae and soil microbial interactions. In: G.J. Bethlenfalvay and R.G. Linderman (eds) *Mycorrhizae in Sustainable Agriculture*, ASA Special Publication 54, Madison, WI, USA, pp.45-70.

Lumini, E., Bosco, M., Puppi, G., Isopi, R., Frattegiani, M., Buresti, E. and Favilli, F. (1994) Field performance of *Alnus cordata* Loisel (Italian alder) inoculated with Frankia and VA-mycorrhizal strains in mine-spoils afforestation plots. *Soil Biol. Biochem.* (in press)

Lynch, J.M. (1976) Products of soil micro-organisms in relation to plant growth. CRC *Critical reviews in Microbiology* **5**: 67-107.

Mayo, K., Davies, R.E. and Motta, J. (1986) Stimulation of germination of spores of *Glomus versiforme* by spore associated bacteria. *Mycologia* **78**: 426-431.

McGonigle, T.P. (1988) A numerical analysis of published field trials with vesicular-arbuscular mycorrhizal fungi. *Funct. Ecol.* **2**:473-478.

McGrinnity, P. and Kapusta, G. (1980) Soybean cyst nematode *Heterodera glycines* and *Rhizobium* strain influences on soybean nodulation and N_2-fixation. *Agron. J.* **72**: 785-789.

Merbach, W. (1982) Untersuchungen über Stickstoff und symbiontische N_2-Fixierung bei Körnerlegumi-nosen. MLU Halle-Wittenberg Diss. B.

Meyer, J.R. and Linderman, R.G. (1986a) Response of subterranean clover to dual inoculation with vesicular-arbuscular mycorrhizal fungi and a plant growth-promoting bacterium, *Pseudomonas putida*. *Soil Biol. Biochem.* **18**: 185-190.

Meyer, J.R. and Linderman, R.G. (1986b) Selective influence on populations of rhizosphere or rhizoplane bacteria and actinomycetes by mycorrhizas formed by *Glomus fasciculatum*. *Soil Biol. Biochem.* **18**: 191-196.

Milto, N.I. (1982) Kluben'kovye bakterii i produktivnost' bobovych rastenij. *Nauka i Technika*

Mosse, B. (1962) The establishment of vesicular-arbuscular mycorrhizae under aseptic conditions. *J. Gen. Microbiol* **27**: 509-520

Mosse, B. and Hepper, C.M. (1975) Vesicular-arbuscular mycorrhizal infections in root organ cultures. *Physiol. Plant Pathol.* **5**: 215.

Nielands, J.B. and Leong, S.A. (1986) Siderophores in relation to plant growth and disease. *Ann. Rev. Plant Physiol.* **37**: 187-208.

Pacovsky, R.S. (1988) Influence of inoculation with *Azospirillum* brasilense and *Glomus fasciculatum* on sorghum nutrition. *Plant and Soil* **110**: 283-287

Pacovsky, R.S. (1989) Metabolic differences in *Zea-Glomus-Azospirillum* symbioses. *Soil Biol.Biochem.* **21**:953-960.

Pacovsky, R.S., Fuller, G. and Paul, E.A. (1985) Influence of soil on the interactions between endomycorrhizae and *Azospirillum* on sorghum. *Soil Biol. Biochem.* **17**: 525-531.

214

Paula, M.A., Reis, V.M. and Doberainer, J. (1991) Interaction of *Glomus clarum* with *Acetobacter diazotrophicus* in infection of sweet potato (*Ipomea batatas*), sugarcane (*Saccharum* spp.), and sweet sorghum (*Sorghum vulgare*). *Biology and Fertility of Soils* 11: 111-115.

Paula, M.A., Urquiaga, S., Siqueira, J.O. and Doberainer, J. (1992) Synergistic effects of vesicular-arbuscular mycorrhizal fungi and diazotrophic bacteria on nutrition and growth of sweet potato (*Ipomea batatas*). *Biology and Fertility of Soils* 14: 61-66.

Pena-Cabriales, J.J. and Alexander, M. (1979) Survival of *Rhizobium* in soils undergoing drying. *Soil Sci. Soc. Am. J.* 43: 5.

Piccini, D. and Azcon, R. (1987) Effect of phosphate-solubilizing bacteria and vesicular-arbuscular mycorrhizal fungi on the utilization of the Bayovar rock phosphate by alfalfa plants using a sand-vermiculite medium. *Plant and Soil* 101: 45-50.

Rabatin, S.C. and Stinner, B.R. (1989) The significance of vesicular-arbuscular mycorrhizal fungi-soil macroinvertebrate interactions in agroecosystems. *Agric. Ecosyst. Environ.* 27: 195-204.

Rambelli, A. (1973) The rhizosphere of mycorrhizae. In: G.C. Marks and T.T. Kozlowski (eds) *Ectomycorrhizae*. Acad.Press, New York, pp. 299-350.

Reinhard, S., Martin, P. and Marschner, H. (1992) Interactions in the tripartite symbiosis of pea (*Pisum sativum* L.), *Glomus* and *Rhizobium* under non-limiting phosphorus supply. *J.Plant Physiol.* 141:7-11.

Rose, S.L. (1980) Mycorrhizal associations of some actinomycete nodulated nitrogen-fixing plants. *Can. J. Bot.* 58: 1449-1454.

Rose, S.L. and Youngberg, C.T. (1981) Tripartite associations in snowbrush (*Ceanothus velutinus*): effects of vesicular-arbuscular mycorrhizae on growth, nodulation, and nitrogen fixation. *Can.J.Bot.* 59: 34-39.

Ruiz-Lozano, J.M. and Azcon, R. (1993) Specificity and functional compatibility of VA mycorrhizal endophytes in association with *Bradyrhizobium* strains in *Cicer arietinum*. *Symbiosis* 15: 217-226.

Russo, R.O. (1989) Evaluating alder-endophyte (*Alnus acuminata - Frankia - Mycorrhizae*) interactions. Acetylene reduction in seedlings inoculated with *Frankia* strain ArI3 and *Glomus intraradices*, under three phosphorus levels. *Plant and Soil* 118: 151-155.

Sanchez-Diaz, M., Pardo, M., Antolin, M., Peña, J. and Aguirreolea, J. (1990) Effect of water stress on photosynthetic activity in the *Medicago-Rhizobium-Glomus* symbiosis. *Plant Science* 71: 215-221.

Sieverding, E. (1990) Ecology of VAM fungi in tropical agrosystems. *Agric. Ecosyst.Environ.* 29:369-390.

Singh, C.S. (1992) Mass inoculum production of vesicular-arbuscular (VA) mycorrhizae: II. Impact of N_2-fixing and P-solubilizing bacterial inoculation on VA-mycorrhiza. *Zentralbl. Mikrobiol.* 147: 503-508.

Smith, S.E. and Bowen, G.D. (1979) Soil temperature, mycorrhizal infection and nodulation in *Medicago truncatula* and *Trifolium subterraneum*. *Soil Biol. Biochem.* 11: 469-473.

Smith, S.E. and Daft, M.J. (1977) Interactions between growth phosphate content and nitrogen fixation in mycorrhizal and non- mycorrhizal *Medicago sativa*. *Austr. J. Pl. Physiol.* 4: 403-413.

Smith, S.E. and Gianinazzi-Pearson, V. (1988) Physiological interaction between symbionts in vesicular-arbuscular mycorrhizal plants. *Ann. Rev. Plant Physiology* 39: 221-244.

Smith, S.E., Nicholas, D.J.D, and Smith, F.A. (1979) Effect of early mycorrhizal infection on nodulation and nitrogen fixation in *Trifolium subterraneum* L. *Austr. J. Plant Physiol.* 6: 305-316.

Subba Rao, N.S., Tilak, K.V.B.R., and Singh, C.S. (1985a) Synergistic effect of vesicular-arbuscular mycorrhizas and *Azospirillum brasiliense* on the growth ow barley in pots. *Soil Biol. Biochem.* 17: 119-121.

Subba Rao, N.S., Tilak, K.V.B.R., and Singh, C.S. (1985b) Effect of combined inoculation of *Azospirillum brasiliense* and vesicular-arbuscular mycorrhiza on pearl millet (Pennisetum americanum). *Plant and Soil* 81: 283-286.

Suslow, T.V. (1982) Role of root colonizing bacteria in plant growth: In: R. Mount and C. Lacey (eds) *Phytopathogenic Prokaryotes*. Vol. 1. Acad. Press, New York, pp. 187-223.

Teixeira, K.R.S., Stephan, M.P. and Doberainer, J. (1987) Physiological studies of *Saccharobacter nitrocaptans* a new acid tolerant N_2-fixing bacterium. 4th International Symposium on Nitrogen Fixation with non-legumes. Final Program abstracts, p.149. Rio de Janeiro.

Tilak, K.V.B.R. and Singh. C.S. (1988) Response of pearl millet (*Pennisetum americanum*) to inoculation with vesicular-arbuscular mycorrhizae and *Azospirillum brasiliense* with different sources of phosphorus. *Current Sci.* 57: 43-44.

Tilak, K.V.B.R., Li, C.Y., Ho, I. (1989) Occurence of nitrogen-fixing *Azospirillum* in vesicular-arbuscular mycorrhizal fungi. *Plant and Soil* 116: 286-288.

Torrey, J.G. (1978) Nitrogen fixation by Actinomycete-nodulated angiosperms. *BioScience* 28: 386-392.

Trabulsi, I. (1980) Vlijanie juznoj gallovoj nematody na aktivnost'azotofiksirujuscich bakterij *Rhizobium japonicum* pri vyrascivanii razlicnych sortov. *Soi.* 8: 171-175.

Trappe, J.M., Molina, R. and Castellano, M. (1984) Reactions of mycorrhizal fungi and mycorrhiza formation to pesticides. *Ann. Rev. Phytopathol.* 22: 331-359.

Vancura, V., Orozco, M.O., Gravová, O., Prikryl, Z. (1989) Properties of bacteria in the hyphosphere of a vesicular-arbuscular mycorrhizal fungus. *Agric. Ecosyst. Environ.* 29: 421-427

Varma, A.K., Singh, K., Lall, V.K. (1981) Lumen bacteria from endomycorrhizal spores. *Curr. Microbiol.* 6: 207-211

Visser, S., Danielson, R.M. and Parkinson, D. (1991) Field performance of *Elaeagnus commutata* and *Shepherdia canadensis* (Elaeagnaceae) inoculated with soil containing *Frankia* and vesicular-arbuscular mycorrhizal fungi. *Can. J. Bot.* 69: 1321-1328.

Wache, H. (1987) Effect of herbicides on symbiotic nitrogen fixation of lucerne (*Medicago sativa* L.) *Zentralbl. Mikrobiol.* 142: 349-355.

Waidyanatha, U.P.S., Yogaratnan, N. and Ariyaratne, W.A. (1979) Mycorrhizal infection effect on growth and nitrogen fixation of *Pueraria* and *Stylosanthes* and uptake of phosphorus from two rock phosphates. *New Phytol.* 82: 147-152.

Impact of Arbuscular Mycorrhizas on
Sustainable Agriculture and Natural Ecosystems
S. Gianinazzi and H. Schüepp (eds.)
© 1994 Birkhäuser Verlag Basel/Switzerland

Micropropagated plants, an opportunity to positively manage mycorrhizal activities

M. Vestberg and V. Estaún[1]

Agricultural Research Centre of Finland, Laukaa Research and Elite Plant Unit, FIN-41340 Laukaa, Finland
[1]Institut de Recerca i Tecnologia Agroalimantàries, Centre de Cabrils s/n, E-08348 Cabrils (Barcelona), Spain

Introduction

Plant biotechnology arouses great interest in both developed and developing countries because of its vast impact on agriculture. In plant biotechnology, the emphasis is on manpower rather than on expensive equipment. Among plant biotechnologies, micropropagation is often cited as the most successful example of a laboratory curiosity which has become an important commercial industry. The fast growing industry of micropropagation was illustrated by O'Riordain (1992), who found a three- to fourfold increase in the number of European commercial and official micropropagation laboratories between 1982 and 1992. Nowadays micropropagation is the most widely and successfully used technology by private companies for the mass production of horticultural plants: ornamentals, fruits, vegetables, plantation crops and spices.

Mycorrhiza is a mutualistic symbiosis formed by most species in the plant kingdom, including almost all the plants currently micropropagated. Mycorrhiza has been shown to positively affect plant growth by improving nutrient uptake, especially phosphorus uptake, to increase the rhizosphere soil volume and to alleviate biotic and abiotic stresses. Mycorrhiza is also ecologically significant and is particularly important in new plantations, in soils with problems (salinity, heavy metal contamination, low nutrient availability) or in sustainable agricultural systems where the non-biological inputs are minimized.

The technology used for micropropagating plants does not take into consideration the existence of this important symbiosis. The media used are devoid of mycorrhizal propagules and therefore the plants obtained from these systems are non-mycorrhizal. Such non-mycorrhizal plants obtained from the micropropagation process eventually become mycorrhizal when they are planted in field soil. However, an early inoculation of these plants with selected mycorrhizal inoculum might improve plant performance and allow lower chemical inputs. This paper presents the results obtained so far in the combination of the two biotechnologies, micropropagation and mycorrhization by arbuscular mycorrhizal fungi (AMF).

Factors affecting the result of mycorrhizal inoculation

There is a range of factors affecting the inoculation result in a micropropagation system. These factors are for example the timing of inoculation (*in vitro* vs *ex vitro*), effect of species and cultivar and fungal isolates (host-fungus interactions), physical and chemical composition of the growth substrate, conditions in growth chamber and greenhouse. In order to achieve an optimal benefit from inoculation of AMF to micropropagated plants, all these factors must be considered.

Timing   Micropropagation involves an *in vitro* and an *ex vitro* stage with different developmental phases in each. However, mycorrhizal inoculation is generally done after roots are formed (exception: experiments with inoculation of non-rooted microcuttings), which limits the suitable stages for inoculation to three:

1. *In vitro* - rooting phase
2. *Ex vitro* - immediately after the rooting phase at the beginning of the acclimatization period
3. *Ex vitro* - after the acclimatization phase, before starting the post acclimatization period under greenhouse conditions

Mycorrhizal inoculation *in vitro*   The rooting phase is normally conducted in agar based media. Germination and early stages of mycelial growth of AMF resting spores have been studied extensively (Mosse, 1959; Green et al., 1976; Koske, 1981) and the results of these studies show that most AMF spores studied germinate readily on media with low nutrient content, and that high nutrient levels, especially phosphate can inhibit spore germination and growth. Arbuscular mycorrhizal symbiosis has been synthesized on agar culture using seedlings and root organ cultures (Mosse and Hepper, 1975; Miller-Wideman and Watrud, 1984; Mugnier and Mosse, 1987; Bécard and Fortin, 1988; Gemma and Koske, 1988; Bécard and Piché, 1989; Chabot et al., 1992).

The right medium for the growth of both the micropropagated plant and the fungus has to be found for each plant-fungus combination. Pons et al. (1983) and Ravolanirina et al. (1989) achieved a functional mycorrhizal symbiosis in *Prunus avium* and *Vitis vinifera*, respectively, in the rooting media. However, such a symbiosis in the rooting phase is not always possible because most plants develop only primary roots at the rooting stage (Barea et al., 1992) on agar media and AMF only invade secondary roots (Brundrett et al., 1985). After transplanting, many

219

used, but only when soil was added to the mixes there was a significant response of the plant to the inoculation. Vestberg (1992b) found that sand fertilized with bone meal was superior to rich peat based substrates in initiating rapid AMF colonisation and sporulation. The survival of arctic bramble was considerably increased by AMF when inoculated to a sand substrate (Fig. 1.). In Boston fern, Ponton et al. (1990) found that one of three peat based mixes studied was better at enhancing growth whether or not the plant was inoculated with AMF. These results agree with those of Estaún (1994) on micropropagated *Prunus* (GF 677 clone), where the effect of the potting mix overcame the effects of the AMF inoculation (Fig. 2). In both cases there was an ideal AMF-potting mix-host combination that gave the best results in enhancing plant growth. Host plant and growing medium are major factors in determining AMF infectivity, symbiosis establishment and regulating its functionality. The addition of soil to the potting mix seems to

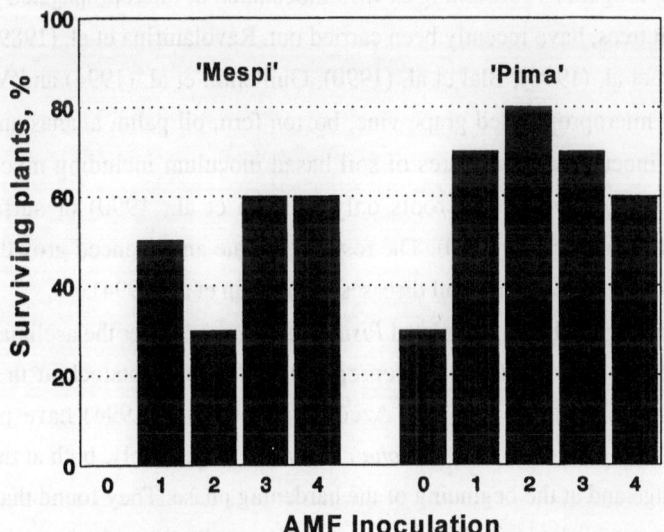

Figure 1. Effect of AMF inoculation (1: *Glomus mosseae* Rothamsted, 2: *G*. sp. V2, 3: *G*. sp. V3 and 4: *G*. sp. V4) or uninoculated (0) on the survival of arctic bramble (*Rubus arcticus* L) 'Pima' and 'Mespi' on sand substrate

increase the positive results of the AMF inoculation. However, even in rich peat based potting mixes there are plant-fungus combinations that can enhance plant growth. The differences found in the efficiency of single fungus inocula for several AMF-host-substrate combinations might indicate the convenience of using mixed inocula of a wide range of fungi that have been shown to be efficient in different situations.

Fertilization AMF play a key role in plant nutrient uptake, especially in phosphorus uptake. The importance and mechanism of this is the same for micropropagated plants as for plants raisedfrom seeds. AMF inoculation allows a considerable lowering of the phosphorus fertilization of highly mycorrhiza dependent micropropagated plants like grape vine (Ravolanirina micropropagated plants also replace the majority of their *in vitro* roots by new ones (Conner and Thomas, 1981).

It can be concluded that the technical aspects of the inoculation of micropropagated plants with an AMF *in vitro* are solved. However, the *in vitro* inoculation is a lengthy process which involves skilled and trained manpower and is not recommended for plants that fall into one of the categories listed above. Therefore, this technique should be used only for specific plants or for research purposes.

Mycorrhizal inoculation *ex vitro* Mycorrhizal inoculation *ex vitro* can be done either directly after the rooting phase *in vitro*, just at the start of the acclimatization period, or after the acclimatization before the start of the hardening phase under normal greenhouse conditions. A great number of investigations concerning *ex vitro* inoculation of micropropagated plants, mainly soft fruits and fruit trees, have recently been carried out. Ravolanirina et al. (1989), Schubert et al. (1990), Ponton et al. (1990), Blal et al. (1990), Guillemin et al. (1994) and Vestberg et al. (1994) inoculated micropropagated grape vine, boston fern, oil palm, ananas and strawberry, respectively. The inocula were mixtures of soil based inoculum including mycorrhizal roots (Schubert et al. 1990), mycorrhizal roots only (Ponton et al., 1990) or surface sterilized mycorrhizal roots (Ponton et al., 1990). The results indicate an enhanced growth and in some cases an increased resistance against soil diseases (Guillemin et al., 1994).

Schubert and Martinelli (1988) inoculated *Pistacia integerrima* after the acclimatization stage. Estaún et al. (1994) also inoculated micropropagated *Prunus* rootstock at the onset of the hardening stage. Vidal et al. (1992) and Azcón-Aguilar et al. (1994) have performed the inoculation with AMF of avocado and *Annona cherimola*, respectively, both at the onset of the acclimatization stage and at the beginning of the hardening phase. They found that although the symbiosis could be established with both techniques, the results showed a better response when the mycorrhization was done at the beginning of the hardening phase. For mycorrhization of apple microcuttings, Uosukainen and Vestberg (1994) suggest a similar inoculation strategy.

Inoculation *ex vitro* is easier than inoculation *in vitro* and should be the method of choice for commercial nurseries. The inoculation at the onset of the acclimatization stage has been studied thoroughly and found satisfacory for many plants. In other plants, however, a delayed inoculation at the hardening phase seems to give better results. Differences in the root growth and the development rate of the plants studied can explain these results. At an early *ex vitro* stage, all micropropagated plantlets and microcuttings still possess a certain degree of heterotrophy. Plants which rapidly become autotrophic are expected to benefit from an early AMF inoculation whilst those with a longer heterotrophic period probably benefit the most from a delayed inoculation. An inoculation protocol should be designed for each plant species, taking into account root growth and development rate, number of transplants after the *in vitro* stage, length of the weaning and hardening periods, objectives of the AMF inoculation (eg. enhanced growth, increased survival, lower fertilization input, increased resistance to biological and abiotic stress).

Fungus-host specificity It is generally accepted that arbuscular mycorrhizal associations lack specificity. Most AMF can form symbiosis with most plants to, at least, some degree. However, the mycorrhizal dependency of the plant may vary with the fungal species or with the soil characteristics. This might be interpreted as a kind of functional host specificity (Clarke and Mosse, 1981, Azcón and Ocampo, 1981, Plenchette et al., 1982, Doud-Miller et al., 1985, Estaún and Hayman, 1987, Fortuna et al., 1992).

When we consider the inoculation of micropropagated plants, the selection of the right fungal partner is essential to achieve a successful and efficient symbiosis. Guillemin et al. (1992), studying three micropropagated pineapple cultivars and five AMF, found that certain cultivar-fungus associations were more efficient than others in enhancing plant growth. Vestberg (1992a) found that three out of six AMF strains assayed with ten strawberry cultivars were significantly better at increasing plant growth and that the colonization rate and sporulation ability were different in the roots of early maturing, late maturing and special cultivars, respectively. Lovato et al. (1994) found that arbuscular mycorrhizal stimulation of growth in micropropagated wild cherry was dependent on the fungus used. Schubert et al. (1990) also found an influence of the endophyte strain on the symbiotic enhancement of plant growth. It has also been discussed whether an isolate mixture or an isolate with a very wide range of hosts, as was suggested by Sieverding (1989), would be preferable. Lovato et al. (1992) stressed the importance of the origin of the fungal strain. Strains isolated from for example acid soils are expected to function best under these conditions. The influence of the fungal strain on the efficiency of symbiosis varies with the plant growing media and with the fertilization rates applied.

Growth substrate The substrates used as growing media for the micropropagated plants are important not only for the development of the symbiosis but for the plant growth. The substrates favoured by growers are peat based, devoid of soil and with inorganic conditioners like perlite or vermiculite. The receptivity to the AMF of these substrates used in potting mixes in commercial nurseries has been little studied. Calvet et al. (1992) found that certain types of peat and composted substrates had a negative effect on the establishment of the arbuscular mycorrhizal symbiosis, although the germination and early mycelial growth were not affected, indicating a biological cause for the inhibition. Vidal et al. (1992) found that the symbiosis could be established in peat-sand mixes although soil-sand mixes were more conducive to AMF root colonization of micropropagated avocado plants. Schubert et al. (1990), working on micropropagated grape vine, found that the AMF colonized the roots in all peat based media

222

Figure 2. Effect of four organic substrates (1: compost A, 2: peat B, 3: compost c and 4: peat D) on the growth of *Prunus* rootstock plants (GF677) inoculated with *Glomus mosseae*

et al., 1989) or apple (Branzanti et al., 1992) in which a phosphorus fertilization of 40 ppm caused the same growth response as the use of AMF at 0 ppm phosphorus in soil based substrate. Schubert et al. (1992) found that kiwi responded most positively to AMF inoculation at an intermediate fertilizer level as compared with unfertilized and highly fertilized (Table I). The controlled-release fertilizers such as Osmocote are of special interest in the mycorrhization of micropropagated plants, because the AMF fungi seem to cope well with this type of fertilizers. This was demonstrated by Williams et al. (1992), who found that mycorrhizal colonization took place within the root system of strawberry plants receiving commercial rates of a controlled-release fertilizer. An Osmocote fertilizer applied at 25% of the minimum recommended commercial rate to mycorrhizal plants was sufficient to produce equivalent dry matter yields as non-mycorrhizal plants receiving the full application of Osmocote (Fig. 3). In another study with micropropagated strawberry, Vosatka et al. (1992) found that the addition of zeolite, which can absorb some nutrients and allows their slow release, enhances plant growth and the effect of mycorrhizal inoculation. Further information concerning the mechanisms of the AMF hyphal phosphorus transport is given by Iver Jakobsen in chapter three of this book.

Potential for use of AMF in commercial micropropagation systems

Most of the micropropagated plants inoculated in *ex vitro* experiments are highly mycorrhiza dependent high-value plants like grape vine (Ravolanirina et al., 1989, Schubert et al., 1990,

Schellenbaum et al., 1991), oil palm (Blal et al., 1990), apple (Branzanti et al., 1992, Uosukainen and Vestberg 1994), plum (Fortuna et al., 1992), pineapple (Guillemin et al., 1992, Lovato et al., 1992) and avocado (Azcon-Aguilar et al., 1992). Therefore, there is a high potential for introducing AMF into the micropropagation system of these plants and of other high-value plants. Salamanca et al. (1992) showed that inoculation with *G. fasciculatum* to

Table I. Shoot fresh weight (g) of micropropagated *Actinidia deliciosa* plants cv Hayward, measured 80 d after inoculation. Plants were inoculated with different amounts of inoculum of *Glomus* strain E_3 and grown in a potted substrate containing different amounts of a compound fertilizer. Values followed by a common letter do not differ significantly at P=0.05. After Schubert et al. (1992)

Fertilizer, g/l	Inoculum, g/pot			
	0	10	20	30
0	2.05b	2.21b	1.66b	2.78b
2	3.10b	5.69ab	6.31ab	6.26ab
4	8.11a	4.89ab	4.96ab	6.00ab
6	6.90ab	4.94ab	5.78ab	5.24ab

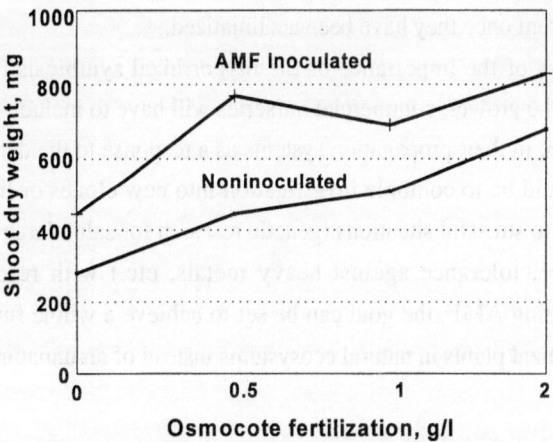

Figure 3. Mean shoot dry weight of mycorrhizal and non-mycorrhizal strawberry plants after five weeks of growth, given different amounts of Osmocote. Data on three mycorrhizal treatments pooled

micropropagated shrub legumes *Anthyllis cytisoides* and *Spartium junceum* shortened their acclimatization process by eight weeks. These shrubs are used in revegetation programs for mediterranean areas, and their shorter propagation cycle is of high value.

Uosukainen and Vestberg (1994) found that AMF inoculated apple plants were more uniform in size, a fact also found by several others working with fruit trees. From a practical point of view the uniformity of the plants is a desirable characteristic in any nursery, allowing for a homogenous classification of the stock. Uosukainen and Vestberg (1994) further estimated that the nursery culture time of AMF inoculated apple was 1/3 to 1/4 shorter than with uninoculated plants. In a large survey including inoculation at the commercial level of micropropagated potato, strawberry, blackberry, apple, rose, ginger and pineapple with optimal AMF strains, Lin et al. (1987) found blackberry and apple the most promising plants, from which it would be feasible and advantageous to produce mycorrhizal nursery stocks.

Concluding remarks and objectives for future research

Micropropagation is of increasing importance as an effective tool to multiply all kinds of agricultural and horticultural plants, or even plants used in regeneration and revegetation programs. Although micropropagation is a well established technique, each plant is a new challenge and specific protocols have to be designed to match its characteristics. Some of these plants have problems during the micropropagation process that could be overcome, at least to some extent by introducing AMF inoculation. Some woody plants, difficult to root (including oak, apple, plum, hydrangea, pear, avocado and other), have been shown to improve their survival when inoculated with AMF. Another problem that could be resolved by the arbuscular mycorrhizal symbiosis is the dormancy that some micropropagated plants (certain *Prunus* and *Malus* rootstocks) present once they have been acclimatized.

Once the awareness of the importance of the mycorrhizal symbiosis for plant growth and development reaches the grower, commercial nurseries will have to include the AMF inoculation as a standard procedure in their propagation systems as a response to the demand. The objectives of future research would be to combine investigation into new clones or lines used to improve plants and adapt them to stressful situations (genetic research for salt tolerance, disease tolerance, water stress tolerance, tolerance against heavy metals, etc.) with research on arbuscular mycorrhizas. By utilizing AMF, the goal can be set to achieve a whole functional plant which functions like mycorrhizal plants in natural ecosystems instead of an unnatural (non-mycorrhizal) plant with bare roots.

References

Azcón-Aguilar, C., Barceló, A., Vidal, M.T. and De la Viña, G. (1992) Further studies on growth and development of micropropagated avocado plants. *Agronomie* 12: 837-840.

Azcón-Aguilar, C., Encina, C.L., Azcón, R. and Barea, J.M. (1994) Effect of arbuscular mycorrhiza on growth and development of *Annona cherimola* micropropagated plants. *Agric. Sci. Finl.* 3: (in press)

Azcón-Aguilar, C. and Ocampo, J.A. (1981) Factors affecting the vesicular-arbuscular infection and mycorrhizal dependency of thirteen wheat cultivars. *New Phytol.* 87: 677-685.

Barea, J.M., Azcón-Aguilar, C. and Azcón, R. (1992) Mycorrhiza and crops. In: I.C. Tommerup (ed) *Advances In plant pathology. Mycorrhiza: A synthesis.* Academic Press. New York.

Bécard, G. and Fortin, J.A. (1988) Early events of vesicular-arbuscular mycorrhiza formation on Ri T-DNA transformed roots. *New Phytol.* 108: 211-218.

Bécard, G. and Piché, Y. (1989) Physiological factors determining vesicular-arbuscular mycorrhizal formation in host and nonhost Ri T-DNA transformed roots. *Can. J. Bot.* 68: 1260-1264.

Blal, B., Morel, C., Ganinazzi-Pearson, V., Fardeau, J.C. and Gianinazzi, S. (1990) Influence of vesicular-arbuscular mycorrhizae on phosphate fertilizer efficiency in two tropical acid soils planted with micropropagated oil palm (*Elaeis guineensis* jacq.). *Biol. Fertil. Soils* 9: 43-48.

Branzanti, B., Gianinazzi-Pearson, V. and Gianinazzi, S. (1992) Influence of phosphate fertilization on the growth and nutrient status of micropropagated apple infected with endomycorrhizal fungi during the weaning stage. *Agronomie* 12: 841-846.

Brundrett, M.C., Piché, Y. and Peterson, R.L. (1985) A developmental study of the early stages in vesicular-arbuscular mycorrhiza formation. *Can. J. Bot.* 63: 184-194.

Calvet, C., Estaún, V. and Camprubi, A. (1992) Germination, early mycelial growth and infectivity of a vesicular-arbuscular mycorrhizal fungus in organic substrates. *Symbiosis* 14: 405-411.

Clarke, C. and Mosse, B. (1981) Plant growth responses to vesicular-arbuscular mycorrhiza. XII. Field inoculation responses of barley at two soil P levels. *New Phytol.* 87: 695-703.

Chabot, S., Bécard, G. and Piché, Y. (1992) Life cycle of *Glomus intraradix* in root organ culture. *Mycologia* 84: 315-321.

Conner, A.J. and Thomas, M.B. (1981) Re-establishing plants from tissue culture: a review. *Pl. Propagator's Soc. Proc.* 31:342-357.

Doud-Miller, D., Domoto, P.A. and Walker, C. (1985) Colonization and efficacy of different endomycorrhizal fungi with apple seedlings at two phosphorus levels. *New Phytol.* 100: 393-402.

Gemma, J.N. and Koske, R.E. (1988) Pre-infection interactions between roots and the mycorrhizal fungus*Gigaspora gigantea*: Chemotropism of germ-tubes and root growth response. *Trans. Br. Mycol. Soc.* 91: 123-132.

Estaún, V., Calvet, C. and Camprubi, A. (1994) Arbuscular mycorrhizae and growth enhancement of micropropagated *Prunus* rootstock in different soilless potting mixes. *Agric. Sci. Finl.* 3: (in press)

Estaún, V., Calvet, C. and Hayman, D.S. (1987) Influence of plant genotype on mycorrhizal infection: Response of three pea cultivars. *Plant Soil* 103: 295-298.

Fortuna, P., Citernesi, S., Morini, S., Giovannetti, M. and Loreti, F. (1992) Infectivity and effectiveness of different species of arbuscular mycorrhizal fungi in micropropagated plants of Mr S 2/5 plum rootstock. *Agronomie* 12: 825-830.

Green, N.E., Graham, S.O. and Schenck, N.C. (1976) The influence of pH on the germination of vesicular-arbuscular mycorrhizal spores. *Mycologia* 68: 929-934.

Guillemin, J.P., Gianinazzi, S. and Trouvelot, A. (1992) Screening of arbuscular mycorrhizal fungi for establishment of micropropagated pineapple plants. *Agronomie* 12: 831-836.

Guillemin, J.P., Gianinazzi, S., Gianinazzi-Pearson, V. and Marchal, J. (1994) Contribution of endomycorrhizas to biological protection of micropropagated pineapple (*Ananas comosus* (L.) Merr) against *Phytophthora cinnamomi* Rands. *Agric. Sci.Finl.* 3: (in press).

Koske, R.E. (1981) *Gigaspora gigantea*: Observations on spore germination of a VA-mycorrhizal fungus. *Mycologia* 73:288-300.

Lin, M.T., Lucena, F.B., Mattos, M.A.M., Paiva, M., Assis, M. and Caldas, L.S. (1987) Greenhouse production of mycorrhizal plants of nine transplanted crops. In: D.M. Sylvia, L.L. Hung and J.H. Graham (eds) *Mycorrhizae in the next decade. Practical applications and research priorities.* 7th NACOM, May 3-8, 1987, Gainesville, Florida. University of Florida, Gainesville, Florida, U.S.A., 281.

Lovato, P., Guillemin, J.P. and Gianinazzi, S. (1992) Application of commercial arbuscular endomycorrhizal fungal inoculants to the establishment of micropropagated grapevine rootstock and pineapple plants. *Agronomie* 12: 873-880.

Lovato, P., Hammatt, N., Gianinazzi-Pearson, V. and Gianinazzi, S. (1994) Mycorrhization of micropropagated wild cherry (*Prunus avium* L.) and common ash (*Fraxinus excelsior* L.). *Agric. Sci. Fin.* 3: (in press).

Miller-Wideman, M.A. and Watrud, L.S. (1984) Sporulation of *Gigaspora margarita* on root cultures of tomato. *Can. J. Microbiol.* 30:642-646.

Mosse, B. (1959) The regular germination of resting spores and some observations on the growth requirements of an *Endogone* sp. causing vesicular-arbuscular mycorrhiza. *Trans. Br. Mycol. Soc.* 42: 273- 286

Mosse, B., and Hepper, C. (1975) Vesicular-arbuscular mycorrhizal infections in root organ cultures. *Physiol. Pl. Path.* 5:215-223.

Mugnier, J. and Mosse, B. (1987) Vesicular-arbuscular mycorrhizal infection in transformed root-inducing T-DNA roots grown axenically. *Phytopath.* 77:1045-1050.

O'Riordain, F. (1990) The European plant tissue culture industry -*Agronomie* 12:743-746.

Plenchette, C., Furlan, V. and Fortin, J.A. (1982) Effects of different endomycorrhizal fungi on five host plants grown in calcined montmorillonite clay. *J. Am. Soc. Hortic. Sci.* 107: 535-538.

Pons, J., Gianinazzi-Pearson, V., Gianinazzi, S. and Navatel, J.C. (1983) Studies of VA-mycorrhizae *in vitro*: mycorrhizal synthsesis of axenically propagated wild cherry (*Prunus avium* L.). *Plant Soil* 71: 217- 221.

Ponton, F., Piché, Y., Parent, S. and Caron, M. (1990) Use of vesicular-arbuscular mycorrhizae in Boston fern production: II. Evaluation of four inocula. *HortSci.* 25: 416-419.

Ravolanirina, F., Gianinazzi, S., Trouvelot, A. and Carre, M. (1989) Production of endomycorrhizal explants of micropropagated grapevine rootstocks. *Agriculture, Ecosystems and Environment* 29: 323-327.

Salamanca, C.P., Herrera, M.A. and Barea, J.M. (1992) Mycorrhizal inoculation of micropropagated woody legumes used in revegetation programmes for desertified Mediterranean ecosystems. *Agronomie* 12: 869-872.

Schellenbaum, L., Berta, G., Ravolanirina, F., Tisserant, B., Gianinazzi, S. and Fitter, A.H. (1991) Influence of endomycorrhizal infection on root morphology in a micropropagated woody plant species *Vitis vinifera* L.). *Ann. Bot.* 68: 135-141.

Schubert, A., Bodrino, C. and Gribaudo, I. (1992) Vesicular-arbuscular mycorrhizal inoculation of kiwifruit (*Actinidia deliciosa*) micropropagated plants. *Agronomie* 12: 847-850.

Schubert, A., and Martinelli, A. (1988) Effect of vesicular-arbuscular mycorrhizae on growth of *in vitro* propagated *Pistacia integerrima*. *Acta Hort.* 441-443.

Schubert, A., Mazzitelli, M., Ariusso, O. and Eynard, I. (1990) Effects of vesicular-arbuscular mycorrhizal fungi on micropropagated grapevines: Influence of endophyte strain, P fertilization and growth medium.*Vitis* 29: 5-13.

Sieverding, E. (1989) Should VAM inocula contain single or several fungal species? *Agric. Ecosystems Environ.* 29: 391-396.

Uosukainen, M. and Vestberg, M. (1994) Effect of inoculation with arbuscular mycorrhizas on rooting, weaning and subsequent growth of micropropagated *Malus* (L.) Moench. *Agric. Sci. Finl.* 3: (in press).

Vestberg, M. (1992a) The effect of vesicular-arbuscular mycorrhizal inoculation on the growth and root colonization of ten strawberry cultivars. *Agric. Sci. Finl.* 1:527-535.

Vestberg, M. (1992b) The effect of growth substrate and fertilizer on the growth and vesicular-arbuscular mycorrhizal infection of three hosts. *Agric. Sci. Finl.* 1:95-105.

Vestberg, M., Palmujoki, H., Parikka, P. and Uosukainen, M. (1994) Effects of arbuscular mycorrhiza on crown rot (*Phytophthora cactorum*) in micropropagated strawberry plants. *Agric. Sci. Finl.* 3: (in press)

Vidal, M.T., Azcón-Aguilar, C. and Barea, J.M. (1992) Mycorrhizal inoculation enhances growth and develpment of micropropagated plants of avocado. *HortSci.* 27:785-787.

Vosatka, M., Gryndler, M. and Prikryl, Z. (1992) Effect of the rhizosphere bacterium *Pseudomonas putida*, arbuscular mycorrhizal fungi and substrate composition on the growth of strawberry. *Agronomie* 12: 859-863.

Williams, S.C.K., Vestberg, M., Uosukainen, M., Dodd, J.C. and Jeffries, P. (1992) Effects of fertilizers and arbuscular mycorrhizal fungi on the *post-vitro* vitro growth of micropropagated strawberry. *Agronomie* 12:851-857.

Soil Monitoring

Early Detection and Surveying of Soil Contamination and Degradation

Edited by
R. Schulin, *ETH, Schlieren, Switzerland*
A. Desaules, *Liebefeld-Bern, Switzerland*
R. Webster and **B. Steiger**, *ETH, Schlieren, Switzerland*

1993. 378 pages. Hardcover
ISBN 3-7643-2956-4
(MV Monte Verità)

Soil pollution is posing increasing hazards to envi-ronmental quality, human health, and economic welfare. As a prerequisite for effective protection and remediation measures, the identification and monitoring of soil pollution has become increasingly important for society in general, and for the responsible governmental authorities in particular.

This book focuses on soil contaminants that threaten the long-term soil fertility and related functions of the soil on a regional scale. The pollutants of most concern were the heavy metals and pe-dosphere. The book assesses the current knowledge and problems being tackled to protect the soil and emphasizes the use mass of balances in soil monitoring. Consequently the connections of soil protection with ecological cycling of elements, air and water pollution are covered as well.

Soil monitoring combines science and technology. This book provides a wide coverage of the field, with each topic covered by a specialist. First results of the Swiss Soil Monitoring Network (NABO) are presented. New methodol-ogy designed for early recognition of soil changes using a mass balance approach and a new view on biomonitoring using soil microbiology are presented. Previously unpublished comparisons of geostatistical and classical estimation of change are also presented in this book.

A synthesis is given which identifies gaps in knowledge and reviews the objectives of soil monitoring. It was found that the primary objective of soil monitoring is to provide sound and relevant information on which political decisions, which may be far reaching, can be based for managing and protecting our environment.

Consequently, this book will provide up-to-date information to ecological scientists and environ-mental agencies working on soil protection and on related topics as environmental cycling, air and water pollution. Political decisionsmakers, their advisers and administrators will find guidance for designing soil monitoring programmes and for formulating legislation on soil protection.

Please order through your bookseller or directly from:
Birkhäuser Verlag AG
P.O. Box 133
CH-4010 Basel / Switzerland
Fax ++41 / 61 721 79 50
Orders from the USA or Canada should be sent to:
Birkhäuser Boston, 333 Meadowlands Parkway,
Secaucus, NJ 07094-2491 / USA
Call Toll-Free 1-800-777-4643

For more information on recent and forthcoming books and journals you can order the Birkhäuser Life Sciences Bulletin, published twice a year and free of charge.

Birkhäuser